T0214523

Lecture Notes in Artificial Intelligence   11221

Subseries of Lecture Notes in Computer Science

LNAI Series Editors

Randy Goebel
  University of Alberta, Edmonton, Canada
Yuzuru Tanaka
  Hokkaido University, Sapporo, Japan
Wolfgang Wahlster
  DFKI and Saarland University, Saarbrücken, Germany

LNAI Founding Series Editor

Joerg Siekmann
  DFKI and Saarland University, Saarbrücken, Germany

More information about this series at http://www.springer.com/series/1244

Maosong Sun · Ting Liu
Xiaojie Wang · Zhiyuan Liu
Yang Liu (Eds.)

Chinese Computational Linguistics and Natural Language Processing Based on Naturally Annotated Big Data

17th China National Conference, CCL 2018
and 6th International Symposium, NLP-NABD 2018
Changsha, China, October 19–21, 2018
Proceedings

 Springer

Editors
Maosong Sun
Tsinghua University
Beijing, China

Zhiyuan Liu
Tsinghua University
Beijing, China

Ting Liu
Harbin Institute of Technology
Harbin, China

Yang Liu
Tsinghua University
Beijing, China

Xiaojie Wang
Beijing University of Posts
  and Telecommunications
Beijing, China

ISSN 0302-9743   ISSN 1611-3349 (electronic)
Lecture Notes in Artificial Intelligence
ISBN 978-3-030-01715-6   ISBN 978-3-030-01716-3 (eBook)
https://doi.org/10.1007/978-3-030-01716-3

Library of Congress Control Number: 2018956472

LNCS Sublibrary: SL7 – Artificial Intelligence

This Springer imprint is published by the registered company Springer Nature Switzerland AG
The registered company address is: Gewerbestrasse 11, 6330 Cham, Switzerland

Preface

Welcome to the proceedings of the 17th China National Conference on Computational Linguistics (17th CCL) and the 6th International Symposium on Natural Language Processing Based on Naturally Annotated Big Data (6th NLP-NABD). The conference and symposium were hosted by Changsha University of Science and Technology located in Changsha City, Hunan Province, China.

CCL is an annual conference (bi-annual before 2013) that started in 1991. It is the flagship conference of the Chinese Information Processing Society of China (CIPS), which is the largest NLP scholar and expert community in China. CCL is a premier nation-wide forum for disseminating new scholarly and technological work in computational linguistics, with a major emphasis on computer processing of the languages in China such as Mandarin, Tibetan, Mongolian, and Uyghur.

Affiliated with the 17th CCL, the 6th International Symposium on Natural Language Processing Based on Naturally Annotated Big Data (NLP-NABD) covered all the topics of NLP, with particular focus on methodologies and techniques relating to naturally annotated big data. In contrast to manually annotated data such as treebanks that are constructed for specific NLP tasks, naturally annotated data come into existence through users' normal activities, such as writing, conversation, and interactions on the Web. Although the original purposes of these data typically were unrelated to NLP, they can nonetheless be purposefully exploited by computational linguists to acquire linguistic knowledge. For example, punctuation marks in Chinese text can help word boundaries identification, social tags in social media can provide signals for keyword extraction, and categories listed in Wikipedia can benefit text classification. The natural annotation can be explicit, as in the aforementioned examples, or implicit, as in Hearst patterns (e.g., "Beijing and other cities" implies "Beijing is a city"). This symposium focuses on numerous research challenges ranging from very-large-scale unsupervised/semi-supervised machine leaning (deep learning, for instance) of naturally annotated big data to integration of the learned resources and models with existing handcrafted "core" resources and "core" language computing models. NLP-NABD 2018 was supported by the National Key Basic Research Program of China (i.e., "973" Program) "Theory and Methods for Cyber-Physical-Human Space Oriented Web Chinese Information Processing" under grant no.2014CB340500 and the Major Project of the National Social Science Foundation of China under grant no. 13&ZD190.

The Program Committee selected 102 papers (69 Chinese papers and 33 English papers) out of 277 submissions from China, Hong Kong (region), Singapore, and USA for publication. The acceptance rate is 36.82%. The 33 English papers cover the following topics:

– Semantics (3)
– Machine translation (6)
– Knowledge graph and information extraction (7)
– Linguistic resource annotation and evaluation (2)

- Information retrieval and question answering (4)
- Text classification and summarization (5)
- Social computing and sentiment analysis (2)
- NLP applications (4)

The final program for the 17th CCL and the sixth NLP-NABD was the result of a great deal of work by many dedicated colleagues. We want to thank, first of all, the authors who submitted their papers, and thus contributed to the creation of the high-quality program that allowed us to look forward to an exciting joint conference. We are deeply indebted to all the Program Committee members for providing high-quality and insightful reviews under a tight schedule. We are extremely grateful to the sponsors of the conference. Finally, we extend a special word of thanks to all the colleagues of the Organizing Committee and secretariat for their hard work in organizing the conference, and to Springer for their assistance in publishing the proceedings in due time.

We thank the Program and Organizing Committees for helping to make the conference successful, and we hope all the participants enjoyed a remarkable visit to Changsha, a historical and beautiful city in South China.

August 2018

Maosong Sun
Ting Liu
Xiaojie Wang
Randy Goebel
Heng Ji

Organization

General Chairs

Sheng Li   Harbin Institute of Technology, China
Changning Huang   Tsinghua University, China
Kaiying Liu   Shanxi University, China

Program Committee

17th CCL Program Chairs

Maosong Sun   Tsinghua University, China
Ting Liu   Harbin Institute of Technology, China
Xiaojie Wang   Beijing University of Posts and Telecommunications, China

17th CCL Area Co-chairs

Linguistics and Cognitive Science
Weiguang Qu   Nanjing Normal University, China
Yulin Yuan   Peking University, China

Fundamental Theory and Methods of Computational Linguistics
Wanxiang Che   Harbin Institute of Technology, China
Yue Zhang   Singapore University of Technology and Design, Singapore

Information Retrieval and Question Answering
Jun Xu   Institute of Computing Technology, CAS, China
Aixin Sun   Nanyang Technological University, Singapore

Text Classification and Summarization
Xipeng Qiu   Fudan University, China
Xiaodan Zhu   Queen's University, Canada

Knowledge Graph and Information Extraction
Xianpei Han   Institute of Software, CAS, China
Ni Lao   Google Inc., USA

Machine Translation
Shujian Huang   Nanjing University, China
Haitao Mi   Ant Financial, USA

Minority Language Information Processing
Xiaobing Zhao   Minzu University of China, China
Celimuge Wu   The University of Electro-Communications, Japan

Language Resource and Evaluation

Muyun Yang	Harbin Institute of Technology, China
Rui Wang	National Institute of Information and Communications Technology, NICT, Japan

Social Computing and Sentiment Analysis

Jia Jia	Tsinghua University, China
Yanyan Zhao	Harbin Institute of Technology, China

NLP Applications

Qingcai Chen	Harbin Institute of Technology, Shenzhen, China
Cui Tao	The University of Texas Health Science Center at Houston, USA

6th NLP-NABD Program Chairs

Maosong Sun	Tsinghua University, China
Randy Goebel	University of Alberta, Canada
Heng Ji	Rensselaer Polytechnic Institute, USA

17th CCL and 6th NLP-NABD Local Arrangements Chairs

Feng Li	Changsha University of Science and Technology, China
Jin Wang	Changsha University of Science and Technology, China
Yang Liu	Tsinghua University, China

17th CCL and 6th NLP-NABD Evaluation Chairs

Ting Liu	Harbin Institute of Technology, China
Wei Song	Capital Normal University, China

17th CCL and 6th NLP-NABD Publication Chairs

Zhiyuan Liu	Tsinghua University, China
Shizhu He	Institute of Automation, CAS, China

17th CCL and 6th NLP-NABD Publicity Chairs

Ruifeng Xu	Harbin Institute of Technology, Shenzhen, China
William Wang	University of California, Santa Barbara, USA

17th CCL and 6th NLP-NABD Advance Lectures Chairs

Xiaojun Wan	Peking University, China
Jiafeng Guo	Institute of Computing Technology, CAS, China

17th CCL and 6th NLP-NABD Sponsorship Chairs

Qi Zhang   Fudan University, China
Jiajun Zhang   Institute of Automation, CAS, China

17th CCL and 6th NLP-NABD System Demonstration Chairs

Tong Xiao   Northeastern University, China
Binyang Li   University of International Relations, China

17th CCL and 6th NLP-NABD Student Seminar Chairs

Yang Feng   Institute of Computing Technology, CAS, China
Dong Yu   Beijing Language and Culture University, China

17th CCL and 6th NLP-NABD Organizers

Chinese Information Processing Society of China

Tsinghua University

Changsha University of Science & Technology, China

Publishers

Lecture Notes in Artificial Intelligence, Springer

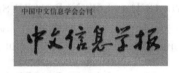

Journal of Chinese Information Processing

Science China

清华大学学报（自然科学版）
Journal of Tsinghua University (Science and Technology)

Journal of Tsinghua University (Science and Technology)

Sponsoring Institutions

Platinum

Gold

SAMSUNG

Silver

LEYAN
TECHNOLOGIES

Bronze

Teksbotics

Contents

Linguistic Resource Annotation and Evaluation

Information Retrieval and Question Answering

Text Classification and Summarization

Social Computing and Sentiment Analysis

NLP Applications

Semantics

Radical Enhanced Chinese Word Embedding

Zheng Chen$^{(\boxtimes)}$ and Keqi Hu

University of Electronic Science and Technology of China,
Section 2-4, Jianshe North Road, Chengdu 610054, China
zchen@uestc.edu.cn

Abstract. The conventional Chinese word embedding model is similar to the English word embedding model in modeling text, simply uses the Chinese word or character as the minimum processing unit of the text, without using the semantic information about Chinese characters and the radicals in Chinese words. To this end, we proposed a radical enhanced Chinese word embedding in this paper. The model uses conversion and radical escaping mechanisms to extract the intrinsic information in Chinese corpus. Through the improved parallel dual-channel network model on a CBOW-like model, the word information context is used together with the Chinese character radical information context to predict the target word. Therefore, the word vector generated by the model can fully reflect the semantic information contained in the radicals. Compared with other similar models by word analogy and similarity experiments, the results showed that our model has effectively improved the accuracy of word vector expression and the direct relevance of similar words.

Keywords: Word embedding · Radical enhanced Chinese word embedding

1 Introduction

Vectorized representation of text is one of the core research areas of natural language processing. One-hot embedding, as the traditional solution, has the advantage of simplifying and efficiency. However, it ignores too much semantic information, failing to meet people's expectations in practical use. Distributed word representation represents a word as a vector in a continuous vector space, makes similar words close to each other. It allows the machine learning model to directly obtain relevant semantic information through text vectors and is conducive to follow-up works. Take the advantage of distributed word representation, also known as word embedding, scholars achieve many excellent results in different natural language processing tasks, such as named entity recognition [1], text classification [2], semantic analysis [3], and question answering system [4]. Among many word-embedding methods [5, 6], Continuous Bag of Words (CBOW) model and Skip-Gram model are most popular ones. They could gain excellent word embedding from large-scale corpus [7].

Although the word embedding method has excellent performance in English, this method that uses the word as the smallest unit of the language does not have a good effect on all languages, especially on Chinese, which is a structurally complex pictographic language. The Chinese characters have semantics itself. In many cases, a single

© Springer Nature Switzerland AG 2018
M. Sun et al. (Eds.): CCL 2018/NLP-NABD 2018, LNAI 11221, pp. 3–11, 2018.
https://doi.org/10.1007/978-3-030-01716-3_1

Chinese character can be a word on its own. And, furthermore, even the sub-word items of Chinese characters, including radicals and components, also contain rich semantic information. For example, the word "智能(intellect)", on the one hand, its semantic information we can learn from the relevant context in the corpus, and on the other hand we can also infer from the individual Chinese characters "智(wisdom)" and "能(ability)". And the Chinese characters "江(large river)", "河(river)", "湖(lake)" and "海(sea)" can be inferred that they are all related to water according to their common radical "氵".

However, as one of the oldest writing system in the human history, Chinese is complicated. At the very beginning, Chinese only have single characters. Subsequently, because of the need of complex expressions, people create a lot of compound characters, which compounded by two, three, or even more characters. Therefore, each part of compound characters need to be simplified, from character to radical. For example, the radical "氵" is actually the reduced form of character "水(water)". If these correspondences cannot be used, the model is difficult to find the semantic relationship of the Chinese character that has common sidelines.

In this paper, we propose a model to jointly learn the embeddings of Chinese words, characters, and sub-character components. By using simplified transformation and radical escaping techniques, the learned Chinese word embeddings can leverage the external context co-occurrence information and incorporate rich internal sub-word semantic information. Both the word similarity experiment and the word analogy experiment prove that this method is real and effective and has better effect than other similar methods.

2   Model

The Radical Enhanced Chinese Word Embedding (RECWE) proposed in this paper is based on the CBOW model. It can effectively synthesizes the Chinese characters and the radicals. The overall structure is shown in Fig. 1.

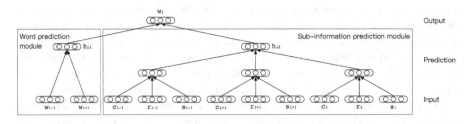

Fig. 1. Radical enhanced Chinese word embedding.

RECWE has dual-prediction modules: word prediction module and sub-information prediction module. The first model is roughly the same as the CBOW model, where w_i is the target word, w_{i-1} and w_{i+1} is the left and right words of the target word in the context, and the h_i^1 is word context vector.

The sub-information prediction module is juxtaposed with the word prediction module. Let c_{i-1}, r_{i-1}, s_{i-1} and c_{i+1}, r_{i+1}, s_{i+1}, corresponding to the context words w_{i-1} and w_{i+1} in the word prediction module, be the "input" Chinese character, radical, and component. h_i^2 is the sub-information context vector. However, because the semantic information contained in Chinese characters and radicals is not as rich as words, using the context vector directly to predict the target words will have a large error. To increase the semantic information, the sub-information prediction module uses the information of target word (c_i, r_i, and s_i in the Fig. 1).

To prediction target words, sub-information prediction module mainly relies on the semantic information contained in the Chinese characters, but not all the words can obtain semantic information through the Chinese characters and the radicals. For example, word "东西(thing)" that represents the physical object, the internal Chinese characters "东(east)" and "西(west)" are only indicative of the position alone and are far from the semantics of the word "东西". In addition, for most of the transliterated words that appear in modern Chinese, such as "苏打(soda)" and "沙发(sofa)", this approach is not suitable. So, for such words, the sub-information prediction module will directly use their word vector to construct h_i^2, instead of disassembling into the Chinese character and the radical.

To make full use of the semantic information of Chinese characters and radicals, before training RECWE, the input text need to perform simplified conversion and radical convert (As shown in Table 1).

Table 1. Radical convert table

Radical	Converted character	Radical	Converted character
艹	艸 (grass)	亻	人 (people)
刂	刀 (knife)	犭	犬 (dog)
灬	火 (fire)	钅	金 (gold)
麦	麥 (wheat)	饣	食 (eat)
礻	示 (show)	月	肉 (meat)
攵	攴 (knock)	罒	网 (net)
扌	手 (hand)	氵	水 (water)
纟	糸 (silk)	耂	老 (old)
牜	牛 (cow)	忄	心 (heart)
衤	衣 (cloth)	王	玉 (jade)
辶	走 (walk)	疒	病 (illness)

Similar to the CBOW model, the objective function of the RECWE model is the log-likelihood function of the conditional probabilities of the two context vectors for the target word w_i, as shown in Eq. 1.

$$L(w_i) = \sum_{k}^{2} \log P(w_i | h_i^k) \tag{1}$$

Where h_i^1 and h_i^2 are the word context vector and the child information context vectors.

The conditional probability $P(w_i|h_i^k)$ for each context vector for the target word w_i can be calculated using the SoftMax function, as shown in Eq. 2.

$$P(w_i|h_i^k) = \frac{\exp(h_i^{kT}\hat{v}_w^i)}{\sum\limits_{j=1}^{N}\exp(h_i^{kT}\hat{v}_w^j)}, k = 1, 2 \tag{2}$$

Where $\hat{v}_w^i$ is the "output" vector of the target word w_i, $\hat{v}_w^j$ is the "output" vector of each word in the input corpora, and N is the length of the input corpora.

The context vector h_i^1 is the average of the "input" vector of each word in the context, obtained by Eq. 3:

$$h_i^1 = \frac{1}{2T}\sum\limits_{-T \leq j \leq T, j \neq 0} v_w^{i+j} \tag{3}$$

Where T is the size of the context window and v_w^{i+j} is the "input" vector of words in the context window.

Similarly, the sub-information context vector h_i^2 is the average of the "input" vectors of the radical and components. The calculation formula is shown in Eq. 4.

$$h_i^2 = \frac{1}{X}\sum\limits_{-T \leq j \leq T, j \neq 0} v_c^{i+j} + v_r^{i+j} + v_s^{i+j} \tag{4}$$

Where v_c^{i+j}, v_r^{i+j}, and v_s^{i+j} are the word, radical, and components vector, and X is the number of v_c^{i+j}, v_r^{i+j}, and v_s^{i+j}.

Thus, for corpus D, the overall log-likelihood function of the RECWE model is shown in Eq. 5.

$$L(D) = \sum\limits_{w_i \in D} L(w_i) \tag{5}$$

3   Experiment

To verify the feasibility and effectiveness of the word embedding trained by our model, in this section, we conduct experiments based on headline data from the news network named "今日头条". We compare our method with word2vec [7], CWE [11], SCWE [8], and JWE [13].

3.1 Parameter Settings

The training corpus is processed through the Ansj segment tool[1], then filter the stop words in the news through HIT and Baidu's stop word vocabulary. All word embedding models use same training parameters, as shown in Table 2. The detail of our implementation can be found at https://github.com/UESTC1010/RECWE.

Table 2. Training parameters

Parameter	Explanation	Value
Size	Word embedding dimension	200
Alpha	Learning rate	0.025
Mincount	The lower bound of low frequency words in corpus	5
Sample	Threshold for down sampling of high frequency words	1e−4
Workers	Number of threads	4
Iter	Number of iterations	5
Window	Windows size	5

To verify the influence of the sub-information of the target word on the word embedding, this experiment adopts three different sub-information superposition modes: The pattern 1 uses the sub-information corresponding to the word context, labeled "p1"; Pattern 2 uses only the sub-information of the target word, labeled "p2"; The pattern three uses the sub-information corresponding to the word context and the target word at the same time, labeled "p3".

3.2 Word Similarity

This task is mainly used to evaluate the ability of word embedding to determine semantically similar word pairs [13]. In this task, two different similar word databases, wordsim-240 and wrodsim-296, which were provided in [11], were selected. They contain 240 Chinese word pairs and 296 Chinese word pairs. Each word pair contains a manually labeled similarity. Wordsim-240 is mainly for semantically related words, while wordsim-296 is for synonyms.

For each word embedding model, the similarity of a word pair is expressed using the cosine distance of the word embedding corresponding to the two words. At the end of the task, we compute the Spearman correlation [14] between the manually labeled similarity and similarity computed by embeddings, which computed as the cosine similarity of word pair's embeddings generated by the model. The evaluation results are shown in Table 3.

From the results, we can see that RECWE outperforms other models in two similar databases. Compared with the SCWE, the CWE has improved word2vec results. However, since only the Chinese character information in the word is used, the result of

---

[1] https://github.com/NLPchina/ansj_seg.

Table 3. Word similarity result

Model	Wordsim-240	Wordsim-296
Word2vec	0.4221	0.4479
CWE	0.4363	0.4750
SCWE	0.4311	0.4648
JWE	0.4467	0.5831
RECWE-p1	0.4962	0.5849
RECWE-p2	0.5011	0.5554
RECWE-p3	0.5290	0.5765

the JWE with the addition of radical information is worse than that of the RECWE. It can also be seen that, in the wordsim-240 database, the RECWE is greatly improved compared to the JWE, especially after the addition of the radical word of the target word. This is mainly because the database contains mostly word pairs that are semantically affiliated, and the RECWE with a radical transformation mechanism can find such relationships well. For example, for the words "淋浴(shower)" and "水 (water)", the RECWE will convert the radical "氵" of character "淋" and "浴" to the character "水" in the pre-processing phase. The intrinsic link between water and water. And then the model can find the inherent connection between "淋浴" and "水". Moreover, comparing the results of the three pattern, it can be found that by adding the sub-information of the target word and word context, the word embedding can be further optimized and the model can achieve better results. However, this feature has little effect on the word sim-296 and may even result in deterioration of the results. Therefore, the specific use depends on the characteristics of the corpus.

3.3 Word Analogy

This task is another technique for measuring the quality of word embedding. It mainly judges whether word embedding can reflect the linguistic regularity between word pairs [14]. For example, given the word relationship group "Beijing-China: Tokyo-Japan", a good word embedding should have the word embedding "$vec(Japan)$" of the word "Japan" close to the embedding produced by the expression "$vec(China) - vec(Beijin) + vec(Tokyo)$". That is, given an analogy relationship "$a - b : c - d$", if the word embedding can find the relationship "$a - b : c - x$" by looking for the embedding x in the vocabulary through formula 6, then it is determined that the word embedding contains this analogy relationship.

$$\arg \max_{x \neq a, x \neq b, x \neq c} \cos\left(\vec{b} - \vec{a} + \vec{c}, \vec{x}\right) \tag{6}$$

This task uses the Chinese word analogy database provided in the literature [11], which contains 1124 group word analogy relationships, each group of analogical relations contains 4 words, all analogy relations are divided into three categories:

"Capital" (group 677), "The provincial capitals" (175 groups) and "people" (272 groups).

The accuracy rate is used as a result indicator. The higher the accuracy rate is, the more the word analogy relationship the word vector contains. The experimental results of the three types of word analogy are shown in Table 4.

Table 4. All kinds of analogy experimental results

Model	Capital	Provincial capitals	People
word2vec	0.11816	0.11428	0.34558
CWE	0.20787	0.14285	0.32720
SCWE	0.21381	0.14022	0.33021
JWE	0.22101	0.16	0.32352
RECWE-p1	0.23632	0.18285	0.38603
RECWE-p2	0.33479	0.2	0.32079
RECWE-p3	0.39606	0.18857	0.44852

From the result, we can see that compared with other word vector models, the RECWE has achieved optimal results in the three types of analogy relationship, indicating that the radical information can enhance the performance of the model in language law. Particularly in the category of "people", the radical transformation mechanism allows the RECWE to find the relationship between word pairs relatively easily based on radicals. For example, RECWE can contact the word "妈妈(mother)" and "姐姐(sister)" through radical "女". Therefore, the accuracy of RECWE compared with the JWE model has greatly improved. For categories such as "Capital" that contain a large number of transliterated words, their internal radicals cannot provide additional semantics, and the promotion is smaller. However, it can be seen from the results of the three patterns of "p1", "p2", and "p3" that using the target information and the sub-information corresponding to the word context can alleviate this short-coming to a certain extent, and the model can be better result.

4 Relate Work

With the continuous development of neural networks and deep learning, word embedding has achieved many achievements in English. However, there has been no breakthrough in how to construct word embedding of Chinese. At the beginning of the study, the researchers tried to directly use the English word embedding model (such as the CBOW and the Skip-gram) to train Chinese word embedding on the Chinese corpora after word segmentation. However, there is obviously a problem with this approach: Most English word embedding model use words as the smallest unit of operation when training, while ignoring the internal morphological in-formation of words. Different from English and other alphabetic characters, Chinese characters are still have a lot of semantic information.

To address this issue and make full use of the semantic information of Chinese words, Chen et al. [11] added Chinese character information to the ordinary CBOW and proposed the CWE. Subsequently, Xu et al. [12] based on the CWE, assigning each character weight in the word to optimize the whole model. However, these models did not use the radicals in Chinese characters and ignored the internal information of all characters.

On the other hand, Sun et al. [8] added radical information to train Chinese character embedding based on the CBOW; Yu et al. [13] designed a multi-granularity word vector model by combining the information of "radical-character" and "character-word" in the JWE. However, these methods simply add the radical information to the model and do not take into account the evolution of the radicals. That is, they do not associate the radicals with the corresponding Chinese characters (such as "氵" to "水"), which makes the information obtained from radicals is very limited, affecting the quality of the word embedding.

5   Conclusion

In this paper, we proposed a Chinese word embedding training model RECWE by effectively using the semantic information Chinese characters and their radicals. Through simplified conversion and radical escaping techniques, RECWE can directly determine the internal semantic relations between Chinese characters based on radicals, allowing words with the same radical to be close to each other in the vector space. Experiments show that our method is more effective than other similar methods in word similarity and word analogy.

Due to the rapid developing of deep learning technology, there're a lot of potential work. About the predict architecture, we will try to employ attention technology in the prediction layer of the model. But more work should to be done on the sematic side, which is to understand Chinese character, and try to employ more features to our model.

Acknowledgements. Financial support for this study was provided by the Fundamental Research Funds for the Central Universities (Grant No. ZYGX2016J198) and Science and Technology Planning Project of Sichuan Province, China (Grant No. 2017JY0080).

References

1. Collobert, R., Weston, J., Bottou, L., et al.: Natural language processing (almost) from scratch. J. Mach. Learn. Res. **12**(Aug), 2493–2537 (2011)
2. Grave, E., Mikolov, T., Joulin, A., et al.: Bag of tricks for efficient text classification. In: Proceedings of the 15th Conference of the European Chapter of the Association for Computational Linguistics, pp. 427–431 (2017)
3. Tang, D., Wei, F., Yang, N., et al.: Learning sentiment-specific word embedding for twitter sentiment classification. In: Proceedings of the 52nd Annual Meeting of the Association for Computational Linguistics (vol. 1: Long Papers), pp. 1555–1565 (2014)

4. Zhou, G., He, T., Zhao, J., et al.: Learning continuous word embedding with metadata for question retrieval in community question answering. In: Proceedings of the 53rd Annual Meeting of the Association for Computational Linguistics and the 7th International Joint Conference on Natural Language Processing (vol. 1: Long Papers), pp. 250–259 (2015)

5. Bengio, Y., Ducharme, R., Vincent, P., et al.: A neural probabilistic language model. J. Mach. Learn. Res. **3**(Feb), 1137–1155 (2003)

6. Mnih, A., Hinton, G.E.: A scalable hierarchical distributed language model. In: Advances in Neural Information Processing Systems, pp. 1081–1088 (2009)

7. Mikolov, T., Chen, K., Corrado, G., et al.: Efficient estimation of word representations in vector space. arXiv preprint arXiv:1301.3781 (2013)

8. Sun, Y., Lin, L., Yang, N., Ji, Z., Wang, X.: Radical-enhanced Chinese character embedding. In: Loo, C.K., Yap, K.S., Wong, K.W., Teoh, A., Huang, K. (eds.) ICONIP 2014. LNCS, vol. 8835, pp. 279–286. Springer, Cham (2014). https://doi.org/10.1007/978-3-319-12640-1_34

9. Li, Y., Li, W., Sun, F., et al.: Component-enhanced Chinese character embeddings. arXiv preprint arXiv:1508.06669 (2015)

10. Collobert, R., Weston, J.: A unified architecture for natural language processing: deep neural networks with multitask learning. In: Proceedings of the 25th International Conference on Machine Learning, pp. 160–167. ACM (2008)

11. Chen, X., Xu, L., Liu, Z., et al.: Joint learning of character and word embeddings. In: IJCAI, pp. 1236–1242 (2015)

12. Xu, J., Liu, J., Zhang, L., et al.: Improve Chinese word embeddings by exploiting internal structure. In: Proceedings of the 2016 Conference of the North American Chapter of the Association for Computational Linguistics: Human Language Technologies, pp. 1041–1050 (2016)

13. Yu, J., Jian, X., Xin, H., et al.: Joint embeddings of Chinese words, characters, and fine-grained subcharacter components. In: Proceedings of the 2017 Conference on Empirical Methods in Natural Language Processing, pp. 286–291 (2017)

14. Mikolov, T., Sutskever, I., Chen, K., et al.: Distributed representations of words and phrases and their compositionality. In: Advances in Neural Information Processing Systems, pp. 3111–3119 (2013)

15. Mikolov, T., Yih, W., Zweig, G.: Linguistic regularities in continuous space word representations. In: Proceedings of the 2013 Conference of the North American Chapter of the Association for Computational Linguistics: Human Language Technologies, pp. 746–751 (2013)

Syntax Enhanced Research Method of Stylistic Features

Haiyan Wu and Ying Liu[✉]

Tsinghua University, Beijing, China
wuhy17@mails.tsinghua.edu.cn, yingliu@mail.tsinghua.edu.cn

Abstract. Nowadays, research on stylistic features (SF) mainly focuses on two aspects: lexical elements and syntactic structures. The lexical elements act as the content of a sentence and the syntactic structures constitute the framework of a sentence. How to combine both aspects and exploit their common advantages is a challenging issue. In this paper, we propose a Principal Stylistic Features Analysis method (PSFA) to combine these two parts, and then mine the relations between features. From a statistical analysis point of view, many interesting linguistic phenomena can be found. Through the PSFA method, we finally extract some representative features which cover different aspects of styles. To verify the performance of these selected features, classification experiments are conducted. The results show that the elements selected by the PSFA method provide a significantly higher classification accuracy than other advanced methods.

Keywords: Style · Lexical and syntactic features
Feature dimension reduction

1 Introduction

In recent years, research on SF has attracted much attention, and many are its applications. Researchers have tried to utilize different ways to study SF, especially statistical methods to text-mine writers' writing style [5,14,18] or a text's potential information [2,7,8,16], achieving quite good performances.

So far, these studies mainly focus on the lexical level and researches have been mainly conducted by calculating the frequency of a limited set of features, such as most commonly used nouns, verbs, adjectives, adverbs, prepositions, conjunctions, high-frequency words, word-length (WL), etc. [1–3,12,15,17]. Nevertheless, it is somewhat inaccurate to understand words solely based on their literal meaning without taking into account their difference nuances. This is especially true in Chinese, where words and their meanings may differ in different context. One attempted solution was to apply the N-Gram model [11,19]

This work is supported by Beijing Social Science Fund (16YYB021) and Project of Humanities and Social Sciences of Ministry of Education in China (17YJAZH056).

M. Sun et al. (Eds.): CCL 2018/NLP-NABD 2018, LNAI 11221, pp. 12–23, 2018.
https://doi.org/10.1007/978-3-030-01716-3_2

and further study texts' style analyzing their syntactic structure [8]. However, due to sparse word collocation, new problems emerged during the calculating process. To address this problem, [13] put forward the concept of word embedding: this method quantifies the words and makes the calculation process easier to analyze, however, it also generates a huge amount of computations. Another solution was formulated by [9] who proposed to analyze SF by using features, such as Sent-Length (SL), questions, declarative sentences, and phrase defaults. Moreover, [4] studied SF and their distribution by calculating the frequency of syntactic structures in a text. Scientific practice shows that good results can be achieved also with other model algorithms applied to the lexicon or combination of the lexicon, or to the syntactic structure of SF.

With the rapid development of deep learning [10,20] achieved excellent results in applying deep learning methods to SF research. However, this does not mean that research in this area has reached a conclusive stage. [21] has pointed out that "there is still little knowledge about the internal operation and behavior of these complex models, or how they achieve such good performance. From a scientific standpoint, this is deeply unsatisfactory". So we still need an efficient way to explore different stylistic features. In this paper, we are going to propose a Principal Stylistic Features Analysis method (PSFA) for an extensive collection of lexical feature sets and syntactic structure features in different styles, which can scientifically and efficiently discover the characteristics of style variation. Meanwhile, PSFA has proved to be effective in various stylistic studies, and that its classification accuracy is significantly higher than other current advanced methods. The main contributions of this work are summarized as following.

1. We take both the lexical and syntactic features into consideration to analyze different text styles.
2. Principal Stylistic Feature Analysis (PSFA) method is proposed to discover typical stylistic features. It helps to mine and visualize internal relationships between features and reveal the principal stylistic characteristic of a text.
3. Experiments show that features selected with the help of PSFA achieve good performance in the stylistic classification task.

2 Related Work

Compared with the previous research work, we have extracted a lexical feature set, which not only includes all the lexical features of the Part-of-speech tags of the Penn Chinese Treebank[1], but also One-Syllable (OS) and Two-Syllable (TS) words, etc. We also have extracted syntactic structure features by analyzing each sentence as a unit, not just as an individual syntactic structure. This is very different from the research work of [4]. Moreover, our method is based on the extrapolation of the sentence components' syntactic structures and their combinations with the lexicon. Its advantage lies in filtering out some highly relevant but redundant words and syntactic structures. This new method helps

---

[1] http://www.cs.brandeis.edu/~clp/ctb/posguide.3rd.ch.pdf.

to reduce repetition between features, it reflects the core SF more concisely [6], and filters out those features with less apparent differences at the beginning, retaining only the most representative features to analyze different text styles Some highly correlated features are filtered out too, and unnecessary features are as well gradually removed. Every step we do is to ensure that the selected elements are the most representative ones.

3   Principal Stylistic Feature Analysis

Figure 1 shows the feature analysis process applied in this paper. Firstly, through *Features Extraction*, we extract the features of the lexical and the syntactic structure. Secondly, we utilize the T-test method to find out which characteristics are different among these three styles. Again, through *Internal Relations Mining*, we dig out the distinctive features of each style. Finally, in order to prove the validity of our method, we filter out as many features as the number of features included in the baseline. The number of features selected by our baseline here is 18, so we filter out 18 features, however, our method can adapt to baselines presenting any number of different features.

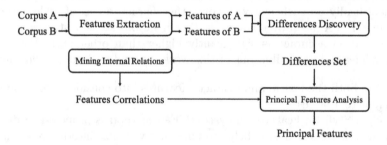

Fig. 1. Process of principal stylistic feature analysis

3.1   Feature Extraction

As in any stylistic analysis, a sentence is made up of words and syntactic structures, and only the combination of these two parts can fully express the complete meaning of the sentence. However, any style is composed of sentences, so for different styles, we can study from the composition of sentences. Our focus is not exclusively on the sentence structure, the lexical is also important. Based on these two considerations, we study the stylistic features from these two aspects: the lexical and syntactic structure. To comprehensively selecting the stylistic elements at lexical-level, we consider not only word types, WL, OS and TS words but also words that can reflect the main stylistic characteristics. In fact, lexical features can reflect the richness of the vocabulary in some ways, while the Pos-word-ratio (PWR) demonstrate how vocabulary is used in a specific writer's work. Moreover, the analysis of Single-word can also reflect the writer's vocabulary richness. Likewise, at the syntactic-level, on the one hand we examine SL, declarative sentences, interrogative sentences and exclamatory sentences, while, on the other, we also explore syntactic structures of verbs, nouns, prepositional, adjectival and adverbial phrases, etc. In the following, we are going to introduce some definitions in a more formal way (Table 1).

Table 1. Penn Treebank part-of-speech (POS) tags

Tag	Description	Tag	Description
AD	Adverbs	IJ	Interjection
AS	Aspect marker	LB	in long bei-construction
BA	in ba-const	LC	Localizer
CD	Cardinal numbers	M	Measure word (including classifiers)
CS	Subordinating conj	MSP	Some particles
DT	Determiner	NR	Proper nouns
ON	Onomatopoeia	OD	Ordinal numbers
SB	in long bei-construction	SP	Sentence-final particle
VC	Copula "是"	VE	"有" as the main verb
CLP	Classifier phrase	DP	Determiner phrase

3.2 Difference Discovery

The t-test[2] is a statistical tool which can be used to detect whether there are differences between two samples. Assume there are two independent samples $X_1 = \{x_i | i = 1 \ldots n_1\}$ and $X_2 = \{x_j | j = 1 \ldots n_2\}$, we can get t-value as following.

$$t = \frac{\bar{X}_1 - \bar{X}_2}{\sqrt{\frac{(n_1-1)s_1^2 + (n_2-1)s_2^2}{n_1 + n_2 - 2}\left(\frac{1}{n_1} + \frac{1}{n_2}\right)}} \tag{1}$$

$\bar{X}_1$ and $\bar{X}_2$ represent the mean the sample X_1 and the sample X_2. S_1 and S_2 respectively represent as the variance of sample X_1 and sample X_2. n_1 and n_2 respectively represents n_1 and n_2 sample sizes. Here, t-test is used to detect which features are significantly different in the two stylistics according to p-value which is calculated from t-value. Here, we select those features with p-value less than 0.01 as the features we want.

3.3 Mining Internal Relations

To excavate the intrinsic relations between the different lexical or syntactic features. We calculate the correlation coefficient (CC) of features[3] between the

---

[2] https://en.wikipedia.org/wiki/Student%27s_t-test.
[3] https://en.wikipedia.org/wiki/Correlation_coefficient.

elements, CC is ranging from -1 to 1. When CC equals to -1, the two features are completely opposites. When, instead, CC equals to 1, the two features are exactly the same. For the CC value between -1 and 1, the absolute value of CC indicates the degree of similarity between features, and the larger the CC value, the higher the similarity; conversely, the lower the value, the lower the feature similarity. Among them, calculation of CC formula is as shown below.

$$\gamma = \frac{\sum_i (x_i - \bar{x}_\iota)(y_i - \bar{y}_\iota)}{\sqrt{\sum_i (x_i - \bar{x}_\iota)^2}\sqrt{\sum_i (y_i - \bar{y}_\iota)^2}} \tag{2}$$

Here, $\bar{x}_\iota$ and $\bar{y}_\iota$ represent the mean of sample X and sample Y. We want to judge whether and how the two samples relate to each other. Clustering[4] CC to judge whether it relates the two features. For example, some words often exist in pairs, such as adjectives and nouns (white skirts), and adverbs and verbs (e.g., lowly climb).

3.4   Principal Feature Analysis

We first assume that there are features A and B, obtained by calculating CC of A and B according to formula 2. Then, through clustering, we obtain the Correlation Cluster Graph G. According to G, we filtered the elements according to the following rule:

- If A and B show high correlation, to reduce redundancy, we choose only one of them. The feature selection between A and B is made in the following way: when A or B associates with other characteristics, we choose the one that shows minimal correlation to other as the candidate feature. In case A or B show little or no relation to other features, A and B are both excluded from selection.
- If A and B show little or no correlation to each other, we need to examine the other characteristics whether associated to A and B respectively, and repeat the first step of the feature selection process.
- Examples of the above two rules will be given in Sect. 4.3.

4   Experiment

To verify the performance of PSFA, we conduct experiments to answer the following questions:

RQ1. Is there any difference between different styles with respect to lexical and syntactic features? If yes, which differences are significant?

RQ2. What is the intrinsic relationship between stylistic features?

RQ3. Can we distinguish different stylistics through the PSFA method?

---

[4] https://en.wikipedia.org/wiki/Hierarchical_clustering.

4.1   Experimental Settings

Datasets. Our experiment is mainly based on Chinese corpus: Novels, News, and Email. Detailed information on the datasets are shown in Table 2.

Table 2. Statistics of evaluation datasets

Dataset	Text size (MB)	Text files#	Sentences#	Docs#
Novel	6.5	10	75,713	2,546
News	16.2	7,479	186,510	8,789
Email	165.5	64,620	441,489	18,148

- **Novels.** *Novels corpus* includes ten selected contemporary novels, such as: Hua Yu's *"To Live"*, Han Han's *"Triple Gate"*, She Lao's *"The Yellow Storm"*, Jingming Guo's *"Never-flowers in never-dream"*, Zhongshu Qian's *"Fortress Besieged"*, Yan Mo's *"Red Sorghum"*, Congwen Shen's *"cities on the border"*, Man Gu's *"Silent Separation/My Sunshine"*, Bihua Li's *"Rouge"*, etc. All of them are well-known novels, and can be downloaded on the Internet.
- **News.** *Sina-News corpus* is made up of news texts crawled from the following website http://news.sina.com.cn/, which includes 7,479 events. Most of them are social news.
- **Email.** *Email corpus*[5] is from TREC 2006 Spam Track Public Corpora. It contains 64,620 emails, and we use 21,766 ham emails. They are from daily communication, and most of them are about sharing the personal experience.

To ensure the purity of texts, we first separate the documents into sentences using pyltp[6], and then, we utilize kits[7] to segment words, part of speech tagging and parse sentences. To eliminate the effects of text length, sentences are combined into batches with around of 500 words which are regarded as documents. Numbers of documents are shown in the Table 2.

4.2   Difference Discovery (RQ1)

With the method of *Differences Discovery* illustrated in Sect. 3.2, we have obtained different stylistic features, as shown in Table 3.

To distinguish between the significance of each kind of stylistic characteristics, for the lower three columns in the middle of Table 3, we calculate the mean of each character, and then draw the following conclusions. Accordingly, for each group, we analyze and explain the difference between two aspects of the lexical and the syntactic structure.

---

[5] https://plg.uwaterloo.ca/~gvcormac/treccorpus06/.
[6] https://github.com/HIT-SCIR/pyltp.
[7] https://nlp.stanford.edu/software/.

Table 3. Differences and Principle features

	Novel & News	Novel & Email	News & Email	Principle
Lexical	Sent-Length	Sent-Length	NT	Start-AD
	One-Syllable	One-Syllable	NN	CS
	LC	Start-AD	LC	IJ
	NN	VC	IJ	VC
	NT	NT	CS	NT
		Start-PN	Sent-Length	Sent-length
		Start-PU	One-Syllable	One-syllable
				NN
	DER	*BA*	*M*	LC
	DEV	*DEV*	*LB*	Declare
			DEC	Start-PU
				Start-PN
Syntax	VP-[DER, VP]	VP-[DER, VP]	QP-[CD, CLP]	IP-[PP, NP]
	QP-[CD, CLP]	IP-[NP, VP]	VP-[PP, VP]	VP-[DER, VP]
	IP-[PP, NP]	VP-[PP, VP]	IP-[PP, NP]	IP-[NP, VP]
		IP-[PP, NP]		VP-[PP, VP]
		DP-[DT, CLP]		DP-[DT, CLP]
				QP-[CD, CLP]
	VP-[VP, AS]	*VP-[VP, AS]*	*VP-[ADVP, VP]*	
	VP-[VP, VP]	*LCP-[NP, LCP]*	*VP-[VP, NP]*	
	VP-[ADVP, VP]	*VP-[NP, VP]*	*VP-[VP, VP]*	
	NP-[NP, NP]	*IP-[ADVP, VP]*	*CP-[IP, SP]*	
	LCP-[IP, LCP]	*PP-[PP, IP]*	*IP-[ADVP, VP]*	
	CP-[IP, SP]	*VP-[ADVP, VP]*	*IP-[IP, VP]*	
	PP-[PP, IP]	*CP-[IP, SP]*		
		VP-[VP, NP]		

1. Italic ones are examples of features which are not selected as the principal features.

Lexical-Level-Analysis of Novel Corpus. Main features are found: OS, LC, NT and DEV. The novel is a literary style that involves characters, events, a complete story and the specific environmental descriptions. Generally speaking, the novels are written to develop a plot, so NT is a quite prominent feature. Besides, novels contain dialogues and brief scene descriptions, as well as many OS words. Characters descriptions are more frequent in novels, and V is commonly used together with AD, DEV and AS to represent a particular state of characters movement. So DEV, AD, and AS are also prominent features of this style. Due to frequent descriptions and variety of themes, word types are more diversified in novels. From the viewpoint of *Difference Discovery*, PWR is quite significant.

Lexical-Level-Analysis of News Corpus. Main features are found: SL, NN, VC and LC. News speak with facts and provide information as authentic as possible. To ensure accuracy, authenticity, conciseness, news often employ time lines and specific figures to state facts, and they are more time-sensitive. Therefore CD NN are prominent words in the news. Moreover, to report facts, long sentences are preferred to shorter ones. Thus, WL and SL are also prominent features in News, IJ too, since IJ serves the purpose to indicate the accuracy of the words reported.

Lexical-Level-Analysis of Email Corpus. These features are found: Start-AD, CS, Declare, and PN. AS communication tool, Emails exchange information between two parties. They are usually written in a narrative tone, for example, asking for information or help, or notifying an event, etc. Emails usually involve NR or PN that are familiar to both the sender and the receiver for discussion. Usually, the register is quite colloquial, and CS are more frequently used in Email. As for the syntactic structures, we have made a similar calculation, and the results are analyzed as follows.

Syntax-Level-Analysis of Novel Corpus. VP-[VP, NP], LCP-[NP, LCP], QP-[CD, CLP] and PP-[PP, IP] are main characteristics. As we all know, compared with News and Email, Novels are more about describing characters, their state, location and actions. Novels frequently show the structures of VP-[VP, NP] and LCP-[NP, LCP], which can better express the characteristics of Novels. Therefore, VP-[VP, NP] and LCP-[NP, LCP] are the most significant syntactic features in Novels. Novels usually contain descriptions of the environment or scenery, so they require quantifier phrases or prepositional phrases, such as QP-[CD,CLP] and PP-[PP,IP], which are also significant syntactic features for Novels.

Syntax-Level-Analysis of News Corpus. There are VP-[VP, AS], NP-[QP, NP], PP-[PP, LCP], IP-[PP, VP], IP-[NP, VP], VP-[PP, VP] and VP-[DER, VP]. News is different from Novels and Emails. In fact, as a medium for reporting facts, its primary task is to describe event-related content truthfully. Therefore, we expect the combined phrase of the verb phrase in the news to be its primary syntactic structure features. After performing the *Difference Discovery*, we found that the News is mainly based on IP and some verb phrases that indicate the state of the action, e.g., VP-[VP, AS]. So IP phrases and VP-[VP, AS] are the main syntactic structure characteristics in News.

Syntax-Level-Analysis of Email Corpus. These features are as the main characteristics in Email: VP-[NP, VP], VP-[ADVP, VP], PP-[PP, NP], IP-[ADVP, VP], IP-[PP, NP], LCP-[IP, LCP], IP[IP, VP], CP-[IP, SP], DP-[DT, CLP]. Email takes narrative utterances to describe events objectively. Therefore, Emails mainly consist of common syntactic structures (IP). As illustrated in Differences Discovery in Sect. 4.2, we found that in Email corpus, IP syntactic structures account for the majority of all structures. Besides, some verbal phrases are used in Emails, in particular when talking about discussion topics of the two sides, the progress made on a project and the extent of this progress. Therefore, verbal phrases are also an essential feature in Email.

4.3   Internal Relation Mining (RQ2)

In the following, using data visualization, we analyze and interpret the internal relations of lexical features among Novels, News, and Email, as shown in Fig. 2. At the same time, we will be filtering out relevant but not distinctive stylistic elements.

Lexical-Level Analysis. From Fig. 2, we find that WL and TS are negatively correlated, while CD and M are positively correlated, and LB and BA are not significantly correlated. WL has a high negative correlation with both TS and OS, as they involve the calculation formula of WL. In the text, if there are many TS words, then WL will become small. The relation between WL and OS is analogous. Since that WL and TS are also highly correlated to other features, OS is selected as the candidate feature to reduce the overall redundancy of feature selection. Moreover, CD and M are positively correlated, because they always exist together to qualify and define the noun. To reduce their correlation, considering the minimum correlation to other characteristics, M is selected as a candidate feature. As for LB and BA, they are independent to other features. That is, their correlation with other features is relatively little, and to balance the characteristics we selected, they are both excluded from selection. Ask and Declare are negatively correlated. In a text, even when many interrogative sentences are present, the corresponding declarative sentences are relatively small, thus they are negatively related. Here we have chosen declarative sentences as the candidate feature. Similar reasons for selection are adopted for PU and Start-PU, AD and Start-AD, PWR and NN, NR and NT are correlated, because, in the News, NT and NR usually appear in pairs to describe an event after comparison with other features, we have selected NN. Also, we have removed some

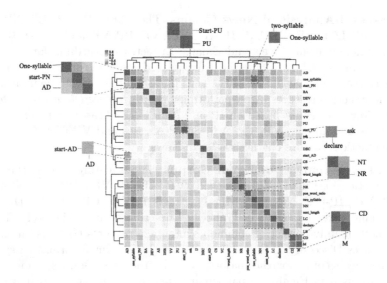

Fig. 2. Correlation analysis and mining internal relations of features

features which showed little relationship with other, such as VV, AS, etc. The remaining, selected lexical features are not in italics, as shown in Table 3.

The selection principle for syntactic structure features and lexical-syntactic structure features is the same as for lexical feature selection, and we'll not repeat its procedure here. Figure 2 shows an example of feature selection. Here, we only give linguistic relevance explanations for these two groups.

Syntax-Level Analysis. We found that IP-[PP, VP] and IP-[PP, NP] are highly correlated. By combining these two structures, it can be expressed locations, actions, and they are frequently used in pairs in articles. Since they are strongly related, and only one of them can be chosen, here we have selected IP-[PP, NP]. PP-[PP, LCP] and LCP-[NP, LCP], since these two phrases are questions and answers. AS PP-[PP, LCP] means "where?", and LCP-[NP, LCP] also means "in what". They are often used in dialogues. Since we can only choose one of them, here, LCP-[NP, LCP] is selected as it is associated with other features. For similar reasons, in the following the pair of phrases IP-[IP, CP] and IP-[IP, SP], DVP-[VP, DEVP]and VP-[DER, VP], IP-[LCP, VP] and LCP-[IP, LCP], NP-[QP, NP] and QP-[CD, CLP], PP-[PP, NP] and VP-[PP, VP], we filter out IP-[IP, CP], DVP-[VP, DEV], IP-[LCP, VP], NP-[QP, NP] and PP-[PP, NP]. The remaining, selected syntactic features are not in italics, as shown in Table 3.

Cross-Filtering of the Lexical and Syntactic Structure. We find that NP-[NP, NP] and NN are highly correlated. In the texts, as NP-[NP, NP] and NN both represent a noun phrase, we can select only one of them and thus NN. M and QP-[CD, CLP] are also highly correlated because they both represent quantifiers, thus only QP-[CD, CLP] is selected. Similarly, DER and VP-[DER, VP] are highly correlated, as they are both adverbs and verbs, and only VP-[DER, VP] is included in the selection. Combinations of VP-[VP, NP] and VP-[NP, VP] can appear in sentence such as "who is doing what", so there is a high correlation between them. These phrases are prevalent in articles, so we also removed them. Most of the time, IP-[ADVP, VP] and VP-[DER, VP] represent the same structure, and only VP-[DER, VP]is included in our selection. There are also a few not syntactic structures, such as VP-[VP, VP], PP-[PP, IP] and IP-[IP, VP]. Furthermore, some words that have no practical meaning, such as DEC, are removed. The remaining, selected lexical-syntactic features are not in italics, as shown in Table 3.

4.4   Verification by Classification (RQ3)

In linguistics, the study of stylistic features is a quite ambitious project. In the second section, we found the lexical and syntactic structural features that can distinguish different styles through the PSFA method, which is to say, we have identified lexical features and syntactic structures with linguistic discrimination. To verify that these lexical and syntactic structural features are distinguishable, we conducted the classification experiment. The classifier we choose is SVM[8]. We

---

[8] http://scikit-learn.org/stable/modules/generated/sklearn.svm.SVC.html.

set up the kernel function as "linear", "C = 0.8", other parameters are applied by default.

To ensure an unbiased comparison, in our analysis, we used only 18 principal features, the same number as the baseline. The remained elements are shown in the last column of Table 3. 2,546 documents are sampled in each corpus to form a balanced dataset. The comparison results are shown in Table 4.

Table 4. Accuracy of classification in 5-fold cross-validation.

	Baseline	Lexical	Syntax enhanced
Accuracy	86.40%	87.73%	**89.16%***

*. significantly better than baselines ($p < 0.05$).

From Table 4, at lexical-level, we found that lexical features selected through the PSFA method are better than the baseline used in classification, which implies that the PSFA method is effective. Similarly, the combination of the lexical and the syntactic features perform better than the lexical features and the baseline too, which further demonstrates that the PSFA approach is effective.

5   Conclusion

In this paper, we have proposed a new method called Principal Stylistic Features Analysis method (PSFA) to study features of three styles, our method combines both the lexical and syntactic features. The PSFA method is not complicated, its starting point is the elemental composition of a sentence: the lexical and syntactic structure. It digs into the natural features of different stylistic combinations of words and syntactic structures, which are necessary for understanding languages. From the viewpoint of statistical analysis, many interesting linguistic phenomena have been found. The PSFA method finally provide some representative features which cover different aspects of stylistics. It can scientifically provide a reasonable explanation for different text styles. In future work, we will pay more attention to comprehensively study the differences of features across different languages from lexicon, syntax, semantics, and context. Deep learning model will help us to find out more characteristic information.

References

1. Ahmad, M., Nadeem, M.T., Khan, T., Ahmad, S.: Stylistic analysis of the 'muslim family laws ordinance 1961'. J. Study Engl. Linguist. **3**(1), 28–37 (2015)
2. Ashraf, S., Iqbal, H.R., Nawab, R.M.A.: Cross-genre author profile prediction using stylometry-based approach. In: CLEF (Working Notes), pp. 992–999 (2016)
3. Bird, H., Franklin, S., Howard, D.: Age of acquisition and imageability ratings for a large set of words, including verbs and function words. Behav. Res. Methods Instrum. Comput. **33**(1), 73–79 (2001)

4. Booten, K., Hearst, M.A.: Patterns of wisdom: discourse-level style in multi-sentence quotations. In: Proceedings of the 2016 Conference of the North American Chapter of the Association for Computational Linguistics: Human Language Technologies, pp. 1139–1144 (2016)
5. Chen, J., Huang, H., Tian, S., Qu, Y.: Feature selection for text classification with Naïve Bayes. Expert Syst. Appl. **36**(3), 5432–5435 (2009)
6. Griffiths, T.L., Steyvers, M., Blei, D.M., Tenenbaum, J.B.: Integrating topics and syntax. In: Advances in Neural Information Processing Systems, pp. 537–544 (2005)
7. Kumar, S., Kernighan, B.: Cloud-based plagiarism detection system performing predicting based on classified feature vectors. US Patent 9,514,417 (2016)
8. Lahiri, S., Vydiswaran, V.V., Mihalcea, R.: Identifying usage expression sentences in consumer product reviews. In: Proceedings of the Eighth International Joint Conference on Natural Language Processing (vol. 1: Long Papers), pp. 394–403 (2017)
9. Liu, Q.: Research on stylistic features of the English international business contract. DEStech Trans. Soc. Sci. Educ. Hum. Sci. (MSIE) (2017)
10. Majumder, N., Poria, S., Gelbukh, A., Cambria, E.: Deep learning-based document modeling for personality detection from text. IEEE Intell. Syst. **32**(2), 74–79 (2017)
11. Mishne, G., et al.: Experiments with mood classification in blog posts. In: Proceedings of ACM SIGIR 2005 Workshop on Stylistic Analysis of Text for Information Access, vol. 19, pp. 321–327 (2005)
12. Niu, X., Carpuat, M.: Discovering stylistic variations in distributional vector space models via lexical paraphrases. In: Proceedings of the Workshop on Stylistic Variation, pp. 20–27 (2017)
13. Pavlick, E., Rastogi, P., Ganitkevitch, J., Van Durme, B., Callison-Burch, C.: PPDB 2.0: better paraphrase ranking, fine-grained entailment relations, word embeddings, and style classification. In: Proceedings of the 53rd Annual Meeting of the Association for Computational Linguistics and the 7th International Joint Conference on Natural Language Processing (vol. 2: Short Papers), pp. 425–430 (2015)
14. Pervaz, I., Ameer, I., Sittar, A., Nawab, R.M.A.: Identification of author personality traits using stylistic features: notebook for PAN at CLEF 2015. In: CLEF (Working Notes) (2015)
15. Ruano San Segundo, P.: A corpus-stylistic approach to dickens' use of speech verbs: beyond mere reporting. Lang. Lit. **25**(2), 113–129 (2016)
16. Santosh, D.T., Babu, K.S., Prasad, S., Vivekananda, A.: Opinion mining of online product reviews from traditional LDA topic clusters using feature ontology tree and sentiwordnet. IJEME **6**, 1–11 (2016)
17. Saparova, M.: The problem of stylistic classification of colloquial vocabulary. **5**(1), 80–82 (2016)
18. Schler, J., Koppel, M., Argamon, S., Pennebaker, J.W.: Effects of age and gender on blogging. In: AAAI Spring Symposium: Computational Approaches to Analyzing Weblogs, vol. 6, pp. 199–205 (2006)
19. Szymanski, T., Lynch, G.: UCD: diachronic text classification with character, word, and syntactic n-grams. In: Proceedings of the 9th International Workshop on Semantic Evaluation (SemEval 2015), United States (2015)
20. Wang, L.: News authorship identification with deep learning (2017)
21. Zeiler, M.D., Fergus, R.: Visualizing and understanding convolutional networks. In: Fleet, D., Pajdla, T., Schiele, B., Tuytelaars, T. (eds.) ECCV 2014. LNCS, vol. 8689, pp. 818–833. Springer, Cham (2014). https://doi.org/10.1007/978-3-319-10590-1_53

Addressing Domain Adaptation for Chinese Word Segmentation with Instances-Based Transfer Learning

Yanna Zhang, Jinan Xu[✉], Guoyi Miao, Yufeng Chen,
and Yujie Zhang

School of Computer and Information Technology, Beijing Jiaotong University,
Beijing, China
{yannazhang, jaxu, gymiao, chenyf, yjzhang}@bjtu.edu.cn

Abstract. Recent studies have shown effectiveness in using neural networks for Chinese Word Segmentation (CWS). However, these models, constrained by the domain and size of the training corpus, do not work well in domain adaptation. In this paper, we propose a novel instance-transferring method, which use valuable target domain annotated instances to improve CWS on different domains. Specifically, we introduce semantic similarity computation based on character-based n-gram embedding to select instances. Furthermore, training sentences similar to instances are used to help annotate instances. Experimental results show that our method can effectively boost cross-domain segmentation performance. We achieve state-of-the-art results on Internet literatures datasets, and competitive results to the best reported on micro-blog datasets.

Keywords: Chinese word segmentation · Domain adaptation
Instance-transferring · Neural network

1 Introduction

Chinese word segmentation (CWS) is a preliminary and important task for many Chinese natural language processing (NLP) tasks. Recently, neural word segmentation has shown promising progress [1–4]. However, these neural network models, mainly trained by supervised learning, rely on massive labeled data. In recent years, large-scale human annotated corpora mainly focus on domains like newswire, the word segmentation performance trained on these corpora usually degrades significantly when the test data shift from newswire to micro-blog texts and Internet literatures [5, 6]. Such a problem is well known as domain adaptation [7]. Usually, the domain of training and testing data is called source and target domain respectively.

There are severe challenges to solving the problem of domain adaptability. On one hand, the In-Vocabulary (IV) word in different domains has different contexts and semantics, which affect the performance of word segmentation on target domain. On the other hand, many domain-related words in target domain that rarely appear in source domains. Therefore, Out-of-Vocabulary (OOV) word recognition becomes an important problem. Take the sentence "普泓 上人 点头，又 看 了 鬼厉 一 眼，转身 便

M. Sun et al. (Eds.): CCL 2018/NLP-NABD 2018, LNAI 11221, pp. 24–36, 2018.
https://doi.org/10.1007/978-3-030-01716-3_3

要 走 了 出去。" for example, the word "鬼厉" is the name of a person that often appears in literatures-related domains while seldom appears in other domains.

Table 1. The People's Daily (PD) is source data, the Micro-blog and "Zhuxian" (ZX) are target data. "Words" is the amount of words in different corpus. "Cover" indicates the percentage of words in target domain that can be covered by source domain.

Datasets	Words	Cover
PD	21653284	–
ZX	96934	80.48%
Micro-blog	421166	70.62%

As listed in Table 1, we count up the numbers of words in different domains and find that about 30% of the words in Micro-blog texts (target domain) are not available in People's Daily (source domain). As a result, it is necessary to develop new methods for addressing the problem of domain adaptability.

Instance-based transfer learning proves to be an excellent fit for alleviating the above two problems [8, 21]. In this paper, we propose a new instance-based transfer learning method, which effectively alleviate the OOV words recognition problem by adding the similar target domain labeled data to the training data. First, we obtain samples from large-scale unlabeled target domain data according to sampling modules, which introduces character-based n-gram embedding to calculate the similarity between two sentences. Second, we train an initial segmentation model with source domain data to annotate samples and then revise the segmentation result with the help of training data. Our proposed method is simple, efficient and effective, giving average 3.5% the recall of OOV words on target domain data.

The contributions of this paper could be summarized as follows.

- The semantic similarity is first introduced to address the problem of adaptation domain for CWS, which is effective to select useful target domain instances (Sect. 3.2).
- We propose a new semantic similarity calculation method with the help of character-based n-gram embedding, which incorporates rich contextual information (Sect. 3.2).
- Training sentences similar to instances are used to construct partial annotated instances, which rectify partial improper annotation caused by word segmentation model (Sect. 3.3).

2   Related Work

Our work focus on domain adaptation for neural word segmentation, mainly using the character-based n-gram contextual information.

Neural Word Segmentation. Most modern CWS methods treated CWS as a sequence labeling problems. There has been a recent shift of research attention in the word segmentation literature from statistical methods to deep learning. Pei et al. [1] used Convolutional Neural Network (CNN) to capture local information within a fixed size window and proposed a tensor framework to capture the information of previous tags. Chen et al. [2] proposed Gated Recursive Neural Network (GRNN) to model feature combinations of context characters. Subsequently, Cai et al. [3] proposed a gated combination neural network which can utilize complete segmentation history. Moreover, Chen et al. [4] adopted an adversarial multi-criteria learning method to integrate shared knowledge from multiple heterogeneous segmentation criteria. However, these neural word segmentation methods rely on a large-scale labeled data which is usually expensive and tends to be of a limited amount.

Transfer Learning in CWS. Transfer learning aims to learn knowledge from different source domains to enhance the word segmentation performance in a target domain. Transfer learning includes domain adaptation, which has been successfully applied to many fields. In particular, several methods have been proposed for solving domain adaptation problem in CWS. Lin et al. [21] proposed a simple yet effective instance-based transfer learning method, which employs a double-selection process to reduce the impact of harmful data in the source. Similar to Lin et al., this paper uses instance-based transfer learning method to solve the domain adaptation problem for segmentation. However, unlike his data selection method, we select samples with abundant target domain feature with the help of use the similarity calculation. Liu et al. [6] considered self-training and character clustering for domain adaptation. Liu et al. [8] proposed a variant CRF model to leverage both fully and partially annotated data transformed from different sources of free annotations consistently. Zhang et al. [9] used domain specific tag dictionaries and only unlabeled target domain data to improve target-domain accuracies. Furthermore, Huang et al. [10] proposed a novel BLSTM-based neural network model which incorporates a global recurrent structure designed for modeling boundary features dynamically. Xu et al. [11] trained a teacher model on source corpora and then use the learned knowledge to initialize a student model, which can be trained on target data by a weighted data similarity method.

The Contextual Information. During CWS, a character or a word usually has different meaning in different positions, which is called the ambiguity of the character or the word. Blitzer et al. [12] indicated that much of ambiguity in character or word meaning can be resolved by considering surrounding words. Inspired by the above, many scholars tried to solve some NLP problems by adding contextual information. Choi et al. [13] used the context-dependent word representation to improve the performance of machine translation. Qin et al. [14] presented a neural model with context-aware character-enhanced embeddings to address implicit discourse relation recognition task. Bao et al. [15] integrated the contextualized character embedding into neural word segmentation to capture the useful dimension in embedding for target domain. Zhou et al. [16] proposed word-context character embeddings (WCC), which contain the label distribution information. Motivated by that, we concatenate the contextual embeddings to calculate the similarity of two sentences.

3   Methods

3.1   Instances-Based Transfer Learning for CWS

Instances-based transfer learning can be regarded as a learning method which continuously increases training data. Our method consists of two main stages. In the first stage, we obtain instances from large-scale unlabeled target domain sentences according to sampling strategy (See Sect. 3.2). In the second stage, we train an initial segmentation model with source domain data to annotate instances and then revise the segmentation result (See Sect. 3.3). These revised labeled instances are added to the training data to retrain the model. Figure 1 illustrates the framework of our instances-based transfer learning training process for CWS. Finally, we can obtain the optimal word segmentation model by continuous iteration, which apply to target domain word segmentation task.

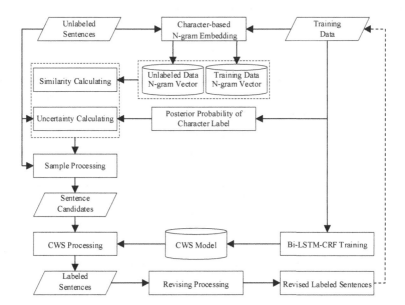

Fig. 1. Framework of instances-based transfer learning training process for CWS.

3.2   Obtaining Unlabeled Target Domain Instances

Character-Based n-gram Embedding. Each character can be represented as a d-dimensional vector by using word2vec [18] toolkit. In order to incorporate the contextual information into vector representation of a character, we define the character-based n-gram embedding. As it is believed that the contextual information from a window size of 5 characters may be sufficient for CWS [19], we employ a max window size of 5 characters to generate the character-based n-gram embedding. For each character x_i in a given input sentence $X = (x_1, x_2, \ldots, x_n)$, it's 5-gram embedding is $e_i^5 = [x_{i-2}, x_{i-1}, x_i, x_{i+1}, x_{i+2}]$, which refers to a concatenate of 5 character embeddings around i-th character. We set the window size to 1, 3 and 5 respectively, and get three kinds of representations of this character.

The Similarity Between Two Sentences. To extract useful samples from the large-scale unlabeled target domain corpus, we propose to calculate the character-based semantic similarity between training data and unlabeled target domain data. Formally, the character-based semantic similarity is defined as follows:

$$\text{Sim}(s, \tilde{s}) = \sum_{j=1}^{3} w_j \, cos < \vec{s}, \, \overrightarrow{\tilde{s}} > \tag{1}$$

where $\vec{s}$ denotes target domain sentence vector and $\overrightarrow{\tilde{s}}$ indicates source domain sentence vector. $cos < \vec{s}, \, \overrightarrow{\tilde{s}} >$ denotes the Cosine distance between two vectors. w_j is the weight of each kind of sentence vector representation. We define the n-gram sentence vector $\vec{s}$ as:

$$\vec{s} = \frac{1}{n} \sum_{i=1}^{n} e_i^T \tag{2}$$

where e_i^T represents the character-based n-gram embedding of the i-th character. Here are three kinds of sentence vectors due to the different values of T, which is window size. n denotes the length of a sentence.

The Uncertainty of Annotation. Higher the uncertainty of an instance's annotation, more useful features the instance contains. In our work, therefore, we choose the instances with higher uncertainty in annotation task.

Additionally, the word segmentation can be represented as a two-class problem according to one factor, that is, whether the character is the right boundary of a word in a sentence. Specifically, the labels B, M, E, S can be divided into two categories: B and M can be grouped together, denoted as N, indicating that the character isn't the right boundary of a word; Also, E and S can be grouped and denoted as Y, indicating that the character is the right border of a word and need to be divided.

We use information entropy to measure the uncertainty of every character label and it is formally denoted as:

$$E(c) = - \sum_{i=N,Y} p_i log_2 p_i \tag{3}$$

where, c is a character in instances, $p_N = p_B + p_M$, $p_Y = p_E + p_S$, p_B represents the posterior probability that c is marked as B. p_M, p_E and p_S are similar to p_B.

Sample Process. As the training data and testing data are sampled from different distributions, CWS models cannot learn enough features for training data. In general, if a sample sentence is more similar with sentences of the training set, the sample sentence contains more OOV words. For example, "为了 还 这份 人情, 冯小刚 这 才 答 应 下来。" in source domain data and "田灵儿 这 才 悻悻然 下来。" in target domain data, "田灵儿" and "悻悻然" are both OOV words. Therefore, we select more similar target domain sentence. According to the above analysis, selecting instances relies on two aspects: The uncertainty of annotation and the semantic similarity between training data and unlabeled sentences. In general, the scoring model for selecting instances is finally defined as:

$$D(s) = \lambda \sum_{i=1}^{n} E(c_i) + (1 - \lambda)maxSim(s, \tilde{s}) \tag{4}$$

where, s represents a sentence in target domain data, n indicates the character number of this sentence, c_i represents the character in a sentence, $\tilde{s}$ is a training sentence, λ is a weight parameter.

3.3 Obtaining Annotated Result of Instances

To annotate instances precisely, we train an initial model on large-scale source domain data firstly, which is used to annotate instances.

Bi-LSTM-CRF Architecture for CWS. CWS task is usually solved by character-level sequence labeling algorithm. Specifically, each character in a sentence is labeled as one of $L = \{B, M, E, S\}$, indicating the begin, middle, end of a word, or a word with a single character. For a given sentence $X = (x_1, x_2, \ldots, x_n)$ containing n characters, the aim of CWS task is to predict label sequence $y^* = (y_1, y_2, \ldots, y_n)$. The Bi-LSTM-CRF architecture (our baseline) for CWS is illustrated in Fig. 2.

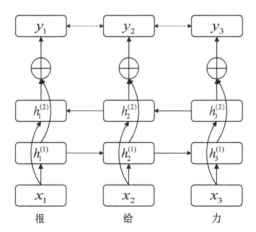

Fig. 2. Bi-LSTM-CRF neural architecture for CWS. Character vector representation are given as input; a bidirectional LSTM produces context-dependent representations; the information is passed through a hidden layer and an output layer. The outputs are confidence scores for CRFs.

Embedding Layer. Similar to other neural models, the first step is to represent characters in distributed vectors. In this work, the vector represent of a character is a concatenation of two parts: character embeddings $e^c(x_i)$ and character-bigram embeddings $e^b(x_i, x_{i+1})$. For each character x_i in a given input sentence $X = (x_1, x_2, \ldots, x_n)$, we regard $e_i = \left[e^c(x_i), e^b(x_i, x_{i+1})\right]$ as the vector representation of the i-th character.

Feature Layers. In order to incorporate information from both sides of the sequence, we use bidirectional Long Short-Term Memory (Bi-LSTM) as feature layers. The update of each Bi-LSTM unit can be described as follows:

$$h_i = \overrightarrow{h_i} \oplus \overleftarrow{h_i} \tag{5}$$

where $\overrightarrow{h_i}$ and $\overleftarrow{h_i}$ are the hidden states at position i of the forward and backward LSTMs respectively; $\oplus$ is concatenation operation.

Inference Layer. Following Lample et al. [17], we employ conditional random fields (CRF) layer to inference labels, which is beneficial to consider the dependencies of adjacent labels. For example, a B (begin) label should be followed by a M (middle) label or E (end) label, and a M label cannot be followed by a B label or S (single) label. Given that y is a label sequence $y_1, y_2, \ldots, y_n$, then the CRF score for this sequence can be calculated as:

$$s(X, y) = \sum_{i=0}^{n} A_{y_i, y_{i+1}} + \sum_{i=1}^{n} P_{i, y_i} \tag{6}$$

where A is a matrix of transition scores such that $A_{i,j}$ represents the score of a transition from the tag i to tag j. y_0 and y_n are the start and end tags of a sentence, that we add to the set of possible tags. A is therefore a square matrix of size $k + 2$. P is the fractional matrix of the Bi-LSTM network's output. The size of P is $n \times k$, where k is the number of labels to be predicted, $P_{i,j}$ corresponds to the j-th label of the i-th word in a sentence.

The output from the model is the tagging sequence with the largest score s(y). A softmax over all possible tag sequences yields a probability for the sequence y:

$$p(y|X) = \frac{e^{s(X,y)}}{\sum_{\tilde{y} \in Y_x} e^{s(X,y)}} \tag{7}$$

Train Strategy. Finally, we directly maximize the log-probability of the correct tag sequence:

$$\log(p(y|X)) = s(X, y) - \log\left(\sum_{\tilde{y} \in Y_X} e^{s(X,\tilde{y})}\right) = s(X, y) - logadd_{\tilde{y} \in Y_X}(X, \tilde{y}) \tag{8}$$

While decoding, the output sequence we predict will have the highest score given by the following:

$$y^* = argmax_{\tilde{y} \in Y_X} s(X, \tilde{y}) \tag{9}$$

We use the Viterbi algorithm to find the optimal labeled sequence during training.

Revising Labeled Instances. Word segmentation systems trained on source domain often suffer a rapid decrease in performance when they are used in target domain. To

mitigate the errors resulting from model segmentation scheme, we revise model segmentation results with the help of training data.

Training sentences similar to instances can be used to get rid of partial implausible
annotation, the three most similar training sentences is therefore selected for every
instance. We employ "Top-N" algorithm to choose them according to semantic similarity calculated in Sect. 3.2. Then we revise the segmentation results by comparing
these three sentences with segmentation instances. Specifically, the same fragments in
the three sentence and the instance are modified to be the same annotation. If the
annotation of a fragment in the three sentences is inconsistent, the most similar sentence shall prevail. The rest of the instance utilize the annotation result of the model.

Table 2. Statistical information of the three datasets.

Dataset		Sentence	Words
PD	Train	441943	21653284
	Dev	15000	394202
	Test	15000	401063
ZX	Train	2373	96934
	Dev	788	20393
	test	1,394	34355
Micro-blog	train	20,135	421,166
	Dev	2052	43697
	Test	8,592	187,877

4 Experiments

4.1 Datasets

Source Domain Data. In this paper, we use the People's Daily[1] (PD) (2014) drawn
from news domain for the source-domain training. We regard the random different 15k
sentences from PD as development and test sets respectively, and the rest are treated as
training sets.

Target Domain Data. The NLPCC 2016 dataset[2] [20] and "Zhuxian"[3] (hereafter
referred to as ZX) are used as the target domain data. The NLPCC 2016 dataset is
selected to evaluate our methods on micro-blog texts. Unlike the popular used newswire dataset, the NLPCC 2016 dataset is collected from Sina Weibo, which consists of
the informal texts from micro-blog with the various topics, such as finance, sports,
entertainment, and so on. ZX is an Internet novel and has a different writing style
comparing to PD. In addition, ZX also contains many novel specific named entity.
Table 2 gives the details of three datasets.

---

[1] https://pan.baidu.com/s/1hq3KKXe.

[2] https://github.com/FudanNLP/NLPCC-WordSeg-Weibo.

[3] This book can be download from https://www.qisuu.la/Shtml812.html.

Target Domain Unlabeled Data. We use micro-blog unlabeled data built from the Internet for free. After filtering special characters and removing duplication, 2.2 million sentences are reserved. For ZX, after some long sentence segmentation, we utilize about 60k sentences the rest of complete ZX novel data as ZX unlabeled data except for ZX test set.

Both recall of out-of-vocabulary words (R-oov) and F1-score are used to evaluate the segmentation performance.

4.2   Hyper-parameter Settings

The hyper-parameters are tuned according to the experimental results. The detailed values are shown in Table 3.

Table 3. Hyper-parameters settings.

Hyper-parameter	Value
Embedding dimension	100
LSTM hidden size	200
LSTM input size	200
Batch size	128
Learning rate	0.01
Dropout	0.5

4.3   Experimental Results

The Value of Parameter λ. We first investigated the impact of λ in formula (9) over segmentation performance. The parameter λ is searched ranging from 0.1 to 1 with a step size of 0.1. We put λ into the formula (9) to calculate the sampling scores and select the most valuable samples. 1000 samples were selected in the experiment. The results of PD to Micro-blog and ZX datasets are shown in Fig. 3. From this figure we can see that setting λ as 0.5 gives the highest F1-score for every corpus. So the parameter λ is set to 0.5.

Effect of Samples Selection. To verify the effectiveness of our sampling strategy, we conduct a comparative experiment which choose different number of samples randomly. We apply the sampling strategy proposed in Sect. 3.2 to the target domain data. The annotation work of unlabeled target domain data is based on the automatically labeled sentences by our baseline model trained with PD corpus. The sentences are automatically annotated and then revised with the help of training data. Experiment results are shown in Fig. 4. It can be seen that with the increase of the number of datasets, the performance becomes better. It shows that our proposed methods can learn continuously knowledge from selected instances. We can see that our method achieves better performance compared to the method with randomly target domain samples, demonstrating that sampling strategy is helpful to improving the domain adaption

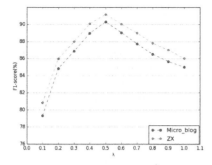

Fig. 3. The results of our methods under different parameter λ on two target domain datasets.

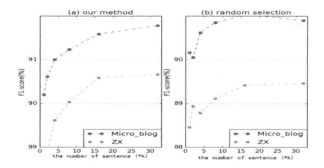

Fig. 4. F1-score on target domain data when adding different numbers of target domain data.

Table 4. The result when 16 K target-domain sentences are added.

Datasets	F1-score	F1-score (unrevised)	R-oov
Micro-blog	91.78	91.40	74.62
ZX	90.66	90.35	83.56

accuracy. And the best result (91.78%) is achieved with 32 K target-domain sentences in micro-blog data set in our experiment.

In addition, to examine whether OOV recognition can benefit from our methods, we also look into the OOV recalls of the ZX dataset. Table 4 show that the proposed samples selection methods can effectively improve the recall of OOV words, which empirically proves its domain adaption ability. The main reason is our method can select sentences similar to the source domain without destroying the original distribution of the training data and alleviate the recognition problem of OOV words.

Effect of Revising Annotation. In order to examine the real effect of the revised methods, we also set a comparison experiment using the baseline segmentation results, as shown in the second column of Table 4. It can be seen that the performance of revised instances is better than the unrevised instances, which indicates that our revised methods can help to correct partially incorrect word segmentation results.

Character-Based and Word-Based n-gram Embedding Comparison. In addition, we add a comparative experiment by using word-based semantic similarity method. The results are shown in Table 5, where the first row shows the performance of our baseline and the second row shows the performance of the method utilizing word-based n-gram embedding, the third row shows the performance of the character-based selection method. Compared the word-based semantic similarity representations, our method performs better with character-based n-gram embedding. Similar conclusion is obtained when adapting from PD to ZX data set. It shows that character-based models have the potential of capturing morpheme patterns, thereby improving generalization ability of word segmentation models. Furthermore, F1-score in character-based experiments increases by 2.19% than our baseline.

Comparisons with State-of-the-Art Models. In this section, we compare our methods with previous advanced methods. As shown in Table 6, our work achieves state-of-the-art results. Although our method is simple, it outperforms the other methods which are very competitive. Since Liu et al. [8] and Zhang et al. [9] used the domain dictionary, our work is not comparable with them. Although contextual information is both used, our method gives better result than Bao et al. [15], showing that our method integrate context more effectively.

Table 5. Comparison with word-based and character-based semantic similarity.

Methods	ZX	Micro-blog
Baseline	88.47	89.90
Word-based	90.35	91.33
Character-based	**90.66**	**91.78**

Table 6. Comparisons with state-of-the-art methods on target domain datasets.

Models	ZX	Micro-blog
Liu et al. [6]	83.99	–
Qiu et al. [5]	90.2	–
Bao et al. [16]	89.35	–
Our work (words-based)	90.35	91.33
Our work (character-based)	**90.66**	**91.78**

5   Conclusion

The performance of CWS can drop significantly when the test domain is different from the training domain. In this paper, we propose a novel instances-based transfer learning method for CWS. We select useful instances containing the higher labeled values to the training set, and the labeled value is calculated by the help of character-based n-gram embedding. The model can be trained by adding training data iteratively to obtain better generalization ability. In our experiments, we evaluated two methods of semantic similarity computation: character-based and word-based. The experimental results on the Micro-blog and ZX dataset fully show that our method is especially effective for segmenting OOV words and enhancing the performance of CWS on different domains.

Acknowledgments. The authors are supported by National Nature Science Foundation of China (Contract 61370130 and 61473294), and the Fundamental Research Funds for the Central Universities(2015JBM033), and International Science and Technology Cooperation Program of China under grant No. 2014DFA11350.

References

1. Pei, W., Ge, T., Chang, B.: Max-margin tensor neural network for Chinese word segmentation. In: The 52nd Annual Meeting of the Association for Computational Linguistics, pp. 293–303, Baltimore, Maryland (2014)
2. Chen, X., Qiu, X., Zhu, C., Huang, X.: Gated recursive neural network for Chinese word segmentation. In: The 53rd Annual Meeting of the Association for Computer Linguistics, pp. 1744–1753 (2015)
3. Cai, D., Zhao, H.: Neural word segmentation learning for Chinese. In: The 54th Annual Meeting of the Association for Computer Linguistics, pp. 409–420 (2016)
4. Chen, X., Shi, Z., Qiu, X., Huang, X.: Adversarial multi-criteria learning for Chinese word segmentation, pp. 1193–1203 (2017)
5. Qiu, L., Zhang, Y.: Word segmentation for Chinese novels. In: AAAI, pp. 2440–2446 (2015)
6. Liu, Y., Zhang, Y.: Unsupervised domain adaptation for joint segmentation and POS-tagging. In: Proceedings of COLING 2012, Posters, pp. 745–754. The COLING 2012 Organizing Committee (2012)
7. Daume, H., Marcu, D.: Domain adaptation for statistical classifiers. J. Artif. Intell. Res. **26**, 101–126 (2006)
8. Liu, Y., Zhang, Y., Che, W., Liu, T., Wu, F.: Domain adaptation for CRF-based Chinese word segmentation using free annotations. In: EMNLP (2014)
9. Zhang, M., Zhang, Y., Che, W., Liu, T.: Type-supervised domain adaptation for joint segmentation and pos-tagging. In: EACL, pp. 588–597(2014)
10. Jiang, W., Huang, L., Liu, Q., Lü, Y.: A cascaded linear model for joint chinese word segmentation and part-of-speech tagging. In: Meeting of the Association for Computational Linguistics, pp. 897–904, 15–20 June 2008, Columbus, Ohio, USA. DBLP (2008)
11. Xu, J., Ma, S., Zhang, Y., Wei, B., Cai, X., Sun, X.: Transfer deep learning for low-resource chinese word segmentation with a novel neural network. In: Huang, X., Jiang, J., Zhao, D., Feng, Y., Hong, Yu. (eds.) NLPCC 2017. LNCS (LNAI), vol. 10619, pp. 721–730. Springer, Cham (2018). https://doi.org/10.1007/978-3-319-73618-1_62
12. Blitzer, J., McDonald, R., Pereira, F.: Domain adaptation with structural correspondence learning. In: EMNLP, pp. 120–128 (2006)
13. Choi, H., Cho, K., Bengio, Y.: Context-dependent word representation for neural machine translation. Comput. Speech Lang. **45**, 149–160 (2016)

14. Qin, L., Zhang, Z., Zhao, H.: Implicit discourse relation recognition with context-aware character-enhanced embeddings. In: The 26th International Conference on Computational Linguistics (COLING), Osaka, Japan, December (2016)

15. Bao, Z., Li, S., Gao, S., Xu, W.: Neural domain adaptation with contextualized character embedding for Chinese word segmentation. In: Huang, X., Jiang, J., Zhao, D., Feng, Y., Hong, Yu. (eds.) NLPCC 2017. LNCS (LNAI), vol. 10619, pp. 419–430. Springer, Cham (2018). https://doi.org/10.1007/978-3-319-73618-1_35

16. Zhou, H., Yu, Z., Zhang, Y., Huang, S., Dai, X.: Word-context character embeddings for chinese word segmentation. In: Conference on Empirical Methods in Natural Language Processing, pp. 760–766 (2017)

17. Lample, G., Ballesteros, M., Subramanian, S., Kawakami, K., Dyer, C.: Neural architectures for named entity recognition, pp. 260–270 (2016)

18. Mikolov, T., Chen, K., Corrado, G., et al.: Efficient estimation of word representations in vector space. arXiv preprint arXiv:1301.3781 (2013)

19. Zheng, X., Chen, H., Xu, T.: Deep learning for Chinese word segmentation and POS tagging. In: Proceedings of the 2013 Conference on Empirical Methods in Natural Language Processing, pp. 647–657. Association for Computational Linguistics (2013)

20. Qiu, X., Qian, P., Shi, Z.: Overview of the NLPCC-ICCPOL 2016 shared task: Chinese word segmentation for micro-blog texts. In: Lin, C.-Y., Xue, N., Zhao, D., Huang, X., Feng, Y. (eds.) ICCPOL/NLPCC -2016. LNCS (LNAI), vol. 10102, pp. 901–906. Springer, Cham (2016). https://doi.org/10.1007/978-3-319-50496-4_84

21. Lin, D., An, X., Zhang, J.: Double-bootstrapping source data selection for instance-based transfer learning. Pattern Recognit. Lett. **34**(11), 1279–1285 (2013)

Machine Translation

Collaborative Matching for Sentence Alignment

Xiaojun Quan[1]([⊠]), Chunyu Kit[2], and Wuya Chen[1]

[1] School of Data and Computer Science, Sun Yat-sen University, Guangzhou, China
`quanxj3@mail.sysu.edu.cn, chenwy58@mail2.sysu.edu.cn`
[2] Department of Linguistics and Translation, City University of Hong Kong,
Kowloon Tong, Hong Kong
`ctckit@cityu.edu.hk`

Abstract. Existing sentence alignment methods are founded fundamentally on sentence length and lexical correspondences. Methods based on the former follow in general the length proportionality assumption that the lengths of sentences in one language tend to be proportional to that of their translations, and are known to bear poor adaptivity to new languages and corpora. In this paper, we attempt to interpret this assumption from a new perspective via the notion of collaborative matching, based on the observation that sentences can work collaboratively during alignment rather than separately as in previous studies. Our approach is tended to be independent on any specific language and corpus, so that it can be adaptively applied to a variety of texts without binding to any prior knowledge about the texts. We use one-to-one sentence alignment to illustrate this approach and implement two specific alignment methods, which are evaluated on six bilingual corpora of different languages and domains. Experimental results confirm the effectiveness of this collaborative matching approach.

Keywords: Sentence alignment · Machine translation

1 Introduction

Sentence alignment has been extensively studied as a first step towards more ambitious natural language processing tasks such as statistical machine translation [2] and cross-language information retrieval [12]. The task is to identify translational correspondence between bilingual sentences in parallel text, also called bitext. The correspondence can then be taken as input to, for example, produce translational correspondence between bilingual words or phrases so as to build a machine translation model. Besides sentence length, lexical clues are also resorted to facilitate sentence alignment, yet they are not necessarily available from scenario to scenario. Thus, a main stream of research in text alignment

The paper was supported by the Program for Guangdong Introducing Innovative and Enterpreneurial Teams (No. 2017ZT07X355).

M. Sun et al. (Eds.): CCL 2018/NLP-NABD 2018, LNAI 11221, pp. 39–52, 2018.
https://doi.org/10.1007/978-3-030-01716-3_4

remains not counting on lexical information but instead on the exploitation of length proportionality between mutual translations. The proportionality follows the observation that longer sentences in one language are likely to be translated into longer sentences in another language, and so are short ones. Statistical evidence also validates that the correlation of sentence lengths between two languages is relatively high [5].

Table 1. Estimation of normal distribution parameters (mean and variance) across corpora of different languages.

		Eng-Chi$_1$	Eng-Chi$_2$	Eng-Fre	Eng-Ger	Eng-Por	Eng-Spa
Mean	Char	0.520	0.386	1.156	1.188	1.108	1.097
	Word	1.066	0.868	1.060	0.940	1.026	1.054
Var	Char	0.045	0.029	0.300	0.180	0.040	0.041
	Word	0.110	0.118	0.050	0.040	0.049	0.051

Sentence length based alignment was initially studied in cognate languages [1,5], and then applied to non-cognate languages [16]. Although the length correlation for non-cognate languages is not as high as that between cognate languages, this length based approach works fairly well. However, a drawback of length-based alignment arises from its insufficient adaptivity to new languages and corpora, because the relied distribution parameters have to be estimated on a bilingual corpus of two specific languages and are not suitable for others, especially unpopular languages. Table 1 illustrates this with distributional parameters estimated on six bilingual corpora, including two English-Chinese corpora and four corpora of English and French, German, Portuguese, and Spanish, covering various domains such as legislation, news and proceedings. Sentence lengths are measured in number of characters and words. The result shows that these parameters vary significantly not only across different language pairs but also across different corpora of the same languages pair. In practice, they need to be specifically customized towards suitable corpora in order to enable the alignment to work effectively.

Inspired by the same proportionality assumption, this paper attempts to go beyond the conventional way and propose the notion of collaborative matching to model length proportionality during alignment. This novel notion takes into account the synergies between sentences, measuring how likely two sequences of sentences should be aligned, instead of treating them separately as in previous works. Based on this notion, two new approaches to one-to-one sentence alignment are developed, one as an approximate solution to exhaustive search and the other built upon differential collaborative similarity. Note that the collaborative matching idea also applies to many-to-many alignment after proper formulation, yet its implementation is much more complicated. That is why we choose to use one-to-one alignment to illustrate our idea in this paper. Besides, we also provide new insights into the measurement of sentence length and present a self-information based approach. The new alignment methods are evaluated on six bilingual corpora of different languages and domains. Experimental results validate their effectiveness regardless whichever sentence length measurements are used, indicating the stability and adaptivity of the new collaborative matching.

2   Problem Statement

This section introduces basic notations and notions of sentence alignment using sentence length, to lay a background for formulating the idea of collaborative matching.

2.1   Notation

A sentence alignment algorithm takes a corpus of bitexts as input, which is comprised of a set of source-language sentences, $\mathcal{S} = \{s_1, s_2, \ldots, s_M\}$, and another set of target-language sentences, $\mathcal{T} = \{t_1, t_2, \ldots, t_N\}$. Its task is to find the translational correspondences between these two sets of sentences. Let $\mathcal{L}_\mathcal{S} = \{l_{s_1}, l_{s_2}, \ldots, l_{s_M}\}$ and $\mathcal{L}_\mathcal{T} = \{l_{t_1}, l_{t_2}, \ldots, l_{t_N}\}$ represent the respective sets of sentence lengths. Let $\mathcal{A} = \{a_1, a_2, \ldots, a_T\}$ be an alignment of $\mathcal{S}$ and $\mathcal{T}$, where $a_i = (\mathcal{S}_i; \mathcal{T}_i)$ for $\mathcal{S}_i \subset \mathcal{S}$ and $\mathcal{T}_i \subset \mathcal{T}$ corresponds to a pairing and T is the number of pairings produced. In the previous work by [1], $\mathcal{S}_i$ and $\mathcal{T}_i$ are also called "bead". For many-to-many alignment type (e.g., 2-2), $\mathcal{S}_i$ and $\mathcal{T}_i$ are each composed of multiple sentences to be merged together as a "bead".

2.2   Sentence Length

As discussed above, there are two natural ways to measure sentence length, namely in number of characters vs. words. Basically, both ways focus on the surface length of sentence and drop other essential information such as word identities. Given that the basic objective of translation is to render the meaning of a text of one language in another language, the equivalence of meaning is the ultimate means for text alignment. Length-based approaches resort to sentence length as a source of "information" about the meaning, assuming that it can be retained to a certain degree after translation.

However, we argue that such "information" can be measured in a different way so as to make the best use of implicit lexical clues indicated by the distribution of words. Our assumption is that if two sentences are translation of one another, they should have a similar distribution of words in their respective texts. Accordingly, this distribution can be used to quantify the amount of "meaning" of a sentence. Specifically, suppose given a word w_i with probability of $Pr(w_i)$, we measure its length in terms of the amount of its self-information $-\log Pr(w_i)$ [4]. Analogically, the length of a sentence is the sum of accumulated self-information over all its words. The rationale of this new length measure can be illustrated by a statistical analysis on an English-Chinese corpus as in Fig. 1, showing that this information based measure gives a higher correlation of sentence length than the other two.

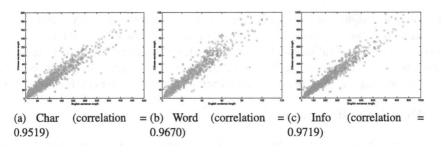

(a) Char (correlation = 0.9519)   (b) Word (correlation = 0.9670)   (c) Info (correlation = 0.9719)

Fig. 1. Correlation analysis of sentence length on an English-Chinese corpus. The length is respectively measured in number of characters and words and in amount of information.

2.3 Sentence Alignment

Most existing approaches to sentence alignment follow the monotonicity assumption that coupled sentences in bitexts appear basically in a similar sequential order in two languages, and employ dynamic programming for global optimization to produce the final output of alignment. Sentence alignment can be formulated as the following optimization

$$\mathcal{A}^* = \arg\max_{\mathcal{A}} \sum_{a_i \in \mathcal{A}} \log P(a_i), \tag{1}$$

where $P(a_i)$ is a probabilistic score measuring the likelihood of aligning sentences $\mathcal{S}_i$ and $\mathcal{T}_i$ in a_i in terms of their lengths. In [5] and [1], $P(a_i)$ is calculated by means of the sentence length ratio in two languages that is assumed to follow a normal distribution, with mean μ and variance σ^2 to be estimated on real data.

3   Collaborative Matching

This section formulates our collaborative matching and illustrates its application to sentence alignment. A simple example is first given to illustrate the basic concept.

3.1   An Illustrative Example

Consider a scenario where we need to decide the translations of given sentences based solely on their lengths. Initially, suppose we are given a source-language sentence of length 4, as well as two candidates of translation with length 5 and 6, respectively. Without any additional knowledge, it is impossible to decide between the two candidates which is the potential translation. If one more sentence is added to each side of the three sets, say with respective length 5, 6, and 4, the answer becomes a bit clearer according to the length proportionality assumption. The source-language sentences (with length 4 and 5) are more likely to be translated into the target-language sentences with length 5 and 6 than that with 6 and 4. If another three sentences are given, say with lengths 6, 7 and 2, there should be little doubt about the correct alignment. Highest possibility is that source-language sentences with lengths {4, 5, 6} are aligned with sentences with lengths {5, 6, 7}, rather than that with {6, 4, 2}.

From this example, we can see that sentences from the same language can work collaboratively during alignment rather than separately as in previous works. The collaboration is in fact a collective reflection of the length proportionality and can be intuitively demonstrated in the Euclidean space, in which the lengths of sentences from one language are assembled into a vector. The vector of true translations, that collectively satisfy the length proportionality better, have smaller angle (i.e., higher similarity) with the vector of source sentences. This gives a strong motivation for collaborative matching to measure how translations and their source texts fit each other.

3.2   Collaborative Similarity

The above example shows a new perspective on the use of length proportionality for sentence alignment. To this end, we propose the notion of collaborative similarity to measure the likelihood of aligning two equal-sized (this restriction will be released later) sequences of sentences in terms of their length vectors. In particular, we propose to use a trigonometric function to estimate the similarity score. Among many possibilities, we use cosine function to produce a score monotonic with the likelihood. Specifically, the collaborative similarity of two sequences of sentences, $\hat{S}$ and $\hat{T}$, is defined as

$$C(\hat{S}, \hat{T}) = \frac{\sum_i l_{\hat{S}_i} l_{\hat{T}_i}}{\sqrt{\sum_i l_{\hat{S}_i}^2} \sqrt{\sum_i l_{\hat{T}_i}^2}}. \tag{2}$$

Then sentence alignment becomes a task of finding a set of aligned sentences with the maximum collaborative similarity from given bitext, as

$$(\hat{S}, \hat{T})^* = \arg \max_{(\hat{S}, \hat{T})} C(\hat{S}, \hat{T}). \tag{3}$$

Once $\hat{S}$ and $\hat{T}$ are aligned one by one sequentially to give the final 1-1 alignment, the remaining sentences will be treated as 1-0/0-1 type of alignment.

The differences of this approach from the previous alignment approaches are as follows. Given the prior knowledge (e.g., normal distribution) about the length proportionality, the previous approaches search for an alignment with the highest possibility. They can be regarded as a "supervised" alignment in the sense that their alignment results are produced subjecting to some prior knowledge. Our approach, however, is unable to align two single sentences straightforwardly but needs to rely on the synergies between sentences, which appears to be "unsupervised" in sense of using little prior knowledge.

3.3   Two Alignment Approaches

In our framework, sentence alignment becomes a task through exhaustive search of all possible alignments for one with the highest collaborative similarity, yet this is impractical due to the heavy computational cost. This section formulates two approximate approaches to resolving this issue.

Near-Optimal Alignment. The first approach employs the divide-and-conquer strategy to find a near optimal solution through the following inference from Eq. 3.

$$
\begin{aligned}
(\hat{\mathcal{S}}, \hat{\mathcal{T}})^* &= \underset{(\hat{\mathcal{S}}, \hat{\mathcal{T}})}{\arg\max} \; \frac{\sum_{i=1}^{\hat{T}} l_{\hat{\mathcal{S}}_i} l_{\hat{\mathcal{T}}_i}}{\sqrt{\sum_{i=1}^{\hat{T}} l_{\hat{\mathcal{S}}_i}^2} \sqrt{\sum_{i=1}^{\hat{T}} l_{\hat{\mathcal{T}}_i}^2}} \\[2mm]
&= \underset{(\hat{\mathcal{S}}, \hat{\mathcal{T}})}{\arg\max} \left(\frac{\sum_{i=1}^{\hat{T}-1} l_{\hat{\mathcal{S}}_i} l_{\hat{\mathcal{T}}_i}}{\sqrt{\sum_{i=1}^{\hat{T}} l_{\hat{\mathcal{S}}_i}^2} \sqrt{\sum_{i=1}^{\hat{T}} l_{\hat{\mathcal{T}}_i}^2}} + \frac{l_{\hat{\mathcal{S}}_{\hat{T}}} l_{\hat{\mathcal{T}}_{\hat{T}}}}{\sqrt{\sum_{i=1}^{\hat{T}} l_{\hat{\mathcal{S}}_i}^2} \sqrt{\sum_{i=1}^{\hat{T}} l_{\hat{\mathcal{T}}_i}^2}} \right) \\[2mm]
&\approx \underset{(\hat{\mathcal{S}}, \hat{\mathcal{T}})}{\arg\max} \left(\frac{\sum_{i=1}^{\hat{T}-1} l_{\hat{\mathcal{S}}_i} l_{\hat{\mathcal{T}}_i}}{\sqrt{\sum_{i=1}^{\hat{T}-1} l_{\hat{\mathcal{S}}_i}^2} \sqrt{\sum_{i=1}^{\hat{T}-1} l_{\hat{\mathcal{T}}_i}^2}} + \frac{l_{\hat{\mathcal{S}}_{\hat{T}}} l_{\hat{\mathcal{T}}_{\hat{T}}}}{\sqrt{\sum_{i=1}^{\hat{T}} l_{\hat{\mathcal{S}}_i}^2} \sqrt{\sum_{i=1}^{\hat{T}} l_{\hat{\mathcal{T}}_i}^2}} \right) \quad (4)
\end{aligned}
$$

That is, the final alignment can be approximately derived from the alignment of the preceding steps and the sentence pair under consideration in the current step. Correspondingly, the alignment can be performed via dynamic programming as follows. Let $W(i,j)$ be the maximum collaborative similarity that can be derived between sentences $s_1, ..., s_i$ and their translations $t_1, ..., t_j$. Following the idea of near-optimal alignment, it is the max of the following three cases:

$$
W(i,j) = \max \begin{cases} W(i-1,j) & + & 0 \\ W(i,j-1) & + & 0 \\ W(i-1,j-1) + & \dfrac{l_{\hat{\mathcal{S}}_{\hat{T}}} l_{\hat{\mathcal{T}}_{\hat{T}}}}{\sqrt{\sum_{i=1}^{\hat{T}} l_{\hat{\mathcal{S}}_i}^2} \sqrt{\sum_{i=1}^{\hat{T}} l_{\hat{\mathcal{T}}_i}^2}} \end{cases} \quad (5)
$$

The first two cases have zero second terms because the second term of Eq. 4 is 0 for a 1-0/0-1 alignment. The dynamic programming procedure is initialized with zero $W(\cdot, 0)$ and $W(0, \cdot)$ and then proceeds to fill every cell of matrix W. The final alignment is retrieved by tracing W backwards starting from W_{MN}.

Differential Collaborative Similarity. The collaborative similarity of two sequences of adjacent sentences from two languages can actually be used to find a set of neatly aligned sentences. Yet the actual alignment is intertwined with 1-0/0-1 type of alignment, making it infeasible to perform the entire alignment by virtue of the collaborative similarity of adjacent sentences. Nevertheless, high-quality initial alignment can be used as certain "benchmark" to measure how likely two new bilingual sentences are to be aligned. To this end, we propose the differential collaborative similarity. Before that, we first define three items associated with the initial alignment $\hat{A}$. Let $d_{\hat{A}} = \sum_{\hat{a}_i \subset \hat{A},\ \hat{S}_i \in \hat{a}_i,\ \hat{T}_i \in \hat{a}_i} l_{\hat{S}_i} l_{\hat{T}_i}$, $q_{\hat{A}} = \sum_{\hat{a}_i \subset \hat{A},\ \hat{S}_i \in \hat{a}_i} l_{\hat{S}_i}^2$ and $r_{\hat{A}} = \sum_{\hat{a}_i \subset \hat{A},\ \hat{T}_i \in \hat{a}_i} l_{\hat{T}_i}^2$. Then, the differential collaborative similarity of two bilingual sentences, s_i and t_j, is defined as

$$w_{ij} = \frac{d_{\hat{A}} + l_{s_i} l_{t_j}}{\sqrt{q_{\hat{A}} + l_{s_i}^2} \sqrt{r_{\hat{A}} + l_{t_j}^2}} - \frac{d_{\hat{A}}}{\sqrt{q_{\hat{A}}} \sqrt{r_{\hat{A}}}}. \tag{6}$$

The rationale behind this measure is that the derived initial alignment is comprised of paired sentences highly reflecting the length proportionality. It is not unreasonable to assume that a new sentences pair should increase or at least preserve the collaborative similarity of an existing initial alignment if they are true translation of each other. This will contribute a positive differential collaborative similarity score. Otherwise, the score is negative. With this new similarity measure, the alignment can be performed by setting each $W(i, j)$ as

$$W(i, j) = \max \begin{cases} W(i-1, j) & + & \gamma \\ W(i, j-1) & + & \gamma \\ W(i-1, j-1) + w_{ij} \end{cases} \tag{7}$$

where γ is a penalty parameter, necessary in general in various sequence alignment tasks and has to be determined empirically. Similar dynamic programming as the above one for near-optimal alignment can be adopted straightforwardly to produce the final alignment using this similarity measure.

4 Evaluation

This section reports the evaluation of our new alignment approaches on six bilingual corpora, with Gale and Church's length based alignment approach as baseline for comparison, which is the most classic length-based alignment algorithm and serves as the foundation of most more advanced approaches and tools.

4.1 Data Sets

The following six bilingual corpora of different languages and domains are used for evaluation.

BLIS. Bilingual Laws Information System (BLIS)[1] is an electronic text database of Hong Kong legislation. BLIS provides English-Chinese bilingual texts of ordinances and subsidiary legislation and organizes the legal texts into a hierarchy of chapters, sections, parts, subsections, paragraphs and subparagraphs. This

---

[1] http://www.legislation.gov.hk.

corpus has been used in a number of previous works on sentence alignment and other machine translation tasks [7,14]. 175 are randomly selected from the original 31,401 bitexts for manual alignment by two experts. Then, sentences are identified based on punctuations, resulting in a set of 1619 1-1 and 71 1-0/0-1 sentence pairs.

LDC. This corpus is comprised of English-Chinese news stories collected via Sinorama Magazine, Taiwan, from 1976 to 2004.[2] Of the original corpus of 365,568 sentence pairs, 27 bitexts of 2041 1-1 and 564 1-0/0-1 pairings are randomly selected for use in our evaluation.

Europarl Parallel Corpus. The original corpus is extracted from the proceedings of the European Parliament with versions in 21 European languages [8]. It is aligned at sentence level between English and other 20 languages. For this evaluation, 4 of the 20 aligned corpora of French, German, Portuguese, and Spanish are chosen. They are the cognate languages with the most widely used bitext data in the field. Each raw corpus includes more than 2 millions sentence pairs. Among them, 5000 pairs are randomly selected from each, with 1000 sentences from two sides are randomly discarded so as to form 1-0/0-1 type of alignment. Noted that there may be also pairs of 1-1 aligned sentences discarded during this process.

In our experiment, sentence length will be firstly measured in number of characters and number of word tokens. For the former, each cognate language character or punctuation is counted as 1, while each Chinese character is 2. For the latter, the length is measured in number of word tokens as segmented by spaces and punctuations. For Chinese, since there are no marked word boundaries, word segmentation needs to be performed. To measure sentence length in self-information of words, it is intuitively more desirable to stem them into their root forms, which, however, appears to be impractical as multiple languages are involved here while most existing stemming works focus only on the English language. For this reason, there will be no stemming performed for all the corpora.

4.2 Adaptivity of the Baseline

This part examines the adaptivity of Gale and Church's algorithm to the situation of inaccurate estimation of parameters. For this purpose, we first derive normal distribution parameters (mean and variance) from the six corpora according to the benchmark alignments and then alter them gradually by a rate of 10% each time. The alignment performance corresponding to such alteration of parameters is shown in Fig. 2. It reveals the effect of different ways of changing the parameters: (a) and (d) resulting from changing mean only, (b) and (e) from changing variance only, and (c) and (f) from changing both. One observation we can make is that the algorithm is significantly more sensitive to the change of mean than that of variance. The possible reason is that the standardization

---

[2] http://www.ldc.upenn.edu/Catalog/catalogEntry.jsp?catalogId=LDC2005T10.

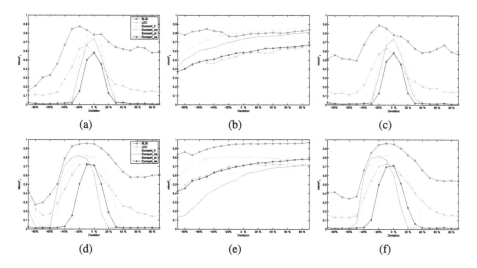

Fig. 2. Adaptivity of Gale and Church's algorithm with different means (a, d), variances (b, e), and both (c, f). Sentence length in a-c and d-f is measured in number of characters and words, respectively.

of normal distribution makes the two parameters have different degrees of sensitivity. These figures also show that when the mean is misestimated by 20% or more, which could happen in practice, the alignment performance becomes disastrously poor. This study shows that the baseline tends to have weak adaptivity to the estimation of parameters.

4.3   Alignment by Collaborative Matching

This part reports the alignment result of near-optimal alignment (NOA) and differential similarity based alignment (DCSA), with sentence length measured in character, word and self-information. The size of the initial alignment for DCSA is set to $\min(M, N)/10$ of each bitext. Alignment performance is reported in precision (P), recall (R), defined as the proportions of correctly aligned pairings in produced pairings and gold standard, respectively, and F-measure (F_1), their harmonic mean, defined as $F_1 = 2PR/(P + R)$. In addition, micro-averaged performance in terms of precision, recall, and F-measure are also computed to measure the overall 1-1 and 1-0 performance. The final alignment results are shown in Tables 2, 3 and 4, from which several findings can be derived. Firstly, Gale and Church's algorithm performs better with sentence length in number of words than in number of characters, and suffers seriously from sentence length in amount of self-information. Secondly, although the non-parametric approach, NOA, achieves competitive performance against the baseline when sentence length is measured in number of characters, it underperforms when the length is measured in terms of words. But its performance is not prone to being affected

Table 2. Alignment performance with sentence length in number of characters.

Dataset	Type	Gale&Church			NOA			DCSA		
		P	R	F_1	P	R	F_1	P	R	F_1
BLIS	1-1	0.827	0.808	0.818	0.970	0.965	0.967	0.975	0.977	0.976
	1-0	0.180	0.380	0.244	0.565	0.732	0.638	0.686	0.676	0.681
	Micro	0.771	0.790	0.780	0.948	0.955	0.952	0.963	0.964	0.964
LDC	1-1	0.604	0.602	0.603	0.823	0.823	0.823	0.856	0.855	0.855
	1-0	0.441	0.452	0.447	0.678	0.681	0.680	0.676	0.681	0.678
	Micro	0.568	0.570	0.569	0.792	0.792	0.792	0.817	0.817	0.817
Ep_fr	1-1	0.847	0.877	0.862	0.689	0.698	0.693	0.875	0.901	0.888
	1-0	0.641	0.432	0.517	0.248	0.220	0.233	0.654	0.481	0.554
	Micro	0.821	0.796	0.808	0.617	0.611	0.614	0.845	0.824	0.834
Ep_de	1-1	0.790	0.814	0.802	0.719	0.729	0.724	0.861	0.880	0.870
	1-0	0.471	0.338	0.394	0.230	0.202	0.215	0.596	0.473	0.527
	Micro	0.748	0.729	0.739	0.642	0.635	0.639	0.823	0.808	0.815
Ep_pt	1-1	0.594	0.619	0.606	0.707	0.723	0.715	0.863	0.888	0.875
	1-0	0.241	0.148	0.183	0.263	0.209	0.233	0.627	0.457	0.529
	Micro	0.553	0.534	0.543	0.642	0.630	0.636	0.831	0.811	0.821
Ep_es	1-1	0.635	0.657	0.646	0.697	0.708	0.703	0.864	0.885	0.874
	1-0	0.302	0.208	0.247	0.266	0.227	0.245	0.639	0.497	0.559
	Micro	0.593	0.576	0.585	0.630	0.622	0.626	0.832	0.815	0.823

by how sentence length is measured. Neither is the performance of DCSA, a stable and consistently outstanding performance across all the datasets.

5   Related Work

Sentence length based alignment follows essentially the length proportionality assumption, which can be straightforwardly observed from Indo-European and non-Indo-European languages [16]. Sentence length, together with other information such as lexical correspondence, has served as the foundation of many successful sentence aligners. For example, [11] developed an aligner with a three-pass procedure, which first aligns bitext using only length information of sentences, from which a set of finely aligned sentence pairs is yielded for training a translation model [2]. Then, it realigns the bitext using both sentence length and the word correspondences derived by the trained model. Hunalign [15] is another aligner developed via a hybrid of sentence length and lexical information. When lexical information is unavailable, it performs a similar initial alignment using sentence length and then automatically generates a lexicon. If a lexicon exists, it yields a rough translation for a source text and then compares it with its true

Table 3. Alignment performance with sentence length in number of words.

Dataset	Type	Gale&Church			NOA			DCSA		
		P	R	F_1	P	R	F_1	P	R	F_1
BLIS	1-1	0.968	0.968	0.968	0.970	0.962	0.966	0.981	0.981	0.981
	1-0	0.568	0.592	0.579	0.520	0.746	0.613	0.684	0.732	0.707
	Micro	0.950	0.952	0.951	0.943	0.953	0.948	0.968	0.970	0.969
LDC	1-1	0.753	0.758	0.756	0.798	0.791	0.794	0.837	0.842	0.839
	1-0	0.558	0.528	0.543	0.636	0.672	0.653	0.691	0.661	0.676
	Micro	0.713	0.709	0.711	0.761	0.765	0.763	0.806	0.803	0.805
Ep_fr	1-1	0.791	0.819	0.805	0.699	0.711	0.705	0.858	0.885	0.871
	1-0	0.528	0.358	0.427	0.257	0.216	0.235	0.617	0.446	0.517
	Micro	0.757	0.735	0.746	0.630	0.621	0.625	0.826	0.805	0.815
Ep_de	1-1	0.804	0.827	0.815	0.710	0.718	0.714	0.852	0.876	0.864
	1-0	0.539	0.394	0.455	0.240	0.215	0.227	0.617	0.451	0.521
	Micro	0.768	0.750	0.759	0.635	0.629	0.632	0.820	0.801	0.810
Ep_pt	1-1	0.775	0.805	0.790	0.703	0.718	0.711	0.859	0.882	0.87
	1-0	0.462	0.300	0.364	0.264	0.213	0.235	0.623	0.471	0.536
	Micro	0.737	0.714	0.725	0.638	0.627	0.632	0.826	0.808	0.817
Ep_es	1-1	0.813	0.841	0.826	0.697	0.709	0.703	0.864	0.883	0.874
	1-0	0.546	0.374	0.444	0.262	0.220	0.239	0.602	0.479	0.533
	Micro	0.779	0.757	0.768	0.630	0.621	0.626	0.826	0.811	0.818

translation to form a similarity matrix. The matrix is then taken as input to a dynamic programming algorithm to produce a final alignment.

Besides internal lexical correspondences derived during alignment, many other works resort to external lexicons. For example, [6] leveraged an external dictionary together with an internally-derived lexicon to build lexical correspondences. [3] introduced a hybrid system for sentence alignment by combining sentence length and an external lexicon, as well as sentence offset information. To take fuller advantage of lexical information, [10] assumed that different words should have differentiated importances in his aligner - Champollion. [9] proposed a revised version of Champollion, to improve its speed without performance loss, by first dividing input bitext into smaller by a length-based approach and aligned fragments and then applying Champollion to derive finer-grained alignment. Another assumption that most approaches to sentence alignment follow is monotonicity assumption, that coupled sentences in bitexts appear in a similar sequential order in two languages. Differently, [13,14] studied the problem of non-monotonic sentence alignment.

Table 4. Alignment performance with sentence length in amount of self-information.

Dataset	Type	Gale&Church			NOA			DCSA		
		P	R	F_1	P	R	F_1	P	R	F_1
BLIS	1-1	0.857	0.837	0.847	0.976	0.967	0.971	0.981	0.981	0.981
	1-0	0.162	0.338	0.219	0.549	0.789	0.647	0.684	0.732	0.707
	Micro	0.797	0.816	0.806	0.950	0.960	0.955	0.968	0.970	0.969
LDC	1-1	0.477	0.470	0.473	0.817	0.816	0.817	0.862	0.866	0.864
	1-0	0.388	0.426	0.406	0.668	0.677	0.673	0.727	0.699	0.712
	Micro	0.456	0.461	0.458	0.785	0.786	0.785	0.833	0.830	0.831
Ep_fr	1-1	0.544	0.572	0.558	0.684	0.696	0.690	0.875	0.898	0.886
	1-0	0.253	0.137	0.178	0.249	0.212	0.229	0.637	0.484	0.55
	Micro	0.514	0.493	0.503	0.616	0.608	0.612	0.841	0.823	0.832
Ep_de	1-1	0.562	0.586	0.574	0.722	0.733	0.727	0.854	0.879	0.866
	1-0	0.226	0.136	0.170	0.243	0.207	0.224	0.626	0.453	0.526
	Micro	0.525	0.506	0.515	0.648	0.640	0.644	0.824	0.803	0.813
Ep_pt	1-1	0.572	0.602	0.587	0.708	0.724	0.716	0.855	0.876	0.865
	1-0	0.217	0.113	0.149	0.266	0.211	0.236	0.611	0.473	0.533
	Micro	0.537	0.514	0.525	0.643	0.631	0.637	0.820	0.803	0.812
Ep_es	1-1	0.540	0.565	0.552	0.716	0.731	0.723	0.865	0.885	0.874
	1-0	0.254	0.146	0.186	0.271	0.218	0.242	0.603	0.475	0.531
	Micro	0.509	0.490	0.499	0.650	0.639	0.644	0.827	0.811	0.819

6   Conclusion

Sentence length has been widely utilized as a fundamental clue in most works
and tools for sentence alignment. While most previous efforts have focused on
resorting to probability theory to leverage this information under the length
proportionality assumption, an unavoidable drawback is poor adaptivity to new
languages and corpora. To find a solution, we establish our methodology on the
notion of collaborative matching, an idea to follow this assumption in a collective manner to treat sentences collaboratively rather than separately during
alignment. It makes the alignment less dependent on prior knowledge or specific languages and corpora and thus tends to be more adaptive. Furthermore,
we have also provided new insights into the measurement of sentence length,
presenting a new one by virtue of word self-information. Based on the idea of
collaborative matching, we proposed two novel alignment methods and evaluated them on six corpora of different languages and domains. Several findings can
be obtained from our experimental results. First, among the two conventional
length measures, the one in number of words appears to be more reliable than
the one in number of characters. Second, the proposed length measure leads
to a stable and competitive performance with our alignment methods but a

rather poor performance with the baseline. Next, one of the proposed alignment approaches, DCSA, has shown the best performance in most cases. Finally and more importantly, while the performance of the baseline is likely to be greatly affected by different measurements of sentence length, languages, or corpora, ours are consistently reliable, confirming the adaptivity of our methods.

References

1. Brown, P.F., Lai, J.C., Mercer, R.L.: Aligning sentences in parallel corpora. In: Proceedings of the 29th Annual Meeting on Association for Computational Linguistics (ACL 1991), pp. 169–176 (1991)
2. Brown, P.F., Pietra, V.J.D., Pietra, S.A.D., Mercer, R.L.: The mathematics of statistical machine translation: parameter estimation. Comput. Linguist. **19**(2), 263–311 (1993)
3. Collier, N., Ono, K., Hirakawa, H.: An experiment in hybrid dictionary and statistical sentence alignment. In: Proceedings of the 17th International Conference on Computational Linguistics - The 36th Annual Meeting of the Association for Computational Linguistics (COLING-ACL 1998), pp. 268–274 (1998)
4. Cover, T.M., Thomas, J.A.: Elements of Information Theory. Wiley-Interscience, Hoboken (1991)
5. Gale, W.A., Church, K.W.: A program for aligning sentences in bilingual corpora. In: Proceedings of the 29th Annual Meeting on Association for Computational Linguistics (ACL 1991), pp. 177–184 (1991)
6. Haruno, M., Yamazaki, T.: High-performance bilingual text alignment using statistical and dictionary information. In: Proceedings of the 34th Annual Meeting on Association for Computational Linguistics (ACL 1996), pp. 131–138 (1996)
7. Kit, C., et al.: Clause alignment for hong kong legal texts: a lexical-based approach. Int. J. Corpus Linguist. **9**, 29–51 (2004)
8. Koehn, P.: Europarl: A parallel corpus for statistical machine translation. In: MT Summit 2005, pp. 79–86 (2005)
9. Li, P., Sun, M., Xue, P.: Fast-champollion: a fast and robust sentence alignment algorithm. In: Proceedings of the 23rd International Conference on Computational Linguistics (COLING 2010): Posters, pp. 710–718 (2010)
10. Ma, X.: Champollion: a robust parallel text sentence aligner. In: LREC 2006, pp. 489–492 (2006)
11. Moore, R.C.: Fast and accurate sentence alignment of bilingual corpora. In: Richardson, S.D. (ed.) AMTA 2002. LNCS (LNAI), vol. 2499, pp. 135–144. Springer, Heidelberg (2002). https://doi.org/10.1007/3-540-45820-4_14
12. Nie, J.Y., Simard, M., Isabelle, P., Durand, R.: Cross-language information retrieval based on parallel texts and automatic mining of parallel texts from the web. In: Proceedings of the 22nd Annual International ACM SIGIR Conference on Research and Development in Information Retrieval (SIGIR 1999), pp. 74–81 (1999)
13. Quan, X., Kit, C.: Towards non-monotonic sentence alignment. Inf. Sci. **323**, 34–47 (2015)
14. Quan, X., Kit, C., Song, Y.: Non-monotonic sentence alignment via semisupervised learning. In: Proceedings of 51st Annual Meeting of the Association for Computational Linguistics (ACL 2013), pp. 622–630 (2013)

15. Varga, D., Németh, L., Halácsy, P., Kornai, A., Trón, V., Nagy, V.: Parallel corpora for medium density languages. In: Recent Advances in Natural Language Processing (RANLP 2005), pp. 590–596 (2005)
16. Wu, D.: Aligning a parallel English-Chinese corpus statistically with lexical criteria. In: Proceedings of the 32nd Annual Meeting on Association for Computational Linguistics (ACL 1994), pp. 80–87 (1994)

Finding Better Subword Segmentation for Neural Machine Translation

Yingting Wu[1,2] and Hai Zhao[1,2(✉)]

[1] Department of Computer Science and Engineering, Shanghai Jiao Tong University, Shanghai, China
wuyingting@sjtu.edu.cn, zhaohai@cs.sjtu.edu.cn
[2] Key Laboratory of Shanghai Education Commission for Intelligent Interaction and Cognitive Engineering, Shanghai Jiao Tong University, Shanghai 200240, China

Abstract. For different language pairs, word-level neural machine translation (NMT) models with a fixed-size vocabulary suffer from the same problem of representing out-of-vocabulary (OOV) words. The common practice usually replaces all these rare or unknown words with a ⟨UNK⟩ token, which limits the translation performance to some extent. Most of recent work handled such a problem by splitting words into characters or other specially extracted subword units to enable open-vocabulary translation. Byte pair encoding (BPE) is one of the successful attempts that has been shown extremely competitive by providing effective subword segmentation for NMT systems. In this paper, we extend the BPE style segmentation to a general unsupervised framework with three statistical measures: frequency (FRQ), accessor variety (AV) and description length gain (DLG). We test our approach on two translation tasks: German to English and Chinese to English. The experimental results show that AV and DLG enhanced systems outperform the FRQ baseline in the frequency weighted schemes at different significant levels.

Keywords: Neural machine translation · Subword segmentation

1 Introduction

Neural Machine Translation [3,14,23] has achieved remarkable performance in recent years. The NMT system is commonly based on a word-level model with a restricted vocabulary of most frequent words. All the infrequent or unseen words are simply replaced with a special ⟨UNK⟩ token at the cost of decreasing translation accuracy. In order to better handle the problems of OOV words, recent work proposed the character-based [9,11,17] or word-character hybrid models [19]. Considering that character is not a sufficiently good minimal unit to compose word representation, it is of great value to find a more meaningful representation unit between word and character, i.e., subword.

The benefits of using variable-length subword representation in NMT are two-fold. First, it can effectively decrease the vocabulary size of the whole training set by encoding rare words with sequences of subword units. Second, it is

© Springer Nature Switzerland AG 2018
M. Sun et al. (Eds.): CCL 2018/NLP-NABD 2018, LNAI 11221, pp. 53–64, 2018.
https://doi.org/10.1007/978-3-030-01716-3_5

also possible to translate or generate unseen words at inference. Currently, BPE [13] has been applied commonly in NMT called subword NMT [22] that enables open-vocabulary translation. It is a frequency-based data compression algorithm that most of the high-frequency words are maintained and the infrequent ones are segmented into subwords. As the widely used BPE segmentation only uses frequency as its measure and does not rely on any morphological or linguistic knowledge, it can be regarded as a computationally motivated method. Some other work segmented words based on morphological information to obtain the representation of OOV words [2,5], which can be called linguistically motivated method. However, the availability of morphological resources for different languages cannot be always guaranteed. Thus, this work will focus on the general computationally motivated subword segmentation for NMT.

The development of Chinese word segmentation has played an important role in the natural language processing tasks [6,7,26]. Zhao and Kit [25] proposed a general unsupervised word segmentation framework for Chinese which consists of two collocative components, decoding algorithm and substring goodness measure. Their work shows that the substring frequency (FRQ) is not the best goodness measure for Chinese word segmentation. Since Chinese is an ideographic language without any explicit word boundaries, we can similarly adapt the reported better measures to the subword segmentation for alphabetic languages such as English by considering their words are written in character sequences without explicit character boundaries. Thus, we regard the BPE style segmentation as a general decoding algorithm, and enable it to work with another two advanced goodness measures, accessor variety (AV) [12] and description length gain (DLG) [16] in the hope of further enhancing the current NMT. We compare the effects of different measures with experiments on the German-English and Chinese-English translation tasks. The evaluations on the test sets prove that the extended BPE-style segmentation methods with the frequency weighted AV and DLG improve the FRQ baseline at different significant levels.

2   Neural Machine Translation

In this paper, we closely follow the neural machine translation model proposed by Bahdanau et al. [3], which is mainly based on a neural encoder-decoder network with attention mechanism [3,20].

The encoder is a bidirectional recurrent neural network (RNN) with gated recurrent unit (GRU) or long short-term memory (LSTM) unit. The forward RNN reads a source sentence $x = (x_1, ..., x_m)$ from left to right and calculates a forward sequence of hidden states $(\overrightarrow{h}_1, ..., \overrightarrow{h}_m)$. Similarly, the backward RNN reads the source sentence inversely and learns a backward sequence of hidden states $(\overleftarrow{h}_1, ..., \overleftarrow{h}_m)$. The hidden states of both directions are concatenated to obtain the annotation vector $h_i = [\overrightarrow{h}_i, \overleftarrow{h}_i]^T$ for each word in the source sentence.

The decoder is a forward RNN initialized with the final state of the encoder. In the decoding phase, a target sentence $y = (y_1, ..., y_n)$ is generated step by step. The conditional translation probability can be formulated as follows.

$$p(y_j|y_{<j}, x) = q(y_{j-1}, s_j, c_j),$$

where s_j and c_j denote the decoding state and the source context at the j-th time step respectively. Here, $q(\cdot)$ is the softmax layer and $y_{<j} = (y_1, ..., y_{j-1})$. The context vector c_j is calculated as a weighted sum of the source annotations according to attention mechanism, where $c_j = \sum_{i=1}^{m} \alpha_{ji} h_i$. The alignment model α_{ji} defines the probability that how well y_j is aligned to x_i, which can be a single layer feed-forward neural network.

3   Unsupervised Subword Segmentation

Extracting substrings inside a word can be regarded as a segmentation process over character sequences, which is similar to splitting words over Chinese character sequences (e.g., Chinese sentences). Therefore, we borrow the idea from Zhao and Kit [25] who proposed a generalized framework for unsupervised Chinese word segmentation including two collocative modules, a decoding algorithm and an alternative goodness measure. They use a top-down decoding method that starts from the sentence level and then searches for the best segmentation for a particular text according to the given goodness function. Their empirical assessment on Chinese used the frequency of substrings as baseline among all goodness measures, which indicates that frequency is not an optimal segmentation criterion for Chinese. Motivated by this, we propose a combined framework of the extended BPE-style segmentation and some advanced substring measures for better subword representation in NMT.

3.1   Generalized BPE Segmentation

The BPE used in subword NMT is a bottom-up method, where the initial state of each word is a sequence of single characters and consecutive substring pairs with highest frequency are merged iteratively to compose subwords. In this section, we introduce a generalized BPE segmentation framework with arbitrary goodness measures besides frequency. The details are described in Algorithm 1.[1]

Given the corresponding segmentation measure, if a substring pair ('a', 'b') has the highest goodness score among all the candidate pairs, every occurrence of this pair will be replaced by a new symbol 'ab'. The merge operations are performed on the whole training corpus and the vocabulary size can be controlled by the number of merge operations N. The final vocabulary size is approximately equal to merge times plus the number of character types in the corpus. According to the merge list of substring pairs learned from training data, we can further apply the merge operation to the development set and test set. Each subword except the end of word is attached to a special token such as "@@" for the sake of restoring the segmented words after translation.

---

[1] The source code has been released at https://github.com/Lindsay125/gbpe.

Algorithm 1. Generalized BPE segmentation

Input: the training corpus D, merge times N, goodness measure g
Output: the segmented text D', merge list V
1: The training corpus D is initialized as a set of character sequences in which every word is split into a sequence of characters, and the merge list V is set empty.
2: Given the current segmentation state on D, calculate the goodness scores of all the distinct successive substring pairs according to the goodness measure g;
3: Search for the highest scored consecutive pair, add it to V and merge all the occurrences of such pairs on D;
4: If the merge times reaches N, the algorithm ends and returns D' and V.
5: Otherwise, go to Step 2.

3.2 Goodness Measures

Given a word W with n characters, its segmentation S can be denoted as a subword sequence $s_1 s_2 \cdots s_m$ $(m \leq n)$. For each merge iteration, every subword s_i (the concatenation of a consecutive pair) is assigned a goodness score $g(s_i)$ for how likely it is an independently translatable item within the whole word. In this study, we examine three types of goodness measures.

Frequency of Substring. The *frequency of substring* (FRQ) serves as the baseline in the Chinese word segmentation system of Zhao and Kit [25]. Its basic idea is to compare the frequency of two partially overlapped character n-grams and then the shorter substring with lower or equal frequency is discarded as a redundant candidate. We define the corresponding goodness score as $g_{FRQ}(s_i)$, which is the count of word types in the vocabulary that contain s_i. For efficiency, only those substrings that occur over once are considered in the candidate list.

Accessor Variety. The *accessor variety* (AV) proposed by Feng *et al.* [12] is a measure to evaluate how likely a substring can be a relatively independent word, which is reported to be good at addressing low frequency words. Given a particular substring, the main idea of AV is that if the type of successive tokens with respect to the corresponding substring increases, it is more likely to be at boundary. Formally, AV is defined as the minimum of $L_{av}(s_i)$ and $R_{av}(s_i)$.

$$g_{AV}(s_i) = AV(s_i) = \min\{L_{av}(s_i), R_{av}(s_i)\},$$

where the left AV $L_{av}(s_i)$ is the number of distinct predecessor tokens of s_i and the right AV $R_{av}(s_i)$ records the number of s_i's different successor tokens.

Description Length Gain. The *description length* of a corpus X is defined as the Shannon-Fano code length for the corpus [15]. It can be calculated as

$$DL(X) = -|X| \sum_{x \in V_X} \hat{p}(x) \log \hat{p}(x) = - \sum_{x \in V_X} c(x) \log \frac{c(x)}{|X|},$$

where V_X is the token vocabulary of X, $c(x)$ is the count of token x in X, $\hat{p}(x) = c(x)/|X|$ and $|X|$ represents the total token count in X.

Kit and Wilks [16] proposed to use the *description length gain* (DLG) for lexical acquisition on word boundary prediction tasks and showed the effectiveness of this goodness measure. The DLG of a particular substring s_i in X is then defined as the description length change while we substitute s_i with an index r and take a note of this operation at the end of X, i.e.,

$$g_{DLG}(s_i) = DLG(s_i) = DL(X) - DL(X[r \rightarrow s_i] \oplus s_i),$$

where $X[r \rightarrow s_i]$ is the resultant corpus by replacing all occurrences of s_i with r throughout X and $\oplus$ represents the concatenation of two strings with a delimiter.

Frequency Weighted Schemes. Note that parallel corpus for NMT typically holds a large vocabulary with noise, all of these statistical goodness measures may be biased by the data noise. Especially for FRQ and AV, which are only based on type statistics, their calculation will be misled by too many word types even though each word type has low frequency. Considering that high frequency words usually correspond to those reliable ones with regular forms, we also introduce three frequency weighted variants,[2]

$$g'_{FRQ}(s_i) = g_{FRQ}(s_i) \sum_{\forall w, s_i \in w} f(w),$$

$$g'_{AV}(s_i) = g_{AV}(s_i) \sum_{\forall w, s_i \in w} f(w), \quad g'_{DLG}(s_i) = g_{DLG}(s_i) \sum_{\forall w, s_i \in w} f(w),$$

where $f(w)$ is the frequency of word w in the corpus. Note that here our FRQ'-BPE will slightly differ from the BPE in [22] whose goodness score is directly computed through counting s_i in the entire corpus.

4 Experiments

4.1 Setup

Our experiments will be performed on two typical language pairs, German to English and Chinese to English. The translation quality is evaluated by the 4-gram case-sensitive BLEU [21] and we use sign test [10] to test the statistical significance of our results.

For German-English task, the evaluations are based on data from TED talks corpora of the IWSLT 2014 evaluation campaign [8]. We tokenize all the training data with the script of Moses[3] and remove sentences longer than 50 words. The training set comprises of about $153K$ sentence pairs with $100K$ German words

---

[2] Though DLG is already frequency weighted as its definition, the proposed extra frequency weight is empirically verified effective from our preliminary experiments.

[3] https://github.com/moses-smt/mosesdecoder/blob/master/scripts/tokenizer.

and $50K$ English words. The development set consists of 6969 sentence pairs which is randomly extracted from training data. The test set is a concatenation of dev2010, dev2012, tst2010, tst2011 and tst2012 which results in 6750 sentence pairs. As German and English share the similar alphabet and have a large overlap of vocabulary, the learning of subwords is performed on the union of the source and target corpora as suggested by Sennrich et al. [22].

For Chinese-English task, we use the News Commentary v12 dataset from the news translation shared task of WMT 2017 [4] which consists of $227K$ parallel sentence pairs. The development set and test set are randomly selected from the whole corpora with 3000 sentence pairs respectively, and the remaining part is served as the training set. Since Chinese and English have totally different character vocabularies, the subword (word for Chinese) learning is performed on the source and target corpora separately and we also try different learning strategies of segmentation on both sides.

During training, we use LSTM units for both encoder and decoder. Each direction of the LSTM encoder and decoder are with 256 dimensions. The word embedding and the attention size are both set to 256. The model is trained using Adam optimizer with the initial learning rate of 0.001. The batch size is set to 32 for German-English task and 64 for Chinese-English task. We train each model for 40 epochs and halve the learning rate every 10 epochs. The training set is reshuffled at the beginning of each epoch. During decoding, greedy search and beam search with size 10 are both performed to optimize the performance.

Table 1. BLEU scores with normal AV and DLG on German-English test set

Method	Merge times		
	$10K$	$20K$	$30K$
FRQ-BPE	27.55	27.94	**28.29**
AV-BPE	27.42	27.96	27.60
DLG-BPE	20.95	20.83	21.03

4.2 Results

Table 1 reports the translation performance after applying FRQ-BPE, DLG-BPE and AV-BPE segmentation on the German-English task with a beam size of 10. From Table 1, we notice that FRQ-BPE and AV-BPE outperform DLG-BPE at different merge times (vocabulary size) with a large margin and FRQ-BPE behaves the best among the three measures. Thus, we consider combining the corpus-level word frequency with the three goodness measures to further enhance the performance as we disscussed in Sect. 3.2.

Table 2 reports BLEU scores with different frequency weighted goodness measures and corresponding merge times on the German-English test set.[4] For

---

[4] "++" indicates that the corresponding BLEU is significantly better than the best score of FRQ'-BPE at the significant level p < 0.01, "+": p < 0.05.

Table 2. BLEU scores with frequency-weighted schemes on German-English test set

N	FRQ′-BPE		AV′-BPE		DLG′-BPE	
	Greedy	Beam	Greedy	Beam	Greedy	Beam
$10K$	27.36	28.86	27.72	29.02	27.56	29.25
$15K$	27.42	28.92	27.56	29.25	27.75	29.30
$20K$	27.51	28.99	27.65	$\mathbf{29.46}^{++}$	$\mathbf{27.76}^{+}$	29.36
$25K$	27.22	28.81	27.69	29.32	27.72	29.34
$30K$	27.06	28.72	27.52	29.09	27.46	29.03

greedy search, both DLG′-BPE and AV′-BPE achieve better performance than FRQ′-BPE and the best score is obtained with DLG′-BPE of $20K$ merge operations. For beam search, AV′-BPE and DLG′-BPE outperform FRQ-BPE′ with improvement of 0.47 and 0.37 BLEU points when $N = 20K$ respectively. Figure 1 plots the variation curves of BLEU scores with different goodness measures against the merge times N when the beam size is 10. From Fig. 1, we observe that the peak value of BLEU with different criteria appears at the $20K$ merge times. In general, AV′-BPE and DLG′-BPE are comparable with each other and both superior to FRQ′-BPE with a margin.

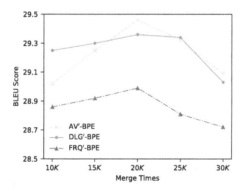

Fig. 1. The curves of BLEU scores with different merge times and segmentation criteria on German-English test set (beam size $= 10$).

We also calculate and plot the average number of segmented words in each sentence and average sentence length increase after subword segmentation with different criteria on German-English task in Fig. 2. Figure 2a shows that, with the same merge times, AV′-BPE and DLG′-BPE maintain more original word forms than FRQ′-BPE within each sentence. The sentence length is an important factor during training as too long sequences will slow down the training and NMT shows a weak performance on long sentences. Figure 2b also shows that both DLG′-BPE and AV′-BPE can better control the sequence length than FRQ′-BPE. For both Fig. 2a and b, the curve of AV′-BPE with best performance is between that of DLG′-BPE and FRQ′-BPE, which means that it is crucial to find the proper granularity for segmentation.

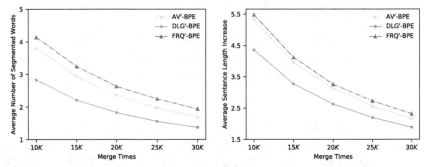

(a) Average number of segmented words in each sentence.   (b) Average sentence length increase.

Fig. 2. Statistics of words and sentence after segmentation.

Table 3. BLEU scores on Chinese-English test set (beam size = 10)

Source	Target		
	FRQ′-BPE	AV′-BPE	DLG′-BPE
FRQ′-BPE	20.07	20.27	20.31
AV′-BPE	19.96	20.23	20.13
DLG′-BPE	20.70	20.45	**20.72**[++]

In Table 3, we report the BLEU scores with different segmentation measures on the test set of Chinese-English task. Here, the subword segmentation learning is performed separately on the source and target sides with $30K$ merge operations which is tuned the same way as German-English task. We also try different goodness measures on each side at the same time. It can be found that DLG′-BPE on the both sides significantly outperforms FRQ′-BPE with 0.65 BLEU points. Figure 3 shows the translation performance comparison of different goodness measures when the source segmentation is fixed. It can be seen that DLG′-BPE does much better at the source segmentation for Chinese than the other two, while AV′-BPE and FRQ′-BPE are on a par with each other.

4.3   Translation Examples

Table 4 shows some translation examples with different segmentation criteria on the test sets of German-English and Chinese-English tasks. We use the "|" token to mark the splitting points within the words (sentences for Chinese).

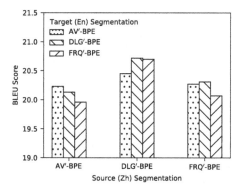

Fig. 3. The bar chart of BLEU scores with different segmentation criteria on Chinese-English test set (beam size = 10).

In the first case, AV′-BPE performs best by translating *"erstaunlich wissbegierige"* to an acceptable alternative *"astonishingly inventive"* compared to the ground truth *"astonishingly inquisitive"*. DLG′-BPE omits the translation of *"wissbegierige"* probably as the segmentation of this word is more confusing than that of the other two methods. FRQ′-BPE behaves not badly in the interpretation of *"erstaunlich"* but generates a non-existing word *"inwalkable"*.

In the second case, DLG′-BPE shows the better translation quality than the other two. The critical difference in splitting phrase "在多大程度上" (*by what a degree, how much*) mostly results in the diverse translation behaviors. Only DLG′-BPE correctly translates it to the correct meaning *"how much"*, while FRQ′-BPE and AV′-BPE give the undesirable answers of *"how else"* and *"what kind of"* respectively.

5 Related Work

Previous work resorted to various techniques to deal with the representation of rare words for NMT. Recently, character-level and other subword-based models have become increasingly popular and achieved great performance for different language pairs. The character-level models often rely on convolutional or recurrent neural networks to encode or decode the character sequences. For instance, Costa-jussà and Fonollosa [11] employed convolutional and highway layers upon characters to form the word embeddings. Ling *et al.* [18] used a bidirectional LSTM to combine character embeddings to word embeddings and generated the target word character by character. Similarly, Ataman and Federico [1] proposed to improve the quality of source representations of rare words by augmenting its embedding layer with a bi-RNN, which can learn compositional input representations at different levels of granularity. Luong and Manning [19] proposed a word-character hybrid model that translates mostly at the word level and consults the character information for rare words with an additional deep character-based LSTM. Different from above methods which still consider

Table 4. German-English and Chinese-English translation examples.

Source	...und wozu wir als erstaunlich wissbegierige Spezies fähig sind .
Reference	...and all that we can be as an astonishingly inquisitive species .
AV'-BPE	...und wo|zu wir als erstaunlich wiss|beg|ier|ige spezies fähig sind .
	...and what we are , as an aston|ish|ingly inven|tive species .
DLG'-BPE	...und wo|zu wir als erstaunlich wis|sbe|gi|erige spezies fähig sind .
	...and what we do as an ast|ound|ing species .
FRQ'-BPE	...und wo|zu wir als erstaun|lich wiss|beg|ier|ige spezies fähig sind .
	...and what we are as an amazing|ly in|walk|able species .
Source	我们应当在多大程度上担心社会不平等呢？
Reference	How much should we worry about inequality ?
AV'-BPE	我们|应当|在多大程度上|担心|社会|不平等|呢？
	What kind of social inequality we should be ?
DLG'-BPE	我们|应当|在|多大程度上|担心|社会|不平等|呢 ？
	How much should we worry about social inequality ?
FRQ'-BPE	我们|应当|在|多|大程度|上|担心|社会|不|平等|呢？
	How else should we worry about social inequality ?

word boundaries, Chung *et al.* [9] introduced a character-level decoder without explicit segmentation while the encoder is still at subword-level. Further, Lee *et al.* [17] proposed the fully character-level NMT model that maps a source character sequence to a target one without any explicit segmentation for both encoder and decoder. At present, the most popular subword-level method is independent of the training procedure and can be regarded as a preprocessing step such as BPE [22] and the wordpieces [24]. The main advantage of this approach is that the original architecture of the word-level model can be reused without increasing much complexity of training or decoding. Here, we focus on the subword segmentation improvement by introducing better substring measures.

6   Conclusion

In this paper, we introduce a generalized subword segmentation framework to enable open-vocabulary translation in NMT. Specifically, we extend the frequency only based BPE segmentation to a general bottom-up case that is capable of incorporating other advanced substring goodness measures, AV and DLG. We empirically compare the effects of different goodness measure schemes on the IWSLT14 German-English and WMT17 News Commentary Chinese-English translation tasks. The experimental results show that frequency weighted DLG'-BPE and AV'-BPE achieves stably better performance than FRQ'-BPE at different significant levels. In general, AV'-BPE shows best performance on the German-English task and DLG'-BPE behaves better on the Chinese-English task. The difference on the best subword segmentation strategies indicates that the choice may be sensitive to specific language pairs to some extent, which deserves further exploration in the future.

Acknowledgments. This paper was partially supported by National Key Research and Development Program of China (No. 2017YFB0304100), National Natural Science Foundation of China (No. 61672343 and No. 61733011), Key Project of National Society Science Foundation of China (No. 15-ZDA041), The Art and Science Interdisciplinary Funds of Shanghai Jiao Tong University (No. 14JCRZ04).

References

1. Ataman, D., Federico, M.: Compositional representation of morphologically-rich input for neural machine translation. arXiv preprint arXiv:1805.02036 (2018)
2. Ataman, D., Negri, M., Turchi, M., Federico, M.: Linguistically motivated vocabulary reduction for neural machine translation from Turkish to English. Prague Bull. Math. Linguist. **108**(1), 331–342 (2017)
3. Bahdanau, D., Cho, K., Bengio, Y.: Neural machine translation by jointly learning to align and translate. In: Proceedings of 3rd International Conference on Learning Representations (2015)
4. Bojar, O., et al.: Findings of the 2017 conference on machine translation. In: Proceedings of the 2nd Conference on Machine Translation, vol. 2: Shared Task Papers, pp. 169–214 (2017)
5. Botha, J., Blunsom, P.: Compositional morphology for word representations and language modelling. In: International Conference on Machine Learning, pp. 1899–1907 (2014)
6. Cai, D., Zhao, H.: Neural word segmentation learning for Chinese. In: Proceedings of the 54th Annual Meeting of the Association for Computational Linguistics, pp. 409–420 (2016)
7. Cai, D., Zhao, H., Zhang, Z., Xin, Y., Wu, Y., Huang, F.: Fast and accurate neural word segmentation for Chinese. In: Proceedings of the 55th Annual Meeting of the Association for Computational Linguistics, pp. 608–615 (2017)
8. Cettolo, M., Niehues, J., Stüker, S., Bentivogli, L., Federico, M.: Report on the 11th IWSLT evaluation campaign. In: The 11th International Workshop on Spoken Language Translation, Lake Tahoe, USA (2014)
9. Chung, J., Cho, K., Bengio, Y.: A character-level decoder without explicit segmentation for neural machine translation. In: Proceedings of the 54th Annual Meeting of the Association for Computational Linguistics (vol. 1: Long Papers), Berlin, Germany, pp. 1693–1703 (2016)
10. Collins, M., Koehn, P., Kučerová, I.: Clause restructuring for statistical machine translation. In: Proceedings of the 43rd Annual Meeting on Association for Computational Linguistics, pp. 531–540 (2005)
11. Costa-jussà, M.R., Fonollosa, J.A.R.: Character-based neural machine translation. In: Proceedings of the 54th Annual Meeting of the Association for Computational Linguistics (vol. 2: Short Papers), Berlin, Germany, pp. 357–361 (2016)
12. Feng, H., Chen, K., Deng, X., Zheng, W.: Accessor variety criteria for Chinese word extraction. Comput. Linguist. **30**(1), 75–93 (2004)
13. Gage, P.: A new algorithm for data compression. C Users J. **12**(2), 23–38 (1994)
14. Kalchbrenner, N., Blunsom, P.: Recurrent continuous translation models. In: Proceedings of the 2013 Conference on Empirical Methods in Natural Language Processing, Seattle, Washington, USA, pp. 1700–1709 (2013)
15. Kit, C.: A goodness measure for phrase learning via compression with the MDL principle. In: Proceedings of the ESSLLI Student Session, pp. 175–187 (1998)

16. Kit, C., Wilks, Y.: Unsupervised learning of word boundary with description length gain. In: Proceedings of the 3rd Conference on Computational Natural Language Learning, pp. 1–6 (1999)
17. Lee, J., Cho, K., Hofmann, T.: Fully character-level neural machine translation without explicit segmentation. Trans. Assoc. Comput. Linguist. **5**, 365–378 (2017)
18. Ling, W., Trancoso, I., Dyer, C., Black, A.W.: Character-based neural machine translation. CoRR abs/1511.04586 (2015)
19. Luong, M.T., Manning, C.D.: Achieving open vocabulary neural machine translation with hybrid word-character models. In: Proceedings of the 54th Annual Meeting of the Association for Computational Linguistics (vol. 1: Long Papers), Berlin, Germany, pp. 1054–1063 (2016)
20. Luong, T., Pham, H., Manning, C.D.: Effective approaches to attention-based neural machine translation. In: Proceedings of the 2015 Conference on Empirical Methods in Natural Language Processing, Lisbon, Portugal, pp. 1412–1421 (2015)
21. Papineni, K., Roukos, S., Ward, T., Zhu, W.J.: BLEU: a method for automatic evaluation of machine translation. In: Proceedings of 40th Annual Meeting of the Association for Computational Linguistics, pp. 311–318 (2002)
22. Sennrich, R., Haddow, B., Birch, A.: Neural machine translation of rare words with subword units. In: Proceedings of the 54th Annual Meeting of the Association for Computational Linguistics, Berlin, Germany, pp. 1715–1725 (2016)
23. Sutskever, I., Vinyals, O., Le, Q.V.: Sequence to sequence learning with neural networks. In: Advances in Neural Information Processing Systems 27, pp. 3104–3112. Curran Associates, Inc. (2014)
24. Wu, Y., et al.: Google's neural machine translation system: bridging the gap between human and machine translation. arXiv preprint arXiv:1609.08144 (2016)
25. Zhao, H., Kit, C.: An empirical comparison of goodness measures for unsupervised Chinese word segmentation with a unified framework. In: Proceedings of the 3rd International Joint Conference on Natural Language Processing, pp. 9–16 (2008)
26. Zhao, H., Utiyama, M., Sumita, E., Lu, B.-L.: An empirical study on word segmentation for Chinese machine translation. In: Gelbukh, A. (ed.) CICLing 2013. LNCS, vol. 7817, pp. 248–263. Springer, Heidelberg (2013). https://doi.org/10.1007/978-3-642-37256-8_21

Improving Low-Resource Neural Machine Translation with Weight Sharing

Tao Feng[1,2], Miao Li[1(✉)], Xiaojun Liu[3], and Yichao Cao[1,2]

[1] Institute of Intelligent Machines, Chinese Academy of Science, Hefei, China
mli@iim.ac.cn
[2] University of Science and Technology of China, Hefei, China
{ft2016,cycao}@mail.ustc.edu.cn
[3] School of Information and Computer, Anhui Agricultural University, Hefei, China
Lxj2442@ahau.edu.cn

Abstract. Neural machine translation (NMT) has achieved great success under a great deal of bilingual corpora in the past few years. However, it is much less effective for low-resource language. In order to alleviate the problem, we present two approaches which can improve the performance of low-resource NMT system. The first approach employs the weight sharing of decoder to enhance the target language model of low-resource NMT system. The second approach applies cross-lingual embedding and source sentence representation space sharing to strengthen the encoder of low-resource NMT. Our experiments demonstrate that the proposed method can obtain significant improvements on low-resource neural machine translation than baseline system. On the IWSLT2015 Vietnamese-English translation task, our model can improve the translation quality by an average of 1.43 BLEU scores. Besides, we can also get the increase of 0.96 BLEU scores when translating from Mongolian to Chinese.

Keywords: Low-resource · Neural machine translation
Weight sharing

1 Introduction

Machine translation is an important part of artificial intelligence, which explores how to use computers to translate one language into the other. In recent years, neural machine translation (NMT) has achieved great success because of the development of deep learning and the availability of large-scale parallel corpus [1,2]. In a variety of language pairs, the performance of neural machine translation has gradually surpassed phrase-based statistical machine translation (SMT) [3,4]. NMT is an end-to-end translation method, which typically consists of an encoder and a decoder [5,6]. More concretely, the encoder network maps the input sequence to a fixed-length vector and the decoder network gets translation from the vector. However, the defect of encoder-decoder framework is that the

© Springer Nature Switzerland AG 2018
M. Sun et al. (Eds.): CCL 2018/NLP-NABD 2018, LNAI 11221, pp. 65–75, 2018.
https://doi.org/10.1007/978-3-030-01716-3_6

encoder only obtains all the information of source sentences through a fixed-length vector. This leads to the poor performance of NMT in long sentences. In order to solve this problem, the attention mechanism was proposed by [7,8]. The attention mechanism can utilize relevant source side information to help predict the current target word, and significantly improve the performance of NMT. Therefore, the encoder-decoder framework with attention has become the mainstream method of the neural machine translation.

However, as a data-driven approach, the performance of NMT is severely affected by the size and the quality of parallel corpus. As the decrease of parallel corpus, the quality of NMT is greatly reduced [9,10]. In this case, the NMT lags behind statistical machine translation on low-resource language pairs. However, the vast majority of language pairs lack a large amount of parallel corpus [11]. Therefore, research on low-resource is valuable.

In this paper, we investigate the usage of the similarity and complementarity between different languages to obtain high-quality context vector and strengthen the decoder for low-resource neural machine translation. Intuitively, we employ multi-task learning framework to build two NMT models, one is a low-resource model (e.g., Vietnamese-English) and the other is a high-resource model (e.g., French-English). These two models share the weights of certain layers. To achieve this goal, we propose two approaches. Inspired by [12], the first approach exploits weight sharing of decoder side between low-resource model and high-resource model to improve the performance of target language model of the low-resource NMT system.

The proposed second approach applies multi-lingual translation system to share word embedding space and sentence representation space between different languages in the source side. The motivation behind this is that we can obtain better context vector for low-resource NMT model with the assistance of high-resource parallel corpus. More concretely, the approach builds upon the recent work on cross-lingual embedding [13]. First, we train the embedding for different source languages on monolingual corpora, and then learn a liner transformation to map the embedding from one space to the other. Therefore, we can align the word embedding space in this way. For sentence representation, not only in order to maintain the independence of each language, but also to map sentence representation into same space, we share the weights of last few layers of the encoder, not all of its layers. In this work, we make following contributions:

- To fully investigate weight sharing in low-resource NMT model, we propose and compare two methods. One attempts to reinforce the decoder side for low-resource model by sharing the weights of decoder layers with the high-resource model, and the other tries to share word embedding space and sentence representation space between source languages, so that we can get high-quality context vector for low-resource model.
- The experiments on Vietnamese-to-English and Mongolian-to-Chinese translations show that our proposed methods significantly outperform the NMT baseline model with attention mechanism.

2   Neural Machine Translation Background

The encoder-decoder NMT model has been proposed in recent years [7,8] and consists of three parts: encoder, attention and decoder. The model takes a source sequence $X = (x_1, x_2, \ldots, x_{T_x})$ as input and generates corresponding translation $Y = (y_1, y_2, \ldots, y_{T_y})$, where x_t and y_t are the symbols of source language and the target language respectively.

Encoder: Given a source sentence X, the encoder builds a continuous representation with recurrent neural networks (RNNs). In NMT model, bi-directional neural networks including a forward RNN and a backward RNN are often implemented. The forward RNN reads the input sentence from left to right: $\overrightarrow{h}_t = \overrightarrow{f}_{enc}(E_x(x_t), \overrightarrow{h}_{t-1})$. Similarly, the backward RNN reads the input sentence from right to left: $\overleftarrow{h}_t = \overleftarrow{f}_{enc}(E_x(x_t), \overleftarrow{h}_{t-1})$, where the E_x is the word embedding matrix and the h_t is a hidden state of RNN at time t. $\overrightarrow{f}_{enc}$ and $\overleftarrow{f}_{enc}$ are some nonlinear functions. In encoder side, the RNN can be a Long Short Term Memory Unit (LSTM) or a Gate Recurrent Unit (GRU).

Attention: The attention mechanism [7,8] was proposed to dynamically compute the context vector of the source end. In general, the current target hidden state is compared with all source states to derive attention weight α_{ts}. Calculate a context vector $c_t = \sum_{s=1}^{T_s} \alpha_{ts} h_s$ as the weighted average of the source states based on the attention weights. Then combine the context vector with the current target hidden state to generate final attention vector a_t. In attention mechanism, the attention weight α_{ts} is used to measure the correlation between the t-th target token and s-th source token, and is calculated as follows:

$$\alpha_{ts} = \frac{exp(score(h_t, h_s))}{\sum_{s'=1}^{T_s} exp(score(h_t, h_{s'}))} \tag{1}$$

where h_t is the hidden state of target at time t and h_s is the hidden state of source at time s. $score()$ is a nonlinear function, usually a feed-forward neural network with a hidden layer.

Decoder: The decoder utilizes recurrent neural networks to predict the target sequence $y = (y_1, y_2, \ldots, y_{T_y})$. Each word y_i is predicted based on recurrent neural network hidden state h_i, the previously predicted word y_{i-1}, and a context vector c_i. Therefore, each conditional probability is calculated as follows:

$$p(y_t | \{y_1, y_2, \ldots, y_{t-1}\}, c) = f(y_{t-1}, h_t, c) \tag{2}$$

where the f is a nonlinear function, usually a multi-layered neural network. However, other architectures such as convolutional neural network or hybrid neural network can be used [5]. In summary, the decoder defines the joint probability for translation y:

$$p(y) = \prod_{t=1}^{T} p(y_t | y_1, \ldots, y_{t-1}, c) \tag{3}$$

where the $y = (y_1, y_2, \ldots, y_{T_y})$.

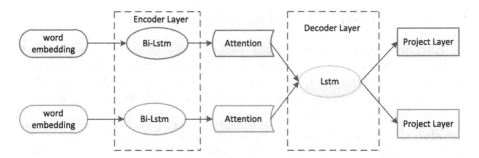

Fig. 1. The framework of SD model. The model consists of three parts, the green curve is a high-resource NMT model, the red curve is a low-resource NMT model, and the yellow curve represents a shared part of the two models. In the training process, the two models independently train the encoder, but share the weights of the decoder. The word embeddings are initialized randomly and updated continuously in iteration. Besides, we investigate the influence of the shared attention mechanism on low-resource NMT model. (Color figure online)

3 Architecture

The low-resource neural machine translation has attracted lots of attention in recent times. Many authors have conducted in-depth research on this issue, especially how to make use of high-resource parallel corpora to assist low-resource NMT [10,14]. It is well known that more high-quality related data can lead to better and more robust network models. For example, the amount of Vietnamese-English corpus is not big enough, but the French-English parallel corpus is large, so we can utilize French-English parallel corpus to improve the performance of Vietnamese-English NMT model. In this paper, we exploit high-resource parallel corpus to enhance the encoder and decoder of low-resource NMT model respectively.

3.1 Strengthen Decoder

In neural machine translation model, the decoder plays an important role in improving fluency for translation system. In essence, the decoder is a recurrent neural network language model that is conditioned on source context in encoder-decoder architecture for NMT. In this section, we aim to exploit the signals from high-resource target side corpora to enhance the decoder of low-resource neural machine translation model, which we refer to as SD model. For detail, we share the weights of decoder between high-resource NMT model and low-resource NMT model to achieve the goal. Given the low-resource parallel corpora $D_L = \{(X^{(n,1)}, Y^{(n,1)})\}_{n=1}^{N_1}$, where the N_1 is not big enough. And we also have large-scale high-resource language pairs $D_H = \{(X^{(n,2)}, Y^{(n,2)})\}_{n=1}^{N_2}$ in which the $N_2 >> N_1$. In the SD model, the neural machine translation is trained with maximum likelihood on the mixed language pairs $\{X^{(n,k)}, Y^{(n,k)}\}_{k=\{1,2\}}^{n=1,\ldots,N_k}$:

$$L(\theta) = \frac{1}{2} \sum_{k=1}^{2} \sum_{n=1}^{N_k} logp(Y^{(n,k)}|X^{(n,k)}; \theta) \tag{4}$$

where θ is parameter of the neural network.

Since the languages utilized by the two models on the decoder are same, we can make full use of the high-resource target language information to improve the performance of the decoder of the low-resource neural machine translation. In addition, we also investigate the influence of the shared attention layer on low-resource neural machine translation model. The Fig. 1 summarizes the general schema of the SD model.

3.2 Strengthen Encoder

Generally speaking, both the source and target word embedding are randomly initialized and then updated as the number of iterators increase in NMT model. This method is feasible for large-scale parallel corpus, but it performs poorly on low-resource corpora. Due to the small size of parallel corpora in low-resource NMT, the translation system can not fully learn the internal structure of sentences and lexical information, which results in the model being unable to generate valid word embedding. This is a major limitation on low-resource neural machine translation system. On the other hand, in many-to-one translation system, it is very significant for low-resource corpora to share sentence representation space with similar languages. For example, it is feasible for low-resource NMT model to share source sentence representation space with high-resource language if the two languages have similar word order. In this section, we use cross-lingual embedding and weight sharing of encoder to alleviate these two issues, which we refer to as SE model.

Share Word Embedding Space. For the first problem, we utilize the cross-lingual embedding to align word embedding space. We train the embedding for each language using monolingual corpora independently, and then learn a liner transformation to map the word embedding from one space to the other. For detail, let X and Y represent the word embedding matrix so that X_i is the word embedding of the i-th entry in the source vocabulary table and Y_j corresponds to the j-th target language embedding. The goal is to find a mapping matrix M that satisfies the following condition:

$$M_* = argmin \sum_{i} \sum_{j} D_{ij}||X_{i*}M - Y_{j*}||^2 \tag{5}$$

where $D_{ij} = 1$ if i-th source language word is aligned with the j-th target language word else 0. In this paper, we follow the cross-lingual embedding method proposed by [13].

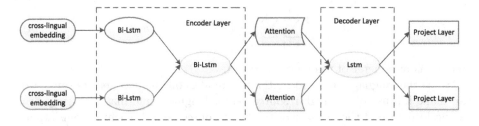

Fig. 2. The framework of SE model. The model can be divided into three parts, the green curve is a high-resource NMT model, the red curve is a low-resource NMT model, and the yellow curve represents a shared part of the two models. In the training process, we utilize cross-lingual embedding to share word embedding space and map the representation of sentences from different languages to same space by sharing weights of the last few layers of the encoder. (Color figure online)

Share Sentence Representation Space. For sentence representation space sharing, we propose a simple and effective way to implement it. As shown in Fig. 2, the proposed method utilizes weight sharing to map representation of sentences from different languages to same space. More concretely, we exploit two independent encoders but sharing the weights of last few layers to extract the high-level representation of the input sentences. Given the input sequence $X = (x_1, x_2, \ldots, x_n)$ and the initial output sequence of the encoder stack $H = (h_1, h_2, \ldots, h_n)$, we calculate H_e as follows:

$$H_e = f \odot H + (1 - f) \odot E(X) \tag{6}$$

where H_e is the final output sequence of the encoder and which will be utilized to calculate the context vector by the decoder and the E is cross-lingual word embedding matrix. f is a gate unit and calculated as follows:

$$f = g(WE + UH + b) \tag{7}$$

where the W and U are the weights of neural network, and the b is bias and they are shared by the two encoders. The following reasons support this approach:

- Through the weight sharing of the last few layers, the two encoders can generate approximate output when the sentences with similar semantics from different languages are used as input. Therefore, the proposed method can get high-quality context vector for low-resource NMT model with the assist of high-resource NMT system.
- There are differences between languages, such as syntax and lexical. In our model, the weights of first few layers of the encoder are not shared, which is to obtain the characteristics of each language for the translation system. Accordingly, we share the weights of last few layers rather than the entire encoder.

4   Experiments

In this section, we describe the data set used in our experiments, data processing, the training details and all the translation results we obtain in experiments.

4.1   Dataset

We evaluate our models on four language pairs: French-English, Vietnamese-English, English-Chinese, Mongolian-Chinese. In our experiments, we translate Vietnamese into English with the help of French-English. Similarly, we translate Mongolian into Chinese with the assistance of English-Chinese.

The Vietnamese-English (133K sentence pairs, 2.7million English words and 3.3 million Vietnamese words) is provided by IWSLT2015 and Mongolian-Chinese (67K sentences pairs, 848K Chinese words and 822K Mongolian words) is provided by CWMT2009. We evaluate our approach on the French-English (2 million sentence pairs, 50 million English words and 52 million French words) translation task of the WMT14 workshop. And we obtain English-Chinese parallel corpus (2 million sentence pairs, 22million English words and 24 million Chinese words) from the WMT17. The Chinese sentences are word segmented using Stanford Word Segmenter. We preserve casing for words and replace those whose frequencies are less than 5 by <unk>. As a result, our vocabulary table size is 17K and 7.7K for English and Vietnamese respectively. And we report BLEU scores on tst2012 and tst2013 for Vietnamese-English translation system. And we make the same treatment for Mongolian-Chinese parallel corpora. Therefore, the size of Chinese and Mongolian vocabulary table is 14K and 12K respectively.

4.2   Training Setup

In our experiment, we exploit encoder-decoder framework with attention mechanism to train NMT model. More concretely, we employ two-layer bi-directional RNN in the encoder, and another two-layer uni-directional RNN in the decoder. All the RNNs use LSTM [15] cells with 600 units, and the dimensionality of word embedding is set to 512. As for attention mechanism, we use the global attention method proposed by [8]. The models are trained using stochastic gradient descent and the maximum length of the sentence is 50. We apply dropout [16] with a probability of 0.25 during training. For all models, the initial learning rate is 0.2, and then it decreases as the number of iterations increases. We initialize all of the parameters of network with the uniform distribution. The maximum value of the gradient is set to 5 in order to solve gradient explosion.

4.3   Results and Analysis

The results of BLEU scores are presented in Table 1. The architecture of baseline system is similar to the one mentioned in Sect. 4.2. However, in order to prevent

Table 1. The performance of proposed method on IWSLT2015 Vietnamese to English tst2012 and tst2013 set and CWMT2009 Mongolian to Chinese test set.

Models	BLEU		
	Vi-En(tst2012)	Vi-En(tst2013)	Mn-Ch
Baseline	20.15	23.07	11.69
SD	20.62	23.59	12.07
SD + share attention	20.48	23.34	11.83
SE	**21.43**	**24.65**	**12.65**

overfitting, we exploit one-layer bi-directional LSTM in the encoder, with 512 units in each cell.

As it can be seen from Table 1, the proposed method obtains very competitive results compared to the baseline system. Our model can reach 21.43 and 24.65 BLEU scores in Vietnamese-English tst2012 and tst2013 set respectively, and we can also achieve 12.65 BLEU scores in Mongolian-Chinese test set, which is much stronger than the baseline system, with improvements of at least 6.3% in all cases, and up to 8.2% in some (e.g. from 11.69 to 12.65 BLEU scores in Chinese to Mongolian). The experiment results show that we can improve the performance of low-resource neural machine translation with the help of high-resource language pairs.

In addition, from Table 1, we can see that the model can increase the 0.47 and 0.52 BLEU scores in Vietnamese-English tst2012 and tst2013 set respectively by only strengthening the decoder of the low-resource NMT model, and the BLEU scores of Mongolian-Chinese is also improved. It reveals that the low-resource NMT system generates a better target side language model than the baseline system by sharing the weights of decoder with high-resource language pairs. However, the performance of low-resource neural machine translation is reduced when we share the weights of attention layer with the high-resource neural machine translation model. Compared with the SD model, the BLEU scores of Vietnamese-English decrease by 0.14 and 0.25 respectively, and the BLEU scores of Mongolian-Chinese also decline. The attention mechanism is used to capture the source side information dynamically, which allows model learning to align between the target language and source language. For different source languages, the alignment matrix obtained by the model is not same. Therefore, sharing attention layer can lead to a decrease in the performance of low-resource neural machine translation system.

5   Related Works

Low-resource neural machine translation has attracted a lot of attention in recent years. [10] presented a transfer learning method to improve the performance of low-resource neural machine translation. Their main idea was to first train a high-resource language pair model, then transfer some of the learned parameters to the low-resource pair to initialize and constrain training. Besides, semi-supervised approach is another way to deal effectively with insufficient resources. [17] explored strategies to train with monolingual data without changing the neural network architecture. They utilized dummy source sentences and synthetic source sentences to construct pseudo-parallel corpora, which brings substantial improvements to neural machine translation. [18] converted a monolingual corpus in the target language into a parallel corpus by copying it, so that each source sentence is identical to its corresponding target sentence. [19] investigated how to utilize the source-side monolingual data in NMT to enhance encoder network. They applied the multi-task learning framework using two NMTs to predict the translation and the reordered source-side monolingual sentences simultaneously. For zero-resource neural machine translation, [20] made attempt to train a source-to-target NMT model without parallel corpora available, guided by an existing pivot-to-target NMT model on a source-pivot parallel corpus. [21] proposed an approach to zero-resource NMT via maximum expected likelihood estimation. Their results revealed that maximum expected likelihood estimation can greatly improve the performance of NMT. [22] introduced automatic encoder to neural machine translation, and proposed a semi-supervised learning method based on bilingual corpus and monolingual corpus. [23] proposed two methods, which are referred to as shallow fusion and deep fusion, to integrate a language model into NMT. The basic idea is to use the language model to score the candidate words proposed by the translation model at each time step or concatenating the hidden states of the language model and the decoder. [24] proposed a finetuning algorithm for multiway, multilingual neural machine translation that enables zero-resource machine translation. In unsupervised Machine Translation, [11] proposed a method to build unsupervised NMT model. They combined unsupervised cross-lingual embedding, denoising auto-encoder and dual learning together to train NMT system in a unsupervised manner. The method proposed in [25] was similar to [11], but the work in [25] was more complete. Although unsupervised machine translation methods are promising, their performance is far lower than supervised machine translation. In multi-task neural machine translation, [26] proposed a method that can simultaneously translate sentences from one source language to multiple target languages. In detail, their models shared source languages representation and separated the modeling of different target language translation.

6   Conclusion

In this paper, we aim to utilize high-resource languages to improve the performance of low-resource neural machine translation. We propose two methods to achieve this goal. One is to exploit the weight sharing of decoder to enhance the target side language model of low-resource NMT system. The other is to enhance the encoder by using cross-lingual embedding and shared sentence representation space.

The experiments show the effectiveness of our proposal, which has significant improvements in the BLEU scores over baseline system. Our model can improve the translation quality on the IWSLT2015 Vietnamese-English translation task. In addition, the proposed approaches in this paper is also effective for Mongolian-Chinese translation.

In the future, we plan to combine unsupervised or semi-supervised methods with our model. Besides, we will verify the approach with more datasets from different domains.

Acknowledgements. The work is supported by the Nation Natural Science Foundation of China under No. 61572462, 61502445.

References

1. Wu, Y., Schuster, M., Chen, Z., et al.: Google's neural machine translation system: bridging the gap between human and machine translation. arXiv preprint arXiv:1609.08144 (2016)
2. Gehring, J., Auli, M., Grangier, D., et al.: Convolutional sequence to sequence learning. arXiv preprint arXiv:1705.03122 (2017)
3. Junczys-Dowmunt, M., Dwojak, T., Hoang, H.: Is neural machine translation ready for deployment? A case study on 30 translation directions. arXiv preprint arXiv:1610.01108 (2016)
4. Bojar, O., Chatterjee, R., Federmann, C., et al.: Findings of the 2016 conference on machine translation. In: ACL 2016 First Conference On Machine Translation (WMT16), pp. 131–198. The Association for Computational Linguistics (2016)
5. Kalchbrenner, N., Blunsom, P.: Recurrent continuous translation models. In: Proceedings of the 2013 Conference on Empirical Methods in Natural Language Processing, pp. 1700–1709 (2013)
6. Sutskever, I., Vinyals, O., Le, Q.V.: Sequence to sequence learning with neural networks. In: Advances in Neural Information Processing Systems, pp. 3104–3112 (2014)
7. Bahdanau, D., Cho, K., Bengio, Y.: Neural machine translation by jointly learning to align and translate. arXiv preprint arXiv:1409.0473 (2014)
8. Luong, M.T., Pham, H., Manning, C.D.: Effective approaches to attention-based neural machine translation. arXiv preprint arXiv:1508.04025 (2015)
9. Koehn, P., Knowles, R.: Six challenges for neural machine translation. arXiv preprint arXiv:1706.03872 (2017)
10. Zoph, B., Yuret, D., May, J., et al.: Transfer learning for low-resource neural machine translation. arXiv preprint arXiv:1604.02201 (2016)
11. Artetxe, M., Labaka, G., Agirre, E., et al.: Unsupervised neural machine translation. arXiv preprint arXiv:1710.11041 (2017)
12. Johnson, M., Schuster, M., Le, Q.V., et al.: Google's multilingual neural machine translation system: enabling zero-shot translation. arXiv preprint arXiv:1611.04558 (2016)
13. Artetxe, M., Labaka, G., Agirre, E.: Learning bilingual word embeddings with (almost) no bilingual data. In: Proceedings of the 55th Annual Meeting of the Association for Computational Linguistics (Volume 1: Long Papers), vol. 1, pp. 451–462 (2017)

14. Nguyen, T.Q., Chiang, D.: Transfer learning across low-resource, related languages for neural machine translation. arXiv preprint arXiv:1708.09803 (2017)
15. Hochreiter, S., Schmidhuber, J.: Long short-term memory. Neural Comput. 9(8), 1735–1780 (1997)
16. Gal, Y., Ghahramani, Z.A.: Theoretically grounded application of dropout in recurrent neural networks. In: Advances in Neural Information Processing Systems, pp. 1019–1027 (2016)
17. Sennrich, R., Haddow, B., Birch, A.: Improving neural machine translation models with monolingual data. arXiv preprint arXiv:1511.06709 (2015)
18. Currey, A., Barone, A.V.M., Heafield, K.: Copied monolingual data improves low-resource neural machine translation. In: Proceedings of the Second Conference on Machine Translation, pp. 148–156 (2017)
19. Zhang, J., Zong, C.: Exploiting source-side monolingual data in neural machine translation. In: Proceedings of the 2016 Conference on Empirical Methods in Natural Language Processing, pp. 1535–1545 (2016)
20. Chen, Y., Liu, Y., Cheng, Y., et al.: A teacher-student framework for zero-resource neural machine translation. arXiv preprint arXiv:1705.00753 (2017)
21. Zheng, H., Cheng, Y., Liu, Y.: Maximum expected likelihood estimation for zero-resource neural machine translation. In: Proceedings of the Twenty-Sixth International Joint Conference on Artificial Intelligence (IJCAI 2017), Melbourne, Australia, pp. 4251–4257 (2017)
22. Cheng, Y., Xu, W., He, Z., et al.: Semi-supervised learning for neural machine translation. arXiv preprint arXiv:1606.04596 (2016)
23. Gulcehre, C., Firat, O., Xu, K., et al.: On using monolingual corpora in neural machine translation. arXiv preprint arXiv:1503.03535 (2015)
24. Firat, O., Sankaran, B., Al-Onaizan, Y., et al.: Zero-resource translation with multi-lingual neural machine translation. arXiv preprint arXiv:1606.04164 (2016)
25. Lample, G., Denoyer, L., Ranzato, M.A.: Unsupervised machine translation using monolingual corpora only. arXiv preprint arXiv:1711.00043 (2017)
26. Dong, D., Wu, H., He, W., et al.: Multi-task learning for multiple language translation. In: Proceedings of the 53rd Annual Meeting of the Association for Computational Linguistics and the 7th International Joint Conference on Natural Language Processing (Volume 1: Long Papers), vol. 1, pp. 1723–1732 (2015)

Identifying Word Translations in Scientific Literature Based on Labeled Bilingual Topic Model and Co-occurrence Features

Mingjie Tian, Yahui Zhao, and Rongyi Cui[✉]

Intelligent Information Processing Lab.,
Department of Computer Science and Technology,
Yanbian University, Yanji 133002, China
iipmjtian@qq.com, {yhzhao,cuirongyi}@ybu.edu.cn

Abstract. Aiming at the increasingly rich multi language information resources and multi-label data in scientific literature, in order to mining the relevance and correlation in languages, this paper proposed the labeled bilingual topic model and co-occurrence feature based similarity metric which could be adopted to the word translation identifying task. First of all, it could assume that the keywords in the scientific literature are relevant to the abstract in the same article, then extracted the keywords and regard it as labels, labels with topics are assigned and the "latent" topic was instantiated. Secondly, the abstracts in article were trained by the labeled bilingual topic model and got the word representation on the topic distribution. Finally, the most similar word between both languages was matched with similarity metric proposed in this paper. The experiment result shows that the labeled bilingual topic model reaches better precision than "latent" topic model based bilingual model, and co-occurrence features enhance the attractiveness of the bilingual word pairs to improve the identifying effects.

Keywords: Topic model · Label · Co-occurrence features · Word translations

1 Introduction

Without any doubt, the Web is growing rapidly, which can be reflected by the amount of online Web content. One challenging but very desirable task accompanying the Web growth is to organize information written in different languages, to make them easily accessible for all users. Recently there has been a new trend that from news to scientific literature, a significant proportion of the world's textual data is labeled with multiple human-provided tags. This trend allows us to understand the content of the document with a more detailed dimension.

The diversity of language has enriched information resources, but differences between languages have inevitably hindered users to use them. Take word translation task as an example, it would be very helpful to leverage the word w_1 in source language for finding the translation w_2 in target language. This work is called word alignment and many research works has been conducted to solve this problem. [1–3] acquired translation candidates based on methods that need an initial lexicon of translations, cognates or similar words to extend additional translations from the context. Some

© Springer Nature Switzerland AG 2018
M. Sun et al. (Eds.): CCL 2018/NLP-NABD 2018, LNAI 11221, pp. 76–87, 2018.
https://doi.org/10.1007/978-3-030-01716-3_7

attempts of obtaining translations using cross-lingual topic models have been made in the last few years. [4] built topics as distributions over bilingual matchings where matching priors may come from different initial evidences such as machine readable dictionary. The main shortcoming of method above is that it introduces external knowledge. [5–7] proposed cross-lingual topic models to solve cross-lingual text classification, machine translation and event detection task, [8] made use of these cross-lingual topic models to obtain word translation in bilingual corpus and proposed word similarity standard *TF-ITF* (term frequency – inverse topic frequency) to improve the matching effects.

We combine the multi-label in scientific literature with the application of topic models word translation task, and propose the Labeled Bilingual Topic Model (LBTM) to solve the implicit meaning of the topics in "latent" topic models, make the topic has defined semantic and be better explained. And we propose a new similarity metric based on co-occurrence features, combine word's feature in topic and corpus level to enhance the translation word's relevance.

2 Methodology

2.1 Latent Dirichlet Allocation

The LDA topic model is a generative model for documents proposed by Blei [9]. It is also known as a three levels Bayesian probability model, which contains three levels of word, topic and document. The LDA topic model introduce Dirichlet prior parameters [10] on the distribution of document-topic and topic-word, which solves the over-fitting problem that occurs when processing the large-scale corpus.

The LDA topic model is a directed probabilistic graphical model [11], as is shown in Fig. 1.

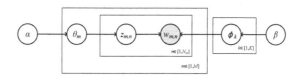

Fig. 1. Probabilistic graphical model of LDA

Given document dataset D contains number of M documents, the m-th document contains number of N_m words, assuming that the number of topics in document dataset D is K totally, the generative process can be described by the following steps:

1. For each document, get the length $N \sim Poission(\xi)$
   a. Choose the distribution over topics $\theta \sim Dir(\alpha)$
2. For each position n ($n = 1$ to N_m) in document
   a. Choose topic $z_n \sim Multinomial(\theta)$
   b. Choose word w_n with probability $p(w_n|z_n, \beta)$

Here, *Poission*(·), *Dir*(·) and *Multinomial*(·) represent possion, dirichlet and multinomial distributions respectively.

The parameters should be estimated during the modelling of the dataset through LDA topic model, the commonly used methods is Variational Bayesian Inference [12], EM algorithm [13] or Collapsed Gibbs Sampling [14, 15]. The method based on Collapsed Gibbs Sampling can effectively sample topics from large-scale document dataset. The parameter estimation process can be considered as the inverse process of the document generation, the parameters are estimated with the documents distribution have been known. The conditional probability of the topic sequence under the known word sequence is calculated as follow.

$$
\begin{aligned}
&p(z_i = k | \overrightarrow{z_{\neg i}}, \overrightarrow{w}) \\
&= \frac{p(\overrightarrow{w}, \overrightarrow{z})}{p(\overrightarrow{w}, \overrightarrow{z_{\neg i}})} = \frac{n_{k,\neg i}^t + \beta_t}{\sum\limits_{v=1}^{V} (n_k^v + \beta_v) - 1} \cdot \frac{n_{m,\neg i}^k + \alpha_k}{\sum\limits_{l=1}^{L} (n_m^l + \alpha_l) - 1}
\end{aligned}
\tag{1}
$$

2.2 Labeled Bilingual Topic Model

The LDA topic model replace the high-dimensional term by the low-dimensional "latent" topic to capture the semantic information of the document. However, each "latent" topic is implicit defined and lack of explanations. We make flexible use of the multi-label data in scientific literature or news such as keywords in the literature to improve the LDA topic model, and propose the Labeled Bilingual Topic Model. Regard the labels in the literature or news as the "topic", make the topics can be explained, instantiate the "latent" topics and assign an explicit meaning. After modelling the document dataset by the model proposed by us, the documents are represented by the meaningful topics. Each word in the document has a probability distribution on the topics, and can be represented as a vector in the vector space to implement word translations identifying.

The Model

Assume that the document dataset consists of M documents, the content of each document is described by two languages L_1 and L_2, the content described by each language is the same. The model has a set of language-independent "common" topics to describe the two languages documents, each "common" topic has two different representations, each corresponding to the each language. The probabilistic graphical model of the Labeled Bilingual Topic Model is shown in Fig. 2.

Where α is the hyper parameter of Dirichlet distribution between document and topics, β_{Lj} is the hyper parameter of Dirichlet distribution between topic and words in language L_j ($j = 1, 2$), Λ is the constraint relationship between document and topics, and the relationship between document and topics are specific, γ is the hyper parameter of the Bernoulli distribution constrained by the document and topics, Λ and γ are given empirically. $w_{m,n}$ represents the n-th word in the m-th document, $z_{m,n}$ is the topic corresponding to the word $w_{m,n}$, the parameter θ_m is the distribution of the m-th

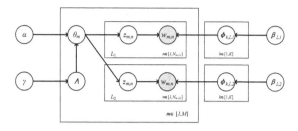

Fig. 2. Probabilistic graphic model of LBTM

document on the topic, the parameter $\phi_{k,Lj}$ is the distribution of the k-th topic on the language L_j's word; Given document dataset D contains number of M documents, the m-th document's language L_j's part contains number of N_m, L_j words, assuming that the number of topics in document dataset D is K totally, the generative process of Labeled Bilingual Topic Model can be described by the following steps:

1. For each "common" topic z, $z = 1, 2, \ldots, K$
   a. For each language L_j, $j = 1$ or 2
      (1) Choose topic's distribution on words $\phi_{z,L_j} \sim Dir(\beta_{L_j})$
2. For the document m in document dataset
   a. Choose document-topic constraint $\Lambda_{m,k} \in \{0, 1\} \sim bernoulli(\gamma)$
   b. Choose the distribution over "common" topics $\theta_m \sim Dir(\alpha)|\Lambda$
3. For each position n ($n = 1$ to N_m) in document
   a. Choose topic $z_{m,n} \sim Multinomial(\theta_m)$
   b. Choose word $w_{m,n,L_j} \sim Multinomial(\phi_{m,n,L_j})$

Estimation

In parameter estimation phase, the Gibbs Sampling method should be modified to fit the bilingual and labeled features. The conditional probability of the topic sequence under the word sequence is extended from monolingual to bilingual. The conditional probability can be calculated as follow.

$$p(z_{i,L_j} = k | \vec{z}_{\neg i, L_j}, \vec{w}_{L_j})$$

$$= \frac{n_{k, \neg i, L_j}^t + \beta_{L_j}^t}{\sum_{v=1}^{V_{L_j}} (n_{k, L_j}^v + \beta_{L_j}^v) - 1} \cdot \frac{\sum_{j=1}^{2} n_{m, \neg i, L_j}^k + \alpha_k}{\sum_{l=1}^{L} (\sum_{j=1}^{2} n_{m, L_j}^l + \alpha_l) - 1} \tag{2}$$

Where $n_{k, \neg i, L_j}^t$ is the number of times word t of language L_j assigned to topic k except t's current assignment, $\sum_{v=1}^{V_{L_j}} n_{k, L_j}^v - 1$ is the total number of words in language L_j assigned to topic k except t's current assignment, V_{L_j} is the vocabulary of language L_j, $n_{m, \neg i, L_j}^k$ is the number of words in language L_j in document m assigned to topic k except t's current assignment, $\sum_{l=1}^{L} \sum_{j=1}^{2} n_{m, L_j}^l - 1$ is the total number of words in all languages in document m except the current word t. Finally, we can obtain word's distribution on the topics φ as follow.

$$\varphi_{t,k,Lj} = \frac{n^t_{k,L_j} + \beta^t_{L_j}}{\sum\limits_{v=1}^{V_{L_j}} (n^v_{k,L_j} + \beta^v_{L_j})} \tag{3}$$

Differences with "Latent" Topic Model

Compare to the multi-lingual models [5–7] derived from "latent" topic model [9], Labeled Bilingual Topic Model takes advantage of the multi-label data in the documents, instantiate the "latent" topics so that the meaning of the topic is no longer "implicit", but "explicit". The differences with the method derived from "latent" topic model are as follows:

1. Topic size K: The determination of the number of topics is one of the difficulties in the LDA topic model, the value of K needs to be choose from the experimental results. The number of topics in LBTM is determined, that is, the number of unique labels in the dataset;
2. Vector representation of the word and document: The topic sampling range for each word in each document at "latent" topic model is $1 \sim K$, documents are represented by the assignment of word and topic in the document, therefore, the value of each topic component in the vector representation of the document and word may not be 0. In LBTM, each document has a constraint with the fixed labels, and the words in the document are also have a constraint with the labels, the topic component value that is constrained to the document which contains the word is not 0, and the rest must be 0.
3. Range of topic sampling: The conditional probability between each word in each document and all topics need to be calculated in each iteration. Due to no constraint between the document and the topic in "latent" topic model, the range of topic sampling for each word in each document is K. In LBTM, there is a fixed constraint relationship between each document and topics, the range of topic sampling of word in a document is a collection of topics (labels) that have a constraint with the document;
4. Sampling computational complexity: Due to the range of sampling, the "latent" topic model needs to calculate the conditional probability between each word and all the topics during each iteration. In LBTM, it is only necessary to calculate the conditional probabilities each word with topics that have fixed constraint with the document. The model proposed by us has advantages in the computational efficiency at sampling process.

3   Similarity Metrics

3.1   Cosine Similarity

The cosine similarity is a measure difference by using the cosine of the angle between two vectors, the smaller the angle between vectors, the more similar the two vectors.

Comparing to Euclidean distance, the cosine similarity focus on the difference in the directions of the two vectors, and the cosine similarity has better robustness to the stretching transformation of the vector. The cosine similarity of words representation on topics are as follow.

$$cos\left(w_{i,L1}, w_{j,L2}\right) = \frac{\varphi_{i,L1} \cdot \varphi_{j,L2}}{||\varphi_{i,L1}|| \, ||\varphi_{j,L2}||} \tag{4}$$

Where $w_{i,L1}$ is i-th word in language L_1 and $w_{i,L2}$ is j-th word in language L_2, and $\varphi_{i,L1}$ is the distribution of i-th word in language L_1 on topics and $\varphi_{j,L2}$ is the distribution of j-th word in language L_2 on topics.

3.2   TF-ITF

Author in [8] borrowed an idea from information retrieval and constructs word vectors over a shared latent topic space. Values within vectors are the *TF-ITF*(term frequency – inverse topic frequency) scores which are calculated in a completely analogical manners as the *TF-IDF* scores for the original word-document space [16]. Given i-th word in language L_j, $n^i_{k,Lj}$ denotes the number of times the i-th word is associated with a topic k. *TF-ITF* score for the i-th word in language L_j and topic k is calculated as follow.

$$TF - ITF_{i,k} = \frac{n^i_{k,Lj}}{\sum\limits_{v=1}^{V_{Lj}} n^v_{k,Lj}} \cdot log \frac{K}{1 + |k : n^i_{k,Lj} > 0|} \tag{5}$$

After words in both languages are represented by *TF-ITF*, the standard cosine similarity metric is then used to find the most similar word vectors from the target vocabulary for a source word vector.

3.3   Co-occurrence Features

In addition to the LBTM to improve the word translation task, we also introduce a similarity metric based on co-occurrence features.

The corpus-based co-occurrence word acquisition method is based on the distributed hypothesis [17], it is based on large-scale corpus and represents the distribution of words in each document as vectors. Finally co-occurrence words is selected by calculating the correlation between the vectors.

If two words in corpus are usually occurred in the same document, the two words can be considered semantically related to each other. The concept of co-occurrence words is mostly applied to the query expansion of information retrieval, when a document is related to the query requirements but does not contain query terms, the query can be expanded by the query co-occurrence word as related information [18]. Method in [19] applied the co-occurrence words to calculate the similarity of cross-lingual documents.

We combined traditional similarity metrics with word co-occurrence to maximize the similarity between the word translation pairs. The similarity combine with co-occurrence features between words in different languages is as follow:

$$sim(w_{i,L1}, w_{j,L2},) = v_{i,L1} \cdot v_{j,L2} + \lambda \, log(m^{i,j} + 1) \qquad (6)$$

Where $w_{i,L1}$ denotes i-th word in language L_1 and $w_{j,L2}$ denotes j-th word in language L_2, $v_{j,L2}$ is the vector representation of word $w_{j,L2}$ and could be topic distributions or TF-ITF. $m^{i,j}$ denotes the number of documents that the words $w_{i,L1}$ and $w_{j,L2}$ co-occurred.

The similarity of the previous item to the right of the equation is the distribution of the word in the topic level, and value of the latter item is the co-occurrence degree between the words at the corpus level. We believe that merging the similarity of different level will take both advantages to improve the word translation precision and it will be proved in the next experiment.

4 Experiments

To verify the validity and feasibility of LBTM and co-occurrence based similarity metric, we carried out word translation identifying experiment. We compare alignment approach with the method proposed in [8], the set of approach contains bilingual topic models and similarity metrics.

4.1 Dataset

The bilingual corpus used in the experiment is the parallel corpus of Chinese-Korean scientific literature. The dataset contains keywords and Abstracts of 2427 Aerospace domain Chinese-Korean parallel scientific literature with the sentence level aligned. In order to reduce data sparsity, we keep only lemmatized nouns and verbs forms for further analysis. Our Chinese vocabulary consists of 20608 terms, while our Korean vocabulary contains 18391 terms. The subset of the 1867 most frequent word translation pairs was used for testing. The sample of parallel corpus is shown in Fig. 3.

4.2 Parameters Setting

The size of topics K, Dirichlet hyper parameters α and β, training and test iterations need to be determined in advance. The topic numbers of LBTM is fixed, and we choose four kinds of topic size to the comparative models. The parameters of LBTM and the "latent" topic model based bilingual model are shown in Table 1.

4.3 Similarity Metrics

We divide the similarity metrics into two types, single kind of similarity method and hybrid method.

Title-一种二元定几何混压式超声速进气道流场控制概念研究
Abstract-针对二元定几何混压式超声速进气道低马赫数时流量系数低加速性能差的问题,提出了一种新的泄流槽流场控制概念,并通过数值仿真,揭示了泄流槽控制激波结构机理及其主要几何参数对进气道性能的影响规律.研究结果表明:采用该流场控制方案可通过泄流槽入口处的波系结构使进气道在低于设计马赫数时的出口总压恢复系数和流量系数相对于原型方案均得到明显提高,而在设计点关闭泄流槽后进气道的性能与原型进气道基本相当,这对改善冲压发动机在低马赫数转级后的加速性能是有利的.
Keywords-航空、航天推进系统;冲压发动机;二元超声速进气道;流场控制;泄流槽

Title- 2차원 고정 형상 혼합-압력식 초음속 흡입구의 유동장 제어에 관한 연구
Abstract-본 논문에서는 2차원 고정 형상 혼합-압력식 초음속 흡입구가 저 마하수(Mach number)에서 흐름 계수의가속 성능이 낮은 문제점을 해결하기 위하여 바이패스 출구로 유동장(flow field)을 제어하는 새로운 기법을 제안하였다.수치해석을 통하여 바이패스 출구가 충격파를 제어하는 메커니즘과 기하학적 파라미터가 흡입구 성능에 주는 영향을 연구하였다.연구 결과를 통하여,제안한 유동장 제어 기법은 바이패스 출구 입구의 파형 구조를 통하여 흡입구의 설계 마하수보다 낮은 유동 속도에서 출구 절대 압력 회복 계수와 흐름 유량 계수가 현저하게 증가하는 것을 확인하였다.바이패스 출구가 닫힌 후, 흡입구의 성능은 원형(Prototype) 흡입구 성능과 같은 특성을 보였으며, 이는 램제트 엔진의 저 마하수 가속 성능에 유리하다는 것을 보여준다.
Keywords-항공 우주 추진 시스템; 램제트 엔진; 2차원 초음속 흡입구; 유동장 제어; 바이패스 출구

Fig. 3. Chinese-Korean parallel scientific literature

Table 1. Comparative experiment parameters setting

Parameters	Comparative model	
	LBTM	Bi-LDA model
K	8718	200/400/600/800
α	50/K	50/K
β	0.01	0.01
Training iterations	1000	1000
Test iterations	100	100

The single similarity method includes cosine similarity, *TF-ITF* proposed in [8] and co-occurrence between words of both language that independent with the topic model's result.

In order to verify that similarity metric combined with co-occurrence features has advantages over the single similarity method, we combined the cos and *TF-ITF* with co-occurrence features, namely cos + co-occurrence and *TF-ITF* + co-occurrence. At the same time, we also compare our metrics with the *TF-ITF* + *TF-IDF* method which obtained the highest result in [8].

4.4 Results and Analysis

Table 2 shows some example of topics produced by LBTM and bilingual "Latent" topic models with K = 200, 400, 600 and 800. Each topic has two representations: first corresponds with the distribution of Chinese words and the second line is associated with Korean words distribution. Words on each line are ranked by probability score in decreasing order. In LBTM, we select topic "涡扇发动机" (turbofan engine), as a fair comparison, we extract the two topics that Chinese word "涡扇发动机" and Korean word "터보팬엔진" (turbofan engine) most frequently appear in each bilingual "Latent" topic models. It can be found that LBTM could catch the words related to the topic "涡扇发动机", even the word "涡扇发动机", 터보팬(turbofan) and 엔진 (engine).

Table 2. Sample of topics

Model	Topic	Words assigned with topic
LBTM	涡扇发动机 터보팬엔진 (TurboFan Engine)	涡扇发动机(turbofan engine) 发动机(engine) 压气机 (compressor) 转速(rotating speed) 高压(high pressure)
		터보팬(turbofan) turbofan(turbofan) engine(engine) 압축 (compression) 엔진(engine)
Bi-LDA Topic 200	86th Topic	湍流(turbulent flow) 湍流模型(turbulent flow model) 对比 (compare) 流动(flow) 机匣(casing)
		난류(turbulence) 모델(model) 결과(result) 평균(average) 비교 (compare)
	42th Topic	工艺(technics) 焊缝(weld seam) 接头(connect) 对接(butt) 焊接接 头(welded joint)
		용접(welding) 접합(joint) 이음(connection) 결과(result) 더블 (double)
Bi-LDA Topic 400	213th Topic	要求(demand) 满足(satisfaction) 需求(requirement) 使用(use) 能 够(can)
		요구(demand) 만족(satisfaction) 대한(about) 충족(satisfy) 요구 사항(requirement)
	361th Topic	变(change) 高(high) 简单(simple) 调节(adjust) 可变(variable)
		가변(variable) variable(variable) 롤링(rolling) 탐색(quest) rolling (rolling)
Bi-LDA Topic 600	372th Topic	吸气(sunction) 对转压气机(Counter rotating compressor) Rotor (rotor) 量(quantity) 效(effect)
		독립(independent) 이중반전(double inversion) 압축기 (compressor) Rotor(Rotor) 샘플(sample)
	87th Topic	发动机(engine) 航空(aviation) 型(type) 工作(working) 加力 (afterburner)
		엔진(engine) 항공기(aircraft) engine(engine) 터보팬(turbofan) 추진력(propulsion)
Bi-LDA Topic 800	771th Topic	压缩(compression) 压缩强度(compressive strength) 记忆合金环 (memory alloy ring) 压缩性能(compression performance) 下降 (descent)
		압축(compression) compression(compression) 파이프(pipe) 실효 성(effectiveness) compressive(compressive)
	562th Topic	加力(afterburner) 起动(start-up) 飞行包线(flight envelope) 调节 规律(regulation law) 加速(Acceleration)
		토대(foundation) 조정(adjust) 비행영역선도(flight area line) off (off) 배기(exhaust)

Table 3 shows the Precision@1 scores (the percentage of words where the first word from the list of translation is the correct one) for all similarity metrics, for different number of topics K in bilingual-LDA and LBTM.

Table 3. Precision@1 scores for experiment

Model	Single metrics			Hybrid metrics			Time
	cos	TF-ITF	co-occur	TF-ITF + TF-IDF	TF-ITF + co-occur	cos + co-occur	Cost (hours)
Topic200	24.64%	24.42%	55.60%	36.31%	36.58%	36.80%	16.62
Topic400	22.28%	22.12%		33.05%	35.99%	37.06%	34.74
Topic600	24.26%	23.03%		34.33%	37.65%	38.35%	50.82
Topic800	25.01%	23.94%		35.62%	38.73%	36.69%	68.64
LBTM	**55.17%**	**50.99%**		**64.69%**	**65.13%**	**68.61%**	**6.06**

LBTM achieves the highest percentage 68.61% in cos + co-occurrence similarity metrics. Compared with the method in [8], LBTM has a higher precision@1 than it in all similarity metrics. We instantiated the "latent" topic means that the topic has a fixed constraint with the documents, it also means that the sampling range of topics assign to the word in the document is fixed. There is a certain relationships between words and topics at the whole corpus level, in other words, the word will must assigned to the one of the certain topic collections. In the vector space model with the topics as the dimensions, the value of each word under these topics are greatly increased, also the probability between word translation pair is increased.

Our model took 6.06 h to train model, with the method in [8] took 16.62 h at least. The reason for saving so much time is that at the training phase, the range of topic sampling in LBTM is given, which is generally the number of keywords (4–6) in per scientific literature, but in comparative experiment, there is no constraint between the document and the topics, the conditional probabilities with all topics should be calculated in whole phase, which increases the computational complexity.

Whether it is LBTM or "latent" topic based bilingual topic model, comparing with the single metrics, the hybrid metrics are greatly improved at least 24%. There are great improvement when cosine similarity or *TF-ITF* combine with the co-occurrence features. Comparing the promotion between co-occurrence and *TF-IDF* to the *TF-ITF*, it could be found the effects are almost the same. Finally, it could be concluded that when the features of the word at corpus level (*TF-IDF* or co-occurrence) are combined with the features of the topic level (Cosine similarity or *TF-ITF*), the combination could play a great synergy. In addition, in the same environment, we can find that whether in single metrics or in hybrid metrics, cosine similarity has better matching result than *TF-ITF*.

5 Conclusion

In this paper, firstly, we combine the multi-label in scientific literature with the application of topic models in word translation identifying, propose the Labeled Bilingual Topic Model (LBTM). Compared with the "latent" topic model, we instantiate the "latent" topics so that the meaning of the topic is no longer "implicit" but "explicit". In the phase of training parameters, due to the given range of topics sampling, in terms of efficiency, LBTM is superior to the "latent" topic model based bilingual LDA model. The result in word translation task indicate that LBTM could reach higher efficiency and precision than "latent" topic model based bilingual LDA model. Secondly, we propose a new similarity metrics that combine bilingual word's co-occurrence features with traditional similarity metrics. Compare with single metrics, the hybrid metrics improved the precision of word translation identifying at least 24%. Limited by the corpus, it can only be applied to the bilingual topic modelling at this stage. If there are multi-lingual multi-labeled parallel corpus in more than two language, we will extend our model to the multi-lingual.

Acknowledgement. This research was financially supported by State Language Commission of China under Grant No. YB135-76.

References

1. Diab, M.T., Finch, S.: A statistical translation model using comparable corpora. In: Proceedings of the 2000 Conference on Content-Based Multi-media Information Access, pp. 1500–1508 (2000)
2. Koehn, P., Knight, K.: Learning a translation lexicon from monolingual corpora. In: Proceedings of the ACL 2002 Workshop on Unsupervised Lexical Acquisition, vol. 9, pp. 9–16. ACL, Stroudsburg (2002)
3. Gaussier, E., Renders, J.M., Matveeva, I., Goutte, C., Déjean, H.: A geometric view on bilingual lexicon extraction from comparable corpora. In: Proceedings of the 42nd Annual Meeting on Association for Computational Linguistics, pp. 526–533. ACL, Stroudsburg (2004)
4. Boyd-Graber, J., Blei, D.M.: Multilingual topic models for unaligned text. In: Proceedings of the Twenty-Fifth Conference on Uncertainty in Artificial Intelligence, pp. 75–82. AUAI Press, Arlington (2009)
5. Ni, X., Sun, J.T., Hu, J., Chen, Z.: Mining multilingual topics from Wikipedia. In: Proceedings of the 18th International World Wide Web Conference, pp. 1155–1156. ACM, New York (2009)
6. Mimno, D., Wallach, H.M., Naradowsky, J., Smith, D.A., McCallum, A.: Polylingual topic models. In: Proceedings of the 2009 Conference on Empirical Methods in Natural Language Processing, pp. 880–889. ACL, Stroudsburg (2009)
7. De Smet, W., Moens, M.F.: Cross language linking of news stories on the web using interlingual topic modelling. In: Proceedings of the 2nd ACM Workshop on Social Web Search and Mining, pp. 57–64. ACM, New York (2009)

8. Vulić, I., De Smet, W., Moens, M.F.: Identifying word translations from comparable corpora using latent topic models. In: Proceedings of the 49th Annual Meeting of the Association for Computational Linguistics: Human Language Technologies: short papers, vol. 2, pp. 479–484. ACL, Stroudsburg (2011)

9. Blei, D.M., Ng, A.Y., Jordan, M.I.: Latent dirichlet allocation. J. Mach. Learn. Res. **3**(Jan), 993–1022 (2003)

10. Qian, X.U., Zhou, J., Chen, J.: Dirichlet process and its applications in natural language processing. J. Chin. Inf. Process. **23**(5), 25–33 (2009)

11. Xu, G., Wang, H.F.: The development of topic models in natural language processing. Chin. J. Comput. **34**(8), 1423–1436 (2011)

12. Fang, A., Macdonald, C., Ounis, I., Habel, P., Yang, X.: Exploring time-sensitive variational Bayesian inference LDA for social media data. In: Jose, J.M., et al. (eds.) ECIR 2017. LNCS, vol. 10193, pp. 252–265. Springer, Cham (2017). https://doi.org/10.1007/978-3-319-56608-5_20

13. Aiping, W., Gongying, Z., Fang, L.: Research and application of EM algorithm. Comput. Technol. Dev. **19**(9), 108–110 (2009)

14. Heinrich, G.: Parameter estimation for text analysis. Technical report (2008)

15. Yerebakan, H.Z., Dundar, M.: Partially collapsed parallel Gibbs sampler for Dirichlet process mixture models. Pattern Recogn. Lett. **90**, 22–27 (2017)

16. Manning, C.D., Schutze, H.: Foundations of Statistical Natural Language Processing. MIT Press, Cambridge (1999)

17. Goodstein, R.L., Harris, Z.: Mathematical structures of language. Math. Gaz. **54**(388), 173 (1970)

18. Bajpai, P., Verma, P.: Improved query translation for English to Hindi cross language information retrieval. Indones. J. Electr. Eng. Inf. **4**(2), 134–140 (2016)

19. Liu, J., Cui, R.Y., Zhao, Y.H.: Cross-lingual similar documents retrieval based on co-occurrence projection. In: Proceedings of the 6th International Conference on Computer Science and Network Technology, pp. 11–15. IEEE (2017)

Term Translation Extraction from Historical Classics Using Modern Chinese Explanation

Xiaoting Wu⬛, Hanyu Zhao⬛, and Chao Che$^{(\boxtimes)}$

Key Laboratory of Advanced Design and Intelligent Computing,
Ministry of Education, Dalian University, Dalian 116622, China
`wuxiaoting2017@163.com`, `hanyuzhao7@163.com`,
`chechao101@163.com`

Abstract. Extracting term translation pairs is of great help for Chinese historical classics translation since term translation is the most time-consuming and challenging part in the translation of historical classics. However, it is tough to recognize the terms directly from ancient Chinese due to the flexible syntactic of ancient Chinese and the word segmentation errors of ancient Chinese will lead to more errors in term translation extraction. Considering most of the terms in ancient Chinese are still reserved in modern Chinese and the terms in modern Chinese are more easily to be identified, we propose a term translation extracting method using multi-features based on character-based model to extract historical term translation pairs from modern Chinese-English corpora instead of ancient Chinese-English corpora. Specifically, we first employ character-based BiLSTM-CRF model to identify historical terms in modern Chinese without word segmentation, which avoids word segmentation error spreading to the term alignment. Then we extract English terms according to initial capitalization rules. At last, we align the English and Chinese terms based on co-occurrence frequency and transliteration feature. The experiment on *Shiji* demonstrates that the performance of the proposed method is far superior to the traditional method, which confirms the effectiveness of using modern Chinese as a substitute.

Keywords: BiLSTM-CRF · Co-occurrence frequency · Transliteration features
Term translation extraction

1 Introduction

Translating outstanding Chinese classics into English is an essential way for Chinese culture promotion. However, at present, only about 0.2% of classical books in China are translated into foreign languages [1]. Speeding up the translation of classics by machine translation is imperative. However, the existing machine translation trained for modern Chinese cannot generate a good translation for historical classics due to the enormous grammatical difference between ancient Chinese and modern Chinese, which can be demonstrated by the example in Table 1. As shown in Table 1, two state-of-the-art machine translation systems, Google and Baidu both give the wrong translation for a sentence from historical book *Shiji*, which is far from correct translation. Besides, it is tough to construct a machine translation system for ancient Chinese because of the lack of ancient Chinese-English parallel corpora. Now it is feasible to carry out targeted

© Springer Nature Switzerland AG 2018
M. Sun et al. (Eds.): CCL 2018/NLP-NABD 2018, LNAI 11221, pp. 88–98, 2018.
https://doi.org/10.1007/978-3-030-01716-3_8

research on the most challenging part of historical classics translation, namely, term translation. This paper extracts the term translation pairs automatically from the bilingual corpus and constructs the term translation dictionary to provide a reference for translators.

Table 1. The translation of the state-of-the-art machine translation systems for an ancient Chinese sentence

Ancient Chinese	固问，语三日，缪公大说，授之国政，号曰五羖大夫。
Google translation	Asked the question, on the 3rd day, Gong Gongda said that he granted the government of the country and the doctor of the 5th class.
Baidu translation	When asked, three days later, Miao Gong said that he gave the national government the number five doctors.
Correct translation	He persisted in his questioning, and they talked for three days. Duke Mu was overjoyed and wanted to hand over the governing of the state to him, entitling him Lord Five Ram Skins.

Accurately identifying terms is the prerequisite for term translation extraction. Nevertheless, it is challenging to extract terms from ancient Chinese directly for three reasons. First, ancient Chinese often omits the context words around the term and uses more flexible grammar; Second, the tagged corpus for ancient Chinese is very limited; Third, there are no efficient algorithms for ancient Chinese word segmentation. Compared to ancient Chinese, term recognition of modern Chinese is more efficient because modern Chinese has a larger scale of tagged corpus, more sophisticated term recognition algorithms [2, 3], and more normative form of language expression. Since most of the terms in ancient Chinese are still preserved in modern Chinese, it is feasible to use the modern Chinese translation of the ancient Chinese to realize the extraction of translated pairs of ancient terms. Moreover, the spread of word segmentation errors can significantly reduce the performance of entity recognition for the traditional Chinese term recognition methods based on words. To this end, this paper proposes a term translation extraction method using character-based sequence model to recognize terms from corresponding modern Chinese explanation instead of ancient Chinese historical books. Specifically, we use the character-based BiLSTM-CRF model [4–6] to identify ancient Chinese terms and extract English terms by the rule of the first letter capitalization. Then extract the Chinese-English term translation pairs according to the co-concurrence frequency [7] and transliteration feature. The framework of term translation extraction is illustrated in Fig. 1.

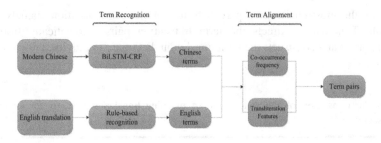

Fig. 1. The framework of term translation extraction method using the character-based model to identify term from modern Chinese instead of ancient Chinese

2   Methodology

According to the framework shown in Fig. 1, this section gives the detailed introduction of each step of the term translation extraction method.

2.1   Recognition of Chinese Terms

Terms in historical classics refer to names, places, official titles and posthumous titles, which are similar to the named entities in modern Chinese. Therefore, extracting terms can be viewed as a name entity recognition(NER) task. Traditional NER methods can be classified into rule based method [8] and statistical machine learning method. The commonly used machine learning methods include Maximum Entropy (ME) [9], SVM [10], CRF [11, 12] and HMM [13]. In recent years, Deep Learning NER method [14] has received extensive attention in the field of natural language processing. Compared to rules-based and statistical machine translation methods, Deep Learning methods have the advantage of generalization ability and less reliance on artificial features.

We employ a character-based BiLSTM-CRF hybrid model to conduct ancient term recognition. The LSTM [15] model has a certain memory function which can effectively solve the problem that the traditional recurrent neural network cannot handle long-distance dependence well. In the sequence labeling task, it is usually necessary to consider the context information at the same time. The unidirectional LSTM only considers the past features, so that the comprehensive feature information cannot be obtained. Therefore, a bidirectional LSTM model was adopted to solve the problem in this paper. The forward LSTM model can record the history information of the sequence and the reverse LSTM model can record the future information. Finally, the outputs of the two models are spliced and used as the final output of the hidden layer. CRF is the most common machine learning model in NER. The objective function of CRF not only considers the input feature, but also takes the label transfer feature into account. In other words, the current prediction label is related to the current input feature and the previous prediction label, and there is a strong interdependence between the prediction label sequences. Adding a CRF layer after the LSTM layer can avoid the occurrence of a label "O" followed by a label "I" when using the "BIO" label strategy [16] for NER. So, the BiLSTM-CRF model can use both the past and the future features learned in the bidirectional LSTM layer, as well as the annotation information learned at the sentence level in the CRF layer.

We use the "BIO" strategy to annotate the corpus. The labeling method is shown in Table 2. "B" denotes the first word of the term, "I" denotes the non-initial word of the term, and "O" denotes the non-term.

Table 2. Example of annotation method

Original sequence	五	大	夫	吕	礼	出	走	逃	亡	到	魏	国	。
Tagged sequence	B	I	I	B	I	O	O	O	O	O	B	I	O

The model structure of Chinese term recognition used in this paper is shown in Fig. 2.

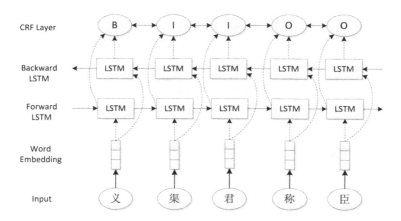

Fig. 2. The structure of the Chinese term recognition model

Given a Chinese sentence, $x = (x_1, x_2, \cdots, x_n)$ indicates the corresponding input word sequence, each layer of BiLSTM-CRF model works as follows:

1. The first layer of the model is the embedding layer, which maps each Chinese character in the sentence into a character vector.
2. The second layer of the model is a bidirectional LSTM layer which automatically extracts sentence features. The vector of each character in the sentence is used as the input for each time step of the bidirectional LSTM. Then, the hidden states output by the forward LSTM $\left(\overrightarrow{h_1}, \overrightarrow{h_2}, \cdots, \overrightarrow{h_n}\right)$ and the reverse LSTM $\left(\overleftarrow{h_1}, \overleftarrow{h_2}, \cdots, \overleftarrow{h_n}\right)$ are bit-spliced as $h_t = \left[\overrightarrow{h_t}; \overleftarrow{h_t}\right] \in R^m$ to obtain a complete hidden state sequence $(h_1, h_2, \cdots, h_n) \in R^{n \times m}$. The probability output matrix of LSTM is defined as P_{n*k},

where k is the number of output labels, $P_{i,j}$ refers to the probability that the i-th word is marked as the j-th tag.

3. The third layer of the model is the CRF layer, where sentence-level sequence annotation is performed. For the tag sequence to be predicted: $y = (y_1, y_2, \cdots, y_n)$, the score of the model for the tag of the sentence x equaling to y is defined as follows:

$$s(x, y) = \sum_{i=0}^{n} A_{y_i, y_{i+1}} + \sum_{i=0}^{n} p_{i, y_i} \tag{1}$$

Where A is the state transition matrix and $A_{i,j}$ represents the probability of the transition from the i-th label to the j-th label. y_0 and y_n are the labels added to the position at the beginning and end of the sentence, respectively. So, A is a square matrix of size k + 2.

Softmax function is employed to get the normalized probability:

$$P(y|x) = \frac{\exp(score(x, y))}{\sum_{y'} \exp(score(x, y'))} \tag{2}$$

Model is trained by maximizing the logarithmic probability of correct label sequence:

$$\log P(y^x|x) = score(x, y^x) - \log(\sum_{y'} \exp(score(x, y'))) \tag{3}$$

The model uses the viterbi algorithm of dynamic programming to obtain the best output label sequence during the prediction process:

$$y^* = \arg \max_{y'} score(x, y') \tag{4}$$

2.2   Recognition of English Terms

This paper uses **Record of the Grand Historian of China** [17] as the English translation, in which all the English terms are capitalized. Consequently, we use the initial capitalization rules to identify English terms. However, the capitalization extraction rule has two problems: (1) the first word of the sentence is extracted as a wrong term and (2) some articles and conjunctions in terms will be missed, such as "the" and "of", which are not capitalized in terms. For the first problem, we do not treat it as a term when the extracted term is at the beginning of the sentence and contains only one word of following part of speech: numeral, preposition, adverb, conjunction, etc. For the second problem, if "the" is followed by a capital word, or "of" is sandwiched between two capital words, they are added to the extracted terms.

2.3   Term Alignment

According to the alignment process, the alignment method based on bilingual parallel corpus can be divided into two categories: (1) the symmetric method, which identifies the terms in two languages, respectively, then the alignment model is used to align the terms in the two sides; (2) the asymmetric method, which recognizes the terms in one language, then find its corresponding translation in another language. The recognition methods used in this paper have good performance on Chinese terms and English terms. Therefore, we use the symmetric method to align terms with co-concurrence frequency and transliteration feature.

Co-occurrence Frequency. To avoid the performance decrease of term alignment caused by word segmentation errors, we adopt a character-based Chinese term recognition method. However, this method can identify many non-terms, which can negatively affect the term alignment and increase the difficulty of alignment. Considering the co-occurrence frequency of terms is helpful for identifying the translation of terms accurately. If a term pair appears more frequently in all the pairs of terms which are related to English term e, then it is more likely to be a correct term pair.

For an English term e, the co-occurrence frequency of term pairs is defined as:

$$F(c_i|e) = \frac{N(c_i, e)}{N_e} \tag{5}$$

Where $N(c_i, e)$ denotes the times that the term pair co-occurs, and N_e denotes all term pairs that contain the English term e.

Transliteration Feature Transliteration is often used in the English translation of historical books. According to two kinds of transliteration of historical classics, we use the method in [18] for reference and use the proportion of the transliteration words in the English terms as transliteration feature values. The transliteration function can be defined as follows:

$$H(c|e) = \frac{N_{pinyin}(c, e) + N_{title}(c)}{len(e)} \tag{6}$$

Wherein, Len(e) indicates the number of the real words in English terms, and the prepositions such as "the", "of", "in", "on" are not included in the counting. The word containing "-" is counted as two words. $N_{pinyin}(c, e)$ denotes the number of Chinese phonetic alphabet in the English term e corresponding to the Chinese term c; $N_{title}(c)$ indicates the number of characters of fixed title in the Chinese term c. We construct a fixed title list manually. The number of fixed title can be determined by querying the list.

Term Alignment For each candidate Chinese term of the English term e, we calculate the co-occurrence frequency F and the transliteration feature value H, which are added to obtain the probability of alignment. Finally, the candidate Chinese term with the largest alignment probability is selected as the translation of the term e.

2.4 Term Translation Extracting Method

Given a bilingual corpus, the algorithm for extracting Chinese and English term translation pairs is shown in Fig. 3.

Input: Chinese-English corpus CE

(1) For every parallel sentence pair $(c_i, e_i), i \in (1, n)$, extract Chinese

terms $A_i = \{t_1^c, t_2^c, \cdots, t_m^c\}$ and English terms $B_i = \{t_1^e, t_2^e, \cdots, t_s^e\}$

from Chinese and English sentence, respectively.

(2) For every English term $t_i^e \in B_i$, calculate the co-occurrence fre-

quency $F(t_j^c | t_i^e)$, $t_j^c \in A_i$.

(3) For every English term $t_i^e \in B_i$, calculate the transliteration feature

value $H(t_j^c | t_i^e)$, $t_j^c \in A_i$.

(4) For each candidate term pair (t_j^c, t_i^e), calculate the alignment prob-

ability: $P(t_j^c | t_i^e) = F(t_j^c | t_i^e) + H(t_j^c | t_i^e)$, and select the term pair

with the greatest probability of alignment P for t_i^e as the final result

to add to the set T_{ce}.

Output: Chinese - English term pairs T_{ce}

Fig. 3. Term translation extraction algorithm

3 Experiment

3.1 Experimental Setup

In the experiment, we used the parallel corpora composed of modern Chinese and English translation of *Shiji* to extract historical term pairs. The modern Chinese we used is *The Chronicle of the Vernacular History* [19], and the English translation is the 1961 version of *Record of the Grand Historian of China* which is written by Burton Watson. Based on the existing ancient Chinese-English bilingual corpus, we find the corresponding modern Chinese for each ancient Chinese sentence to construct the modern Chinese-English bilingual parallel corpora. The training set contains 5060 sentence pairs which are selected from *Annals of Qin, the Basic Annals of the First Emperor of the Qin, the Basic Annals of Hsiang Yu, the basic Annals of Emperor Kao-tsu and the Basic Annals of Empress Li*. 1085 sentence pairs from *the Hereditary House of the Marquis of Liu* and *the Tiereditary House of Prime Minister Chen* are used as a test set. And 287 term translation pairs are manually extracted from test set as standard answer.

3.2 Evaluation Metrics

The experiment results are evaluated by precision (P), recall (R), and F1 measure. Suppose N_{gold} is the number of term pairs in our corpus, $N_{extracted}$ is the number of the term pairs that extracted by our method, and $N_{correct}$ is the number of correct term pairs extracted by our method. P, R, F1 are defined as follows, respectively.

$$P = \frac{N_{correct}}{N_{extracted}} \times 100\% \tag{7}$$

$$R = \frac{N_{correct}}{N_{gold}} \times 100\% \tag{8}$$

$$F1 = \frac{2 \times P \times R}{P + R} \tag{9}$$

3.3 Experimental Result and Analysis

Feasibility Analysis. To verify the feasibility of using modern Chinese to replace ancient Chinese, we conducted term translation extracting method based on BiLSTM-CRF to extract the term pairs from the ancient Chinese-English corpus and the modern Chinese-English corpus, respectively. The results comparison is shown in Table 3.

Table 3. Result comparison of the term translation extraction method using different corpora

Corpus	P	R	F1
Ancient Chinese-English	67.4%	53.5%	59.7%
Modern Chinese-English	**80.8%**	**79.4%**	**80.1%**

From Table 3, it can be seen that the method of extracting terms from modern Chinese significantly improved the P, R, and F1 values compared with the method of directly extracting terms from ancient Chinese. The main reason is that the ancient Chinese often adopts complex and flexible sentence structures such as passive, inversion, and omit structure. When the corpus is relatively small, it is tough to learn accurate and comprehensive features using the BiLSTM-CRF model. Therefore, the performance of the term translation extraction method using ancient Chinese is poor. For example, the sentence "四人至，客建成侯所。" can be translated into modern Chinese as "四个人来了，就住在建成侯的府第中为客。" (When the four arrived, they were entertained as guests at the house of Lü Ze). The sentence "客建成侯所" reflects a more common structure of the object ahead of the predicate in ancient Chinese. In addition, due to the richness of lexical meaning of words in ancient Chinese, a noun "客" expresses the meaning of "to be a guest in …". It is indeed difficult for LSTM to learn similar features in ancient Chinese, so the term "建成侯" was not identified in this sentence. In modern Chinese, this complex sentence structure has been transformed into a simpler and more regular structure of "subject-predicate-object", which is more conducive to the study of characteristics. Therefore, "建成侯" can be successfully identified in modern Chinese corpus.

According to the statistics of experimental corpus, about 96% of the ancient Chinese terms are completely preserved in modern Chinese, which means that the majority of term translation pairs can be extracted by this method. However, for some ancient Chinese terms that have not been completely preserved in modern Chinese, such as interchangeable words, we cannot obtain their translations. For example, the modern Chinese translation "宁昌" corresponding to ancient term "甯昌", therefore, we can not extract the enlish translation of "甯昌" through the method based on modern Chinese explanation.

The Effect of Term Recognition on Term Alignment The quality of the term recognition directly affects the effect of term alignment. So, we compared BiLSTM-CRF with CRF and LSTM-CRF to explore the effect of different term recognition methods on the term pair extraction. The results of the experiment are shown in Table 4.

Table 4. Result comparison of different term recognition methods

Method	Term recognition			Term alignment		
	P	R	F1	P	R	F1
CRF	77.3%	70.3%	73.6%	67.3%	61.2%	64.1%
LSTM-CRF	84.4%	79.4%	81.8%	72.5%	68.5%	70.4%
BiLSTM-CRF	**89.3%**	**87.8%**	**88.5%**	**80.8%**	**79.4%**	**80.1%**

Term recognition is the basis for the term alignment, and higher term recognition recall rate ensures that as many term as possible are identified, which lays the foundation for the term alignment. From Table 4, we can see that the higher the recall rate of the term recognition, the better the result of term alignment.

Comparison with Traditional Term Alignment Methods In this paper, the traditional IBM model4 alignment model was used as the baseline. The IBM model4 method employed Jieba[1] to perform word segmentation for modern Chinese and then aligned the term between modern Chinese and English by GIZA++[2]. The results of the experiment are shown in Table 5.

Table 5. Result comparison of different term translation extraction methods

Method	P	R	F1
IBM model4	67.1%	54.2%	60.0%
Our method	**80.8%**	**79.4%**	**80.1%**

---

[1] https://github.com/fxsjy/jieba.

[2] https://codeload.github.com/moses-smt/giza-pp/zip/master.

Table 5 shows that the proposed method outperforms the traditional term alignment model IBM Model4 with a large margin. This is due to the following two aspects:

Firstly, traditional alignment methods often have high requirements on word segmentation performance. Obtaining a good word segmentation result is a prerequisite for extracting correct translation terms. Nevertheless, the most used word segmentation methods have a very poor performance for ancient Chinese, which will decrease the performance of term alignment. For example, in the segmentation result of Jieba "留/侯张良/,/他/的/祖先/是/韩国/人", "侯张良" is segmented as a word, which will align "侯张良" with "Zhang Liang" wrongly. On the other hand, our method identifies Chinese terms at the character level, which avoids the problem that the word segmentation errors decrease the accuracy of term recognition.

Secondly, our method can effectively alleviate the problem of low recognition performance of the Chinese term. Because the BiLSTM-CRF model extracted some non-terminological vocabularies when recognizing Chinese terms, such as "小子", "中间" and "五谷", this result in the vast number differences between Chinese and English terms identified in a sentence pair, which makes it more difficult in aligning terms. However, the recall rate of our term recognition model is 89.3%, which guarantees that we can identify most of the terms and lays the foundation for the extraction of term pairs. For Chinese terms and English terms extracted from a sentence pair, we calculate the co-occurrence frequency and transliteration feature values between each Chinese term and each English term to determine the final alignment result, which reduces the effect of non-terminology words on the extraction of term pairs.

4   Conclusions

In order to solve the difficulty of term translation in the process of translating ancient Chinese classics, this paper proposes a term translation extracting method using multi-features based on BiLSTM-CRF to extract historical term pairs from modern Chinese-English parallel corpora instead of ancient Chinese-English parallel corpora. Our method not only avoids word segmentation error spreading to the term alignment, but also solves the difficulty of extracting historical term translations directly from ancient Chinese corpus.

However, our method can only identify and align terms preserved in modern Chinese. In the future, we will explore the method which can deal with the terms that are not retained in modern Chinese.

Acknowledgements. This work is supported by the National Natural Science Foundation of China (No. 61402068) and Support Program of Outstanding Young Scholar in Liaoning Universities (No. LJQ2015004).

References

1. Huang, Z.X.: English translation of cultural classics and postgraduate teaching of translation in Suzhou university. Shanghai J. Trans. **1**, 56–58 (2007). (in Chinese)
2. Wang, B.: Translation pairs extraction from unaligned Chinese-English bilingual corpora. J. Chin. Inf. Process. **14**(6), 40–44 (2000). (in Chinese)
3. Yang, P., Hou, H.X., Jiang, Y.P., Jian, Shen, Z., D.U.: Chinese-Slavic Mongolian named entity translation based on word alignment. Acta Scientiarum Naturalium Universitatis Pekinensis **52**(1), 148–154 (2016). (in Chinese)
4. Lample, G., Ballesteros, M., Subramanian, S., Kawakami, K., Dyer, C.: Neural Architectures for Named Entity Recognition, pp. 260–270 (2016)
5. Zeng, D., Sun, C., Lin, L., Liu, B.: LSTM-CRF for drug-named entity recognition. Entropy **19**(6), 283 (2017)
6. Dong, C., Zhang, J., Zong, C., Hattori, M., Di, H.: Character-based LSTM-CRF with radical-level features for Chinese named entity recognition. In: Lin, C.-Y., Xue, N., Zhao, D., Huang, X., Feng, Y. (eds.) ICCPOL/NLPCC-2016. LNCS (LNAI), vol. 10102, pp. 239–250. Springer, Cham (2016). https://doi.org/10.1007/978-3-319-50496-4_20
7. Li, X., Che, C., Liu, X., Lin, H., Wang, R.: Corpus-based extraction of Chinese historical term translation equivalents. Int. J. of Asian Lang. Proc. **20**(2), 63–74 (2010)
8. Zhou, K.: Research on named entity recognition based on rules. Hefei University of Technology (2010). (in Chinese)
9. Hai, L.C., Ng, H.T: Named entity recognition: a maximum entropy approach using global information. In: International Conference on Computational Linguistics (2002)
10. Li, L., Mao, T., Huang, D., Tang, Y.: Hybrid models for chinese named entity recognition. In: Proceedings of the Fifth Sighan Workshop on Chinese Language Processing, pp. 72–78 (2006)
11. Şeker, G.A., Eryiğit, G.: Extending a CRF-based named entity recognition model for Turkish well formed text and user generated content 1. Semant. Web **8**(5), 1–18 (2017)
12. Sun, L., Guo, Y., Tang, W., et al.: Enterprise abbreviation prediction based on constitution pattern and conditional random field. J. Comput. Appl. **36**(2), 449–454 (2016)
13. Patil, N.V., Patil, A.S., Pawar, B.V.: HMM based named entity recognition for inflectional language. In: International Conference on Computer, Communications and Electronics (2017)
14. Wang, G.Y.: Research of chinese named entity recognition based on deep learning. Beijing University of Technology (2015). (in Chinese)
15. Greff, K., Srivastava, R.K., Koutník, J., Steunebrink, B.R., Schmidhuber, J.: LSTM: a search space odyssey. IEEE Trans. Neural Netw. Learn. Syst. **28**(10), 2222–2232 (2017)
16. Settles, B., Craven, M.: An analysis of active learning strategies for sequence labeling tasks. In: Conference on Empirical Methods in Natural Language Processing (2008)
17. Watson, B.: Record of the grand historian of China. J. Asian Stud. **22**(2), 205 (1961)
18. Che, C., Zheng, X.J.: Sub-word based translation extraction for terms in Chinese historical classics. J. Chin. Inf. Process. **30**(3), 46–51 (2016). (in Chinese)
19. Sixty professors in Taiwan's 14 institutions. The Chronicle of the Vernacular History. New World Press (2007). (in Chinese)

Research on Chinese-Tibetan Neural Machine Translation

Wen Lai, Xiaobing Zhao[✉], and Xiaqing Li

National Language Resource Monitoring and Research Center of Minority
Languages, Minzu University of China, Beijing 100081, China
Lavine_Lai@126.com, nmzxb_cn@163.com,
xiaqing0614@foxmail.com

Abstract. At present, the research on Tibetan machine translation is mainly
focused on Tibetan-Chinese machine translation and the research on Chinese-
Tibetan machine translation is almost blank. In this paper, the neural machine
translation model is applied to the Chinese-Tibetan machine translation task for
the first time, the syntax tree is also introduced into the Chinese-Tibetan neural
machine translation model for the first time, and a good translation effect is
achieved. Besides, the preprocessing methods we use are syllable segmentation
on Tibetan corpus and character segmentation on Chinese Corpus, which has a
better performance than the word segmentation on both Chinese and Tibetan
corpus. The experimental results show that performance of the neural network
translation model based on the completely self-attention mechanism is the best
in the Chinese-Tibetan machine translation task and the BLEU score is
increased by one percentage point.

Keywords: Neural machine translation · Tibetan · Syntactic tree
Attention

1 Introduction

Machine translation, studies on how to use computers to achieve the automatic
translation between natural languages, is one of the important research directions in
artificial intelligence and natural language processing (Liu 2017). Natural language
processing is a discipline that crosses computer science and linguistics. Based on
characteristics of this discipline, the system of machine translation can be divided into
two categories, which are the rule-based methods and the corpus-based methods.
Among them, corpus-based methods can be divided into statistics-based methods and
example-based methods (Zhao 1900). In recent years, with the development of internet
technology and the improvement of computing speed of computer, machine translation
has achieved fruitful results both in academia and industry area.

Since the advent of the neural network in the 1940s, it has experienced a period of rise
- low tide - rise. Until 2006, Hinton et al. solved the historic problem of neural net-works
(Hinton et al. 2006), and the related researches of deep learning and neural network
returned to people's attention again. Since then, with the deepening of theoretical
research and the improvement of computing speed of computers, neural net-works have

M. Sun et al. (Eds.): CCL 2018/NLP-NABD 2018, LNAI 11221, pp. 99–108, 2018.
https://doi.org/10.1007/978-3-030-01716-3_9

made great breakthroughs in various fields of artificial intelligence, such as computer vision, machine learning, and speech recognition. Researches about natural language processing have also made a rapid progress along with this tide.

In 2012, the Hinton research group participated in the ImageNet image recognition contest and won the championship, which opened the prelude of deep learning in various fields of artificial intelligence. Neural machine translation (NMT) is also a machine translation method that is gradually emerging at this stage. The main processes of neural machine translation are as follows: Firstly, it uses deep neural net-works (RNN, CNN, etc.) to encode the source language into word-embedding. Secondly, the word-embedding generates the target language by decoding.

Tibetan is a kind of pinyin character, and its syllables are composed of 34 vowel consonants, and then Tibetan words are composed of syllables (Wei 2015). A single character in a Tibetan text is a unit, and it is separated by a syllable separator " " between characters (Cai 2016). Based on the characteristics of Tibetan language, the statistical machine translation model is mainly used in the research on Tibetan trans-lation model (Dong et al. 2012; Luo et al. 2010; Hua et al. 2014a, b), and the relevant theoretical research has basically stopped at the stage of word processing and other corpus pre-processing (Cai et al. 2011; Pang et al. 2015; Xiang et al. 2011; Hua et al. 2014a, b; Nuo et al. 2011). Overall, compared with other rich languages, the research on Tibetan-Chinese or Chinese-Tibetan machine translation are obviously behind. There are few researches focus on neural network model in Tibetan corpus (Li et al. 2017a, b). Besides, Tibetan texts are all word segmentation pre-processed in traditional Tibetan machine translation tasks (Guan 2015). In this article, the traditional method of Tibetan word segmentation is completely abandoned, and Tibetan texts are directly divided into syllables. It gets a better result than Tibetan word segmentation.

The study of Chinese-Tibetan machine translation has a far-reaching significance in promoting Chinese-Tibetan technological and cultural exchanges and promoting the development of educational and cultural undertakings. At present, most of the research related to Tibetan machine translation is focused on Tibetan-Chinese machine trans-lation. There are few researches on Chinese-Tibetan machine translation, and the existing Chinese-Tibetan machine translation is mainly based on statistical machine translation methods. For the related research on machine translation, there is no neural machine translation models used in Chinese-Tibetan corpus, Therefore, in this paper it is innovative to apply the neural network model to Chinese-Tibetan corpus for the first time.

In this paper, five common neural network machine translation models and the latest syntax-tree-based neural network machine translation model are used in Chinese-Tibetan machine translation tasks, and the final translation results are analyzed in detail. The experimental results show that the application of neural network machine trans-lation model on Chinese-Tibetan machine translation tasks has a good performance, and the method of syllable segmentation on Tibetan corpus has a better translation performance than the method of word segmentation on Tibetan corpus in Chinese-Tibetan machine translation tasks. Meanwhile, it shows that the syntax tree is intro-duced in the neural network machine translation model, which can improve the translation performance.

2   Neural Machine Translation Models

2.1   Seq2Seq

The Seq2Seq model was presented in 2014, and two articles published by the Google Brain team (Sutskever et al. 2014) and the Yoshua Bengio team (Cho et al. 2014) illustrate the basic idea of the model. The basic idea of solving the problem of the Seq2Seq model is to map an input sequence to an output sequence through one or more deep neural network models, which is commonly known as LSTM — Long short-term memories network (D'Informa-tique et al. 2001), and this process consists of two parts of encoding input and decoding output.

2.2   RNNSearch

In 2015, RNNSearch machine translation model was proposed by Bahdanau et al. (2014). Based on the encoder-decoder structure, the attention mechanism is added to this model, it is used in natural language processing tasks for the first time, and translation performance is greatly improved.

2.3   Fairseq

Fairseq machine translation model was presented by the Facebook team in May 2017 (Gehring 2017). The traditional method of sequence to sequence learning is to map an input sequence to a variable length output sequence through one or more layers of RNN neural network. In the Fairseq model, a structure is introduced based on con-volutional neural networks (CNNs) entirely. Compared with the recurrent neural network model, all calculations of the element sequence are completely parallel while Fairseq model is in training, the number of nonlinear sequences is fixed and independent of the length of the input sequence.

The research shows that in the same environment, the training time of Fairseq model is 9 times faster than the translation model based on RNN network, and its accuracy is also higher than that of the model based on RNN network.

2.4   Transformer

Transformer machine translation model was proposed by the Google team in June 2017 (Vaswani et al. 2017). In the traditional neural network machine translation model, the neural network is mostly used as the model basis of Encoder-Decoder. This model is based on the attention mechanism and completely abandons the inherent mode of the neural machine translation model without any neural network (CNN or RNN) structure. Experiment results show that this model can run fast in parallel, which greatly improves the training speed of the model while improving performance of machine translation.

The Transformer model performs well in natural language processing tasks such as syntactic parsing and semantic understanding, which is also a breakthrough in the system of natural language processing for decades.

2.5 Syntax-NMT

The syntax-based machine translation model introduces syntactic information into machine translation model, and then modeling the language structure. The research shows that the syntax model can reordering long-distance sentences effectively (Xue et al. 2008). Statistical machine translation methods based on linguistic syntax trees can be further divided into three types: string to tree model, tree to string model, and tree to tree model (Xiong et al. 2008).

Recent years, the syntax-based neural machine translation model is gradually favored by scholars (Chen et al. 2017; Eriguchi et al. 2017; Li et al. 2017a, b; Aharoni et al. 2017). In the NMT case, the syntax information is introduced, it will be easier for the encoder to incorporate long distance dependencies into better representations, which is especially important for the translation of long sentences.

The research shows that the syntax-based neural machine translation model outperform the sequential attentional model as well as a stronger baseline with a bottom-up tree encoder and word coverage in Chinese-English machine translation task.

3 Experimental Setup

3.1 Experimental Corpus

In this paper, we use the Tibetan-Chinese comprehensive evaluation corpus of the 13th National Machine Translation Symposium (CWMT 2017 in china, http://ee.dlut.edu.cn/CWMT2017/index.html). This corpus is processed into Tibetan-Chinese sentence pairs, which contains word segmentation, syllable segmentation and some alignment process. This corpus is shown in following Table 1.

Table 1. Experimental corpus

Corpus	Department	Corpus-Area	Scale (sentence pairs)
QHNU-CWMT2013	Qinghai Normal University (in China)	Government	33145
QHNU-CWMT2015	Qinghai Normal University (in China)	Government	17194
XBMU-XMU	Artificial intelligence institute of Xiamen University (in China) Institute of language (technology), Northwestern University of Nationalities (in China)	Synthesize	52078
XBMU-XMU-UTibent	Institute of language (technology), Northwestern University of Nationalities (in China) Tibet University Artificial intelligence institute of Xiamen University (in China)	Government law	24159
ICT-TC-Corpus	Institute of Computing Technology, Chinese Academy of Sciences (in China)	News	30004

3.2   Corpus Preprocessing

In this paper, Chinese-Tibetan bilingual parallel corpus is preprocessed and then divided into a training set, (141601 sentence pairs), a development set (1000 sentence pairs) and a test set (1000 sentence pairs). Preprocessing tasks include: character segmentation and syllable segmentation on Tibetan corpus, word segmentation and character segmentation on Chinese corpus. For efficient training, we also filter out the sentence pairs whose source or target lengths are longer than 50. Details are shown as Table 2.

Table 2. Sentences and words in corpus

Language	Sentence pairs	Words	Characters
Tibetan	139535	16742	15201
Chinese	139535	23384	4932

3.3   Corpus Preprocessing

In our experiment, to reflect the performance of neural machine translation, phrase-based statistical machine translation model Nitutrans (Xiao et al. 2012) developed by natural language processing laboratory in northeastern university (in china) is used. The syntax tree-based neural machine translation model framework Syntax-NMT developed by Nanjing University's natural language processing laboratory is used. (Chen et al. 2017). Chinese and Tibetan corpus are processed by BPE (Sennrich et al. 2015). The five neural machine translation models used in this paper are consistent in the basic parameter settings, the vocabulary of the sub word table is set to 32000, and the number of training iterations is 200000. Because each model has its own structure, it is difficult to achieve consistent in terms of performance of parameters. In addition, with the language characteristics of the Chinese-Tibetan bilingual corpus, hyperparameters are adjusted to achieve maximum of translation performance on each model. Bilingual evaluation understudy (BLEU) is used as evaluation index in this paper (Papineni et al. 2002).

4   Experimental Results

4.1   Corpus According to Character Segmentation and Word Segmentation

To verify the translation performance of the character segmentation (Tibetan syllable segmentation and Chinese character segmentation) and the word segmentation (Tibetan word segmentation and Chinese word segmentation) on Chinese-Tibetan corpus. Among them, Tibetan word segmentation tool TIP-LAS is used in the Tibetan word segmentation (Li et al. 2015). THU-LAC software opened by Tsinghua university is used to conduct Chinese word segmentation (Li and Sun 2009). The experimental results of the Chinese-Tibetan translation based on the statistical machine translation model are shown in Table 3. The experimental results of the Chinese-Tibetan translation based on the neural network translation model are shown in Table 4.

Table 3. Corpus according to character segmentation and word segmentation (SMT model)

Model	Corpus processing	BLEU
Niutrans	Character	70.65*
Niutrans	Word	69.08

Table 4. Corpus according to character segmentation and Word segmentation (NMT model)

Model	Corpus processing	BLEU
Transformer	Character	71.69*
Transformer	Word	71.25

The experimental results show that in Chinese-Tibetan machine translation task, whether SMT model or NMT model, the performance of character segmentation on corpus is obviously better than that of word segmentation on corpus. The reasons are as follows: Firstly, reduce the granularity of corpus can improve the performance of machine translation; Secondly, based on the language characteristics of Tibetan language, the granularity after segmentation is larger unit, which decreased the performance of machine translation.

4.2   Syntax Tree-Based Machine Translation Models

To verify the translation performance of the syntax-based machine translation models. Niutrans is used in our syntax-based statistic machine translation and Syntax-awared-NMT is used in our syntax-based neural machine translation. We compared the seq2seq model with the Syntax-awared-NMT model and both model have a same framework of neural network. And Berkeley parser tools are used (Petrov et al. 2006) to generate the Chinese syntax tree. Experimental results are show in Table 5.

Table 5. Syntax tree-based models

Model	Corpus processing	BLEU
Niutrans-syntax	Word	64.53*
SEquation2Seq	Word	56.12
Syntax-awared-NMT	Word	61.45

The experimental results show that when the syntax tree is introduced in the seq2seq model, the translation performance has improved. In the Chinese-Tibetan machine translation task, the tree-to-string model based on statistical machine translation is better than the syntax tree-based neural machine translation model, this may be due to the small-scale corpus. The reasons are as follows: Firstly, adding prior knowledge such as linguistics can improve the performance of machine translation; Secondly, syntactic tree information, which can well parse source and target languages and improve machine translation performance.

4.3   Different Neural Networks with the Same Structure

To verify the performance of different neural networks with the same model structure, experiments were conducted in RNNSearch and Fairseq models respectively. Both RNNSearch and Fairseq models are models based on the neural network and attention mechanism. The only difference is that RNNSearch is a model based on cyclic neural networks, whereas Fairseq is a model based on convolutional neural networks. The experimental results are shown in Table 6.

Table 6. Different neural networks with the same structure

Model	Framework	Corpus processing	BLEU
Fairseq	CNN + Attention	Character	18.48
RNNSearch	RNN + Attention	Character	67.71

The experimental results show that there is obvious difference in performance of the same translation structure with different neural networks. Based on its frame-work, Fairseq translation model has a pool performance in Chinese-Tibetan translation tasks. The reasons are as follows: Convolutional neural networks and recurrent neural networks are different in network structure. Convolutional neural networks exhibit strong performance in image processing and computer vision, but natural language processing is a sequence of texts, which is difficult to achieve better results in convolutional neural networks.

4.4   Different Neural Machine Translation Models in Chinese-Tibetan Task

To verify the performance of different neural machine translation models on Chi-nese-Tibetan translation, based on the same corpus, three neural machine translation models are used in this experiment. Niutrans is the SMT system we use in this experiment. The experimental results are shown in Table 7, and Sample translations for each model are shown in Table 8.

Table 7. Different neural networks with the same structure

Model	Framework	Corpus processing	BLEU
Niutrans	Phrased-based	Character	70.65
		Word	69.08
Seq2seq	RNN	Character	63.07
	RNN	Word	56.12
RNNSearch	RNN + Attention	Character	67.71
		Word	67.20
Niutrans-Syntax	SMT + Syntax	Word	64.53
Syntax-NMT	NMT + Syntax	Word	61.45
Fairseq	CNN + Attention	Character	18.48
		Word	18.37
Transformer	Attention	Character	71.69*
		Word	71.25

Table 8. Sample translations for each model

Example 1	
original text	灾害给经济造成重大损失，
Reference translation	གནོད་འཚེ་ཡིས་དཔལ་འབྱོར་ཐད་ལ་གྱོན་གུན་ཚབས་ཆེན་བཟོས།
Niutrans	གནོད་འཚོས་དཔལ་འབྱོར་ལ་གྱོང་གུན་ཚབས་ཆེན་བཟོས་པ་དང་།
Niutrans-syntax	ཀྱི་གནོད་འཚོའི་དཔལ་འབྱོར་ལ་གྱོང་གུན་ཚབས་ཆེན་བཟོས་པ།
Seq2Seq	དཔལ་འབྱོར་ལ་གྱོང་གུན་བཟོས་པར་གྱོང་གུན་ཆེན་པོ་བཟོས་ཡོད།
RNNSearch	གནོད་འཚོའི་དཔལ་འབྱོར་ལ་གྱོང་གུན་ཚབས་ཆེན་བཟོས་པ།
Syntax-NMT	གནོད་སྐྱོན་ལ་བརྟེན་ནས་དཔལ་འབྱོར་ལ་གྱོང་གུན་ཚབས་ཆེན་བཟོས་པ་དང་
Fairseq	བསྐྱར་དུ་འཛུགས་སྐྲུན་བྱ་དགོས།
Transformer	དཔལ་འབྱོར་ལ་གནོད་སྐྱོན་ཚབས་ཆེན་བཟོས་ཡོད་ལས།
Example 2	
original text	给群众生活带来很大困难。
Reference translation	མང་ཚོགས་ཀྱི་འཚོ་བར་དཀའ་ངལ་ཧ་ཅང་ཆེན་པོ་བཟོས།
Niutrans	མང་ཚོགས་ཀྱི་འཚོ་བར་དཀའ་ངལ་ཧ་ཅང་ཆེན་པོ་བྱུང་།
Niutrans-syntax	མང་ཚོགས་ཀྱི་འཚོ་བར་དཀའ་ངལ་ཧ་ཅང་ཆེན་པོ་བཟོ་ཟིན་ཡིན།
Seq2Seq	མང་ཚོགས་ཀྱི་འཚོ་བར་དཀའ་ངལ་ཧ་ཅང་ཆེན་པོ་བྱུང་ཡོད།
RNNSearch	མང་ཚོགས་ཀྱི་འཚོ་བར་དཀའ་ངལ་ཧ་ཅང་ཆེན་པོ་ཡོད།
Syntax-NMT	མང་ཚོགས་ཀྱི་འཚོ་བ་ལ་དཀའ་ངལ་ཧ་ཅང་ཆེན་པོ་ཡོང་།
Fairseq	སྐྱིད་བདེ་འཛུགས་སྐྲུན་ཧ་ཅང་དུ་གཏོང་བ།
Transformer	མང་ཚོགས་ཀྱི་འཚོ་བར་དཀའ་ངལ་ཧ་ཅང་ཆེན་པོ་ཡོད།

5   Conclusion

In this paper, five influential neural machine translation models are used in Chinese-Tibetan translation tasks, which are Seq2Seq, RNNSearch, Fairseq, Syntax-NMT, and Transformer. Through comparison, findings are as following:

1. In Chinese-Tibetan translation tasks, there is no obvious difference in the translation performance between most of the machine translation models based on neural network and the traditional statistical machine translation model, even the translation performance of most of the neural machine translation models is better than that of the traditional statistical machine translation model.
2. In Chinese-Tibetan translation task, the translation performance of character segmentation processing on corpus (Tibetan syllable segmentation, Chinese Character segmentation) is better than that of word segmentation processing on corpus;
3. The translation performance of the Transformer model based on the completely self-attention mechanism is the best in Chinese-Tibetan translation tasks.
4. Introducing the syntax tree based on the neural machine translation model can improve translation performance on Chinese-Tibetan machine translation tasks.
5. There is significant difference in translation performance of different neural machine translation models with the same structure.

Acknowledgement. This work is supported by the National Science Foundation of China (61331013).

References

Liu, Y.: Recent advances in neural machine translation. J. Comput. Res. Dev. **54**(6), 1144–1149 (2017). (in Chinese)

Zhao, T.: Machine Translation Theory. Harbin Institute of Technology Press (1900). (in Chinese)

Hinton, G.E., Osindero, S., Teh, Y.-W.: A fast learning algorithm for deep belief nets. Neural Comput. **18**(7), 1527–1554 (2006)

Wei, S.: Research on Tibetan - Chinese online translation system based on phrases. Dissertation, Northwest University for Nationalities (2015). (in Chinese)

Cai, Z., Cai, R.: Research on the distribution of tibetan character forms. J. Chin. Inf. Process. **30**(4), 98–105 (2016). (in Chinese)

Dong, X., Cao, H., Jiang, T.: Phrase based Tibetan - Chinese statistical machine translation system. Technol. Wind **17**, 60–61 (2012). (in Chinese)

Luo, X.: Research on syntax-based Chinese-Tibetan statistical machine translation system. Dissertation, Xiamen University (2010). (in Chinese)

Hua, Q.: Research on some key technologies of machine translation based on tree-to-string in tibetan language. Dissertation, Shanxi Normal University (2014a). (in Chinese)

Cai, R.: Research on large-scale Sino-Tibetan bilingual corpus construction for natural language processing. J. Chin. Inf. Process. **25**(6), 157–162 (2011)

Pang, W.: Research on the construction technology of Tibetan-Chinese bilingual corpus of corpus based on web. Dissertation, Minzu University of China (2015). (in Chinese)

Xiang, B., Zhang, G.: Research on the translation of han names in Chinese-Tibetan machine translation. J. Qinghai Normal Univ. (Nat. Sci.) **27**(4), 88–90 (2011)

Hua, G.: Tibetan verb researching in Chinese Tibetan machine translation. Dissertation, Qinghai Normal University (2014b). (in Chinese)

Nuo, M., Wu, J., Liu, H., Ding, Z.: Research on phrase translation extraction for Chinese-Tibetan machine translation. J. Chin. Inf. Process. **25**(3), 112–118 (2011)

Li, Y., Xiong, D., Zhang, M., Jiang, J., Ma, N., Yin, J.: Research on Tibetan-Chinese neural machine translation. J. Chin. Inf. Process. **31**(6), 103–109 (2017a)

Guan, Q.: Research on Tibetan segmentation for machine translation. Electron. Test 11x, 46–48 (2015)

Sutskever, I., Vinyals, O., Le, Q.V.: Sequence to sequence learning with neural networks. In: Advances in Neural Information Processing Systems (2014)

Cho, K., et al.: Learning phrase representations using RNN encoder-decoder for statistical machine translation. Comput. Sci. (2014)

D'Informatique, D.E., Ese, N., Esent, P., et al.: Long short-term memory in recurrent neural networks. EPFL **9**(8), 1735–1780 (2001)

Bahdanau, D., Cho, K., Bengio, Y.: Neural machine translation by jointly learning to align and translate. Comput. Sci. (2014)

Gehring, J., Auli, M., Grangier, D., et al.: Convolutional Sequence to Sequence Learning (2017)

Vaswani, A., et al.: Attention is all you need. In: Advances in Neural Information Processing Systems (2017)

Xue, Y., Li, S., Zhao, T., Yang, M.: Syntax-based reordering model for phrasal statistical machine translation. J. Commun. Test **29**(1), 7–14 (2008)

Xiong, D., Liu, Q., Lin, S.: A survey of syntax-based statistic machine translation. J. Chin. Inf. Process. **22**(2), 28–39 (2008)

Chen, H., et al.: Improved neural machine translation with a syntax-aware encoder and decoder, pp. 1936–1945 (2017)

Eriguchi, A., Tsuruoka, Y., Cho, K.: Learning to Parse and Translate Improves Neural Machine Translation (2017)

Li, J., et al.: Modeling Source Syntax for Neural Machine Translation, pp. 688–697 (2017b)

Aharoni, R., Goldberg, Y.: Towards String-to-Tree Neural Machine Translation (2017)

Xiao, T., et al.: NiuTrans: an open source toolkit for phrase-based and syntax-based machine translation. In: Proceedings of the ACL 2012 System Demonstrations. Association for Computational Linguistics (2012)

Sennrich, R., Haddow, B., Birch, A.: Neural machine translation of rare words with subword units. arXiv preprint arXiv:1508.07909 (2015)

Papineni, K., et al.: BLEU: a method for automatic evaluation of machine translation. In: Proceedings of the 40th Annual Meeting on Association for Computational Linguistics. Association for Computational Linguistics (2002)

Li, Y., et al.: TIP-LAS: an open source toolkit for Tibetan word segmentation and POS tagging. J. Chin. Inf. Process. **29**(6), 203–207 (2015)

Li, Z., Sun, M.: Punctuation as implicit annotations for chinese word segmentation. Comput. Linguist. **35**(4), 505–512 (2009)

Petrov, S., et al.: Learning accurate, compact, and interpretable tree annotation. In: Proceedings of the 21st International Conference on Computational Linguistics and the 44th Annual Meeting of the Association for Computational Linguistics. Association for Computational Linguistics (2006)

Knowledge Graph and Information Extraction

Metadata Extraction for Scientific Papers

Binjie Meng[1,2], Lei Hou[3], Erhong Yang[1,2(✉)], and Juanzi Li[3]

[1] Beijing Advanced Innovation Center for Language Resources,
Beijing Language and Culture University, Beijing, China
`mllrose@126.com, yerhong@126.com`
[2] School of Information Science, Beijing Language and Culture University,
Beijing, China
[3] Department of Computer Science and Technology, Tsinghua University,
Beijing, China
`greener2009@gmail.com, lijuanzi2008@gmail.com`

Abstract. Metadata extraction for scientific literature is to automatically annotate each paper with metadata that represents its most valuable information, including problem, method and dataset. Most existing work normally extract keywords or key phrases as concepts for further analysis without their fine-grained types. In this paper, we present a supervised method with three-stages to address the problem. The first step extracts key phrases as metadata candidates, and the second step introduces various features, i.e., statistical features, linguistics features, position features and a novel fine-grained distribution feature which has high relevance with metadata categories, to type the candidates into three foregoing categories. In the evaluation, we conduct extensive experiments on a manually-labeled dataset from ACL Anthology and the results show our proposed method achieves a +3.2% improvement in accuracy compared with strong baseline methods.

Keywords: Metadata extraction · Scientific literature
Fine-grained distribution · Classification

1 Introduction

As the number of scientific literature increases quickly, getting access to the core information of scientific papers easily and fast is becoming more and more important. With these core information, we can improve both the quality and efficiency of information retrieval, literature search engine and research trend prediction. In this paper, we aim at the mining of the core information of scientific, and refer to it as **metadata extraction**.

Metadata extraction is a complicated task and poses the following challenges:

- It is hard to determine whether an extracted item from a scientific paper is representative (i.e., belongs to the metadata) or not.
- To the best of our knowledge, there is not a public labeled dataset, even an effective and widely accepted annotation rules.

© Springer Nature Switzerland AG 2018
M. Sun et al. (Eds.): CCL 2018/NLP-NABD 2018, LNAI 11221, pp. 111–122, 2018.
https://doi.org/10.1007/978-3-030-01716-3_10

- Although all scientific papers follow a common writing rule (e.g., organized as the motivation, background, innovation, contrast, solution and experimental results), the metadata may be flexible enough to appear in any section, making the metadata extraction very challenging.

Traditional metadata extraction [5] is to extract a set of controlled vocabularies with a fixed schema, which greatly depends on the hand-crafted extraction rules and lacks flexibility. Krishnan et al. [7] extract scientific concepts via an automatic manner from paper titles only, which results in a large amount of loss of useful information.

In this paper, we divide metadata into 3 categories: Problem, Method and Dataset, which we believe can represent the main contents of one scientific together. We define the **metadata extraction** as the mining of key phrases belonging to these 3 categories. To extract these information, we construct a manual labeled dataset based on the ACL history data, and propose an supervised domain-specific framework.two

Our model can be divided into three phases. In Phase One, we feed the full paper context into Segphrase [9] to extract key phrases, named as metadata candidates. In Phase Two, we represent all metadata candidates using our proposed novel features, including semantic features and section-leveled tfidf features. In phrase Three, we feed the features into a classifier to predict which category they belong to (Problem, Method, Dataset and Not-Metadata).

To sum up, our contributions are as follows:

- We firstly put forward a pioneering task—metadata extraction—for scientific papers, which is very meaningful to many NLP applications.
- We construct a manual-labeled dataset for metadata extraction, hoping that it can push the development of related researches.
- Based on our dataset, we propose a three-phased supervised domain-specific framework to extract metadata.
- We conduct experiments on our dataset and demonstrate that our framework outperforms the baseline methods for metadata extraction of scientific papers.

The rest of this paper is organized as follows. Section 2 presents problem formalization. In Sect. 3 we present our detailed approaches. The experimental results are introduced in Sect. 4. Section 5 reviews the related literature. Finally, we conclude our work in Sect. 6.

2   Problem Formalization

In this section, we formalize the problem of metadata extraction for scientific papers. Before that, we first introduce some related basic concepts.

Definition 1 (Scientific Paper). *Conceptually, a scientific paper is a report of intellectual work within several key integrant sections, including standardized argumentation structure, which varies slightly in different subjects. Formally, a paper p can be represented as a collection of sections $p = \{s_p, s_m, s_e, s_c\}$ with* *the subscript denoting problem, method, experiment, related work and conclusion respectively. Each section s is a word sequence $s = \langle w_1, w_2, \ldots, w_{|W|}\rangle$ with each word w chosen from the vocabulary V.*

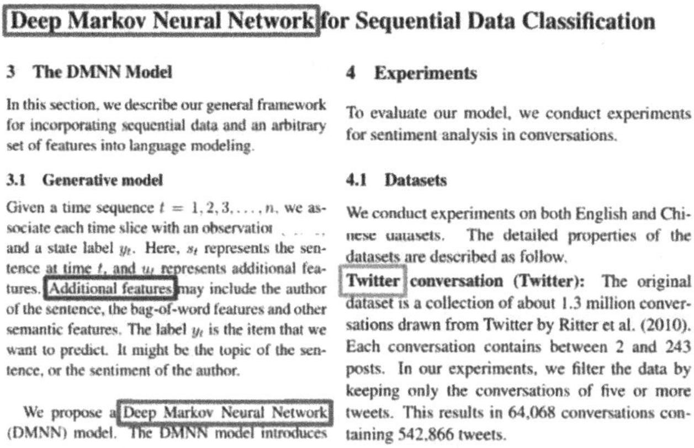

Deep Markov Neural Network for Sequential Data Classification

3   The DMNN Model

In this section, we describe our general framework for incorporating sequential data and an arbitrary set of features into language modeling.

3.1   Generative model

Given a time sequence $t = 1, 2, 3, \ldots, n$, we associate each time slice with an observation . . . . and a state label y_t. Here, s_t represents the sentence at time t, and u_t represents additional features. Additional features may include the author of the sentence, the bag-of-word features and other semantic features. The label y_t is the item that we want to predict. It might be the topic of the sentence, or the sentiment of the author.

We propose a Deep Markov Neural Network (DMNN) model. The DMNN model introduces

4   Experiments

To evaluate our model, we conduct experiments for sentiment analysis in conversations.

4.1   Datasets

We conduct experiments on both English and Chinese datasets. The detailed properties of the datasets are described as follow.

Twitter conversation (Twitter): The original dataset is a collection of about 1.3 million conversations drawn from Twitter by Ritter et al. (2010). Each conversation contains between 2 and 243 posts. In our experiments, we filter the data by keeping only the conversations of five or more tweets. This results in 64,068 conversations containing 542,866 tweets.

Fig. 1. An example of part of scientific paper

For example, Fig. 1 presents an example of scientific paper which contains method and experiment sections. Normally, a scientific paper aims at one or more research tasks, proposes specific methods, designs experiments for validation and achieves some conclusions, and we call these information as metadata.

Definition 2 (Metadata). *Metadata[1] is data that provides information about other data, and can be grouped for different purposes, including descriptive (e.g., title and author), structural (e.g., pages), and administrative (e.g., owner). Previously, metadata about scientific papers usually includes author, publisher, venue, title, abstract and so on. As mentioned above, we focus on more informative and fine-grained metadata in this paper. Formally, the metadata of paper p is described as a triple $md_p = (problem, method, dataset)$, and each element is composed of several continuous words, which can be also called key phrase.*

Also in Fig. 1, "Deep Markov Neural Network" and "Twitter" denote the method and dataset of the example paper. Note that not all key phrases are metadata, e.g., "Additional features". Papers could have various kinds of metadata, and the above schema, i.e., $(problem, method, dataset)$, plays a very important role in understanding papers or inspiring follow-up research. Therefore, we propose the following problem:

Definition 3 (Metadata Extraction for Scientific Paper). *Given a collection of scientific papers $P = \{p_1, p_2, \ldots, p_n\}$, our goal is to extract the metadata*

---

[1] https://en.wikipedia.org/wiki/Metadata.

of each paper, i.e., md_{p_i} for p_i. In particular, we formalize the task as a classification problem. For each paper p_i, we first separate it into disjoint sections $p_i = \{s_{p_i}, s_{m_i}, s_{e_i}, s_{c_i}\}$ and extract several key phrases $K_i = \{k_1, k_2, \ldots, k_m\}$ meanwhile. Then we characterize each key phrase k_i with well-designed syntax and structure features, and finally build a classifier to determine whether it is a kind of metadata (i.e., problem, method and dataset) or not.

Note that our dataset is constructed from papers in PDF format, and section information is not totally identical to the original papers. Even in the original papers, the section organization is also various. Thus the section schema is manually defined as above.

3   The Proposed Approach

In this section, we describe the proposed method for scientific metadata extraction in details. Figure 2 shows the framework in a pipeline paradigm, involving three major steps:

– **Candidate Generation.** Perform a key phrase extractor on the paper collection P in which the abstract of each paper is excluded to get the key phrase set $K = \{k_1, k_2, \ldots, k_m\}$ for each paper $p \in P$, which serves as the candidate metadata mentions.
– **Feature Quantification.** For each key phrase, we characterize it with two types of features, basic features and fine-grained distribution features. The basic features focus on the statistics, position and syntax of the phrase, and the fine-grained distribution features first group paper full text into several disjoint sections and evaluate the importance in different sections for each key phrase. Two types of features are concatenated to represent each key phrase for the following classification.

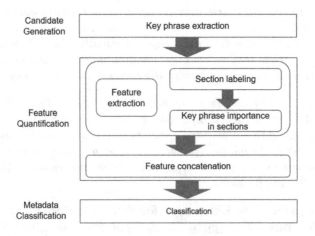

Fig. 2. The framework of the proposed approach

- **Metadata Classification.** Based on the above feature representation, build a classifier to classify extracted key phrases into one of the metadata categories, i.e., problem, method, dataset or not metadata (NOM).

In the above framework, key phrase extraction is to segment the paper text into a sequence of cohesive content units and select representative concept or process phrase as the candidate metadata mentions. In this paper, we emply a distant-supervised phrase segmentation algorithm named SegPhrase [9] with its released implementation[2], and other similar alternatives could also be applied. As for the final classifier, we try several typical classification models and decide to use the random forest (see the experiment part for details). Therefore, the core component of the framework is the design and quantification of different features, and we will demonstrate the details according to feature types, namely basic features and fine-grained distribution features.

3.1   Basic Features

From the perspective of the syntactic and grammar, the metadata phrase obviously and distinctively differs from other normal phrases. In this study, each key phrase is characterized by a set of features in terms of position, statistics and syntax are evaluated. Specifically, the following seven features are calculated for each key phrase $k \in K$:

- **(1) Frequency.** It is the number of times that key phrases appear within one paper, which is applied to evaluate the importance and possibility of a key phrase being metadata. It is based on a straightforward assumption that metadata is more frequently mentioned as compared with normal phrases.
- **(2) Length and (3) Max Word Length.** Intuitively, the length of phrase indicates the information it contains, and thus key phrase length and max word length are applied in this study. The number of characters of each key phrase is defined as key phrase length, and the length of longest word within key phrase is max word length.
- **(4) Leading Letter Capitalized.** Capitalizing the first letter of phrases is a typical and popular way in scientific paper writing to emphasize the importance of these words, highlight the author?s points or ideas, and attract readers? attention. If words in key phrases in accordance with this situation, such key phrases tends to be innovation and informative so that it is likely be related to be metadata. For instance, all the leading letters in each words of "Deep Markov Neural Network" are capitalized, which is labeled as "Method" in our dataset.
- **(5) In Title and (6) In Abstract.** In general, the title and abstract are a condensation or summarization of a paper, which contains the most important and useful information. Based on above backgrounds, whether a key phrase appears in title or abstract reflects its significance for a paper, and thus is related to its probability of being metadata. For example, the metadata phrase "Deep Markov Neural Network" emerges in title of the paper.

---

[2] https://github.com/shangjingbo1226/SegPhrase.

- **(7) Lexical Cohesion.** Words in metadata phrases are usually consistent in lexicon, and thus we define lexical cohesion for quantification, which is computed as follows:

$$w_n \times ((1 + \lg f_k \frac{f_k}{\sum w_c}))$$

where w_n is the number of constituent word one key phrase contains, f_k is the frequencies of appearance in one paper for each key phrase, w_c is the frequencies of each constituent word of one key phrase in the whole paper content.

3.2 Fine-Grained Distribution Features

In this subsection, we focus on the distribution of metadata phrases across the full paper content. We first use a manually labeled sample to demonstrate our assumption, based on which we introduce our proposed feature quantification.

Assumption 1. *Different types of metadata follow different distributions across the full paper content, and metadata and their more correlated sections tend to have consistent semantics, e.g., a key phrase labeled with "dataset" has higher probability of appearance in "experiment" and "introduction" as compared with "related work" and "problem formalization".*

To verify the above assumption, we randomly select 30 papers to statistics the distribution of key phrases in four predefined sections. Table 1 depicts the occurrences of several key phrases in different sections. These key phrases clearly show that the distribution of each key phrase among four sections is closely linked with its semantic meaning. Take "neural network" as an example, the highest occurrence located in section "Model" and not once in section "Experiment", which is in accordance with our expectation. As a contrast, an NLP task named "sarcasm detection" is referred as many as 11 times in section "Problem", highest among other sections. However, in Experiment section, experiment setting and experiment results will regularly be explained, and problem description commonly will not be mentioned. And the figure about "sarcasm detection" clearly supports this convention.

Table 1. Examples on phrase distributions in different sections

Phrase	Section			
	Conclusion	Experiment	Model	Problem
Training set	0	20	1	0
Neural network	1	0	11	7
Hidden state	0	0	7	10
Sarcasm detection	3	0	2	11
Stochastic gradient descent	1	2	7	2

This assumption inspires us that the section-level distributions of candidate phrases are good indicators for metadata classification. To generate such fine-grained distribution features, we first segment a full paper several disjoint sections as defined in Sect. 2, and then quantify the importance of each candidate for all sections as the final feature vectors. In the following part, we will introduce the section segmentation and importance measurement.

Section Segmentation. Scientific literature usually follows an acknowledged logical structure to organize contents (e.g., using sections, subsections). Each section has a different focus from others. To achieve our goal of metadata extraction, the problem, method and experiment sections are necessary. Besides, a paper normally includes introduction, related work and conclusion. All the six sections constitute the section schema. As mentioned after problem definition in Sect. 2, we need to group the paper content into the above sections due to the paper organization is usually various or even missing. To complete the segmentation, we apply a CRF-based parsing tool [13] to attach a section label for each sentence, and then group these sentences into sections. Note that the key phrases extracted by SegPhrase [9] are within one sentence not cross sentences, thus one key phrase mention corresponds to only one section label and key phrase mention with ambiguous section labels does not exist.

Importance Measurement. Various methods can be applied to measure the importance of a given key phrase in different sections. In this paper, we employ the most classical tf-idf measurement. Particularly, we treat all the sentences with the same section labels as a virtual documents, and a paper could be split into at most six documents, based on which the tf-idf is calculated. Note that the importance value is set be 0 if a section does not contain any sentence.

4 Experiment

In this section, we evaluate the proposed framework on a manually created dataset. We first briefly introduce the dataset and experiment settings, then present the detailed experimental results, and finally investigate the feature contributions and some other method details.

4.1 Dataset

To the best of our knowledge, there is no existing benchmark dataset which can be directly used for our metadata extraction evaluation. As such, we crawled the papers in PDF format from ACL Anthology[3], and employed GROBID tool[4] to extract the textual content. Then we randomly selected 30 papers, from which

---

[3] http://aclweb.org/anthology/.
[4] http://grobid.readthedocs.io/en/latest/.

1,947 key phrases were extracted. Finally, we invited graduated students to manually grouped them into predefined metadata schema as defined in Sect. 2, i.e., "problem", "method", "dataset" and NOM. They were required to firstly read the corresponding paper to understand the "problem" and "method". After annotation, over 75% of the phrases (i.e., 1,478 key phrases) were labeled as NOM, leading to an unbalanced classification problem. This results make sense since a paper normally does not contain too many metadata. Note that key phrase labeled with NOM was excluded in experiments to keep balance among the remaining three groups. Detailed statistics are presented in Table 2.

Table 2. Dataset statistics

Metadata	Problem	Method	Dataset	NOM	Total
Numbers	253	165	46	1483	1947

4.2 Experiment Setting

Baseline Methods. We use **MESP** to denote the proposed method of Metadata Extraction for Scientific Papers, and compare it with the following baseline methods to verity its effectiveness:

- **Section Importance Ranking (SIR):** An unsupervised method that treats the most important key phrases in corresponding sections as results. In experiment, we use the section importance introduced in Sect. 3.2 to rank the key phrases in *problem, method* and *experiment* sections, and select the top 5 as results.
- **MESP-B:** The simplified version of our proposed method that only uses basic features in Sect. 3.1.
- **MESP-I:** The simplified version of our proposed method that only uses the section importance features and in Sect. 3.2.
- **MESP-P:** The fusion version of our proposed method that uses the section importance features in Sect. 3.1 and the fine-grained distribution features in Sect. 3.2.

Comparison Metrics. To quickly have a basic knowledge about one scientific paper, only a few key phrases identified as correct metadata are better than many key phrases considered to be metadata, of which only several key phrases are accurate metadata. Therefore, "accuracy" is a most important and effective indicator than others, which is the comparison metric in this experiment. Let $\hat{y}_i$, y_i denote the predicted and corresponding true values of the i-th sample, then the accuracy over n samples is defined as

$$accuracy(y, \hat{y}) = \frac{1}{n} \sum_{i=0}^{n-1} \mathbb{I}(\hat{y_i} = y_i)$$

where $\mathbb{I}(x)$ is the indicator function.

Implementation Details. To find the most appropriate classifier, we tried six typical methods, i.e., Logistic Regression, Nearest Neighbors, Decision Tree, Na?ve Bayes, Adaboost, and Random Forest. In experiment, we used their python implementation in sklearn [12], and searched the best hyper-parameter space via gridsearhcv tool [12] based on standard train/test split and 5-fold cross validation.

4.3 Performance Results

As depicted in Table 3, Random Forest achieves the best performance among all classifiers previously mentioned. Random Forest is an ensemble of several weak learners, which can make full use of the data samples and the features of each sample while training. It implies that metadata extraction is a non-trivial problem which requires extracting different features in diverse ways instead. Besides, metadata phrases are located much closer with each other than NOM phrases in vector space, and that is why Nearest Neighbor classifier achieves good performance.

Table 3. Performance of various classifiers

Classifier	Accuracy
Nave bayes	51.80%
Random forest	67.63%
Nearest neighbor	63.36%
Adaboost classifier	62.28%
Decision tree	54.61%
Linear regression	57.54%

To validate the effectiveness of the proposed framework and designed features, we compare **MESP** with baseline methods using Random Forest and summarize the results in Table 4. From the results, we can see that **MESP** performs best among other baseline methods and achieves a considerable improvement of +3% over the strongest baseline in the accuracy because it is fully equipped with all the grammar, statistics, position and distribution features. The result of **MESP-B** demonstrates that the basic features are fairly good for the extraction task. **MESP** can further improves the performance, which verifies that section importance is a good feature that describes the distribution of key phrases. However, metadata extraction for scientific paper is such a complicated problem which requires much more other information from different perspectives, and this is why **SIR**, which distinguishes metadata merely by section importance value, performs poorest among all methods. Besides, the huge gap of results between **SIR** and other methods demonstrates that superviesed methods are more effective than un-supervised ones in this task.

Table 4. Performance comparison with baselines

Method	SIR	MESP-B	MESP-I	MESP
Accuracy	10.89%	64.44%	58.58%	67.63%

5 Related Work

Two lines of researches are related to the metadata extraction in the current paper, i.e., the definition of metadata and keyword extraction for scientific papers.

5.1 The Definition of Metadata

Metadata could be simply interpreted as *data about data*, and is used for providing information about other data. For metadata definition of scientific literature, the most widely accepted standard is *the Dublin Core*. It defines 15 elements[5] for resource description. From its scope, we can see that it is only used for building unified services for digital libraries and does not understand the deeper semantics within paper content. The metadata in this study is entirely different from the Dublin Core standard. Our metadata definition aims to understand the problem, method and experiment of scientific papers, and could be used for tasks that requires deep semantics, such as "hot ideas" detection in a scientific field [2,4] and literature summarization [1]. For a similar goal, Simone Teufel et al. propose Argumentative Zoning (AZ) based on scientific papers, which tags the rhetorical structure of scientific literature on sentence level [14], and they further extend the AZ scheme to 15 fine-grained categories for annotation [15]. However, they do not provide any corresponding annotation tools, and manually labeling costs a lot of time and human resources. Therefore, automatic metadata extraction is necessary.

5.2 Keyword Extraction for Scientific Papers

Keyword extraction is to extract representative words that occur frequently in texts while other semantic words apart from stop words seldom occurring. Due to its significance to many downstream tasks, the problem attracts numerous research attention, e.g., Liu et al. [9] study the extraction of phrases that capture the main topics discussed in a document from large corpus. However, there is normally no categorization for these keywords during the extraction.

In recent years, many researches turn their attention to the scientific literature. Li et al. extracts keywords in experiment-related graphs for chemistry metadata extraction [8]. Gupta et al. using dependency parsers to extract key

---

[5] *The metadata of papers in the Dublin Core include: Title, Creator, Subject, Description, Contributor, Publisher, Date, Type, Format, Identifier, Source, Relation, Reference, Is referenced By, Language, Rights and Coverage.*

phrases for three focus related categories from each sentence for the analysis of dynamics of research focus of scientific papers [3]. Pan et al. employs a widely-used linguistic pattern introduced by Justeson and Katz [6], i.e., $((A|N)^+|((A|N)^*(NP)?)(A|N)^*))N$, to extract terminology concepts [11]. However, academic key phrases are more ambiguous and hard to type based on frequencies. Tsai et al. uses noun-phrase chunking to extract concept mentions and local textual features and annotating concept mentions iteratively for the identifying scientific concepts [16], but the extracted concepts are too coarse so that they cannot satisfy the fine-grained demands in some analysis tasks. Krishnan et al. introduces an unsupervised method for extraction of representative concepts from scientific literature based on titles scientific paper [7]. However, titles are not sufficient to provide fine-grained metadata for key phrase extraction and categorization [10], and thus we consider the full text of scientific paper for the metadata extraction.

Overall, the problem of metadata extraction for scientific papers is a nontrivial task due to it heavily depends on latent semantic information. Therefore, we address this task in a three-stage framework as described in Sect. 3. The metadata candidates are generated using a widely-accepted tool [9], and our focus is mainly on how to type the extracted metadata. Particularly, we design a novel fine-grained distribution feature based on preliminary data observation, whose effectiveness is verified in the experiments.

6   Conclusion

In this paper, we put forward a novel metadata extraction task on scientific papers and propose a three-stage framework for solution. In particular, we introduce an especially important fine-grained distribution feature for metadata typing. Besides, we perform extensive evaluation on a manually labeled dataset to validate the effectiveness of the proposed framework as well as the designed features. The results show that the performance of our method outperforms the strong baselines by an average accuracy of 3.2%. Our future work is to studying other representative and essential metadata for scientific papers and extending a discipline-independent extractor.

Acknowledgement. This research project is supported by the Major Project of the National Language Committee of the 13rd Five-Year Research Plan in 2016 (ZDI135-3); supported by the Fundamental Research Funds for the Central Universities, and the Research Funds of Beijing Language and Culture University (17YCX148).

References

1. D'Avanzo, E., Magnini, B.: A keyphrase-based approach to summarization: the LAKE system at DUC-2005. In: Proceedings of DUC (2005)
2. Griffiths, T.L., Steyvers, M.: Finding scientific topics. Proc. Nat. Acad. Sci. **101**(Suppl. 1), 5228–5235 (2004)

3. Gupta, S., Manning, C.: Analyzing the dynamics of research by extracting key aspects of scientific papers. In: Proceedings of 5th International Joint Conference on Natural Language Processing, pp. 1–9 (2011)
4. Hall, D., Jurafsky, D., Manning, C.D.: Studying the history of ideas using topic models. In: Proceedings of the Conference on Empirical Methods in Natural Language Processing, pp. 363–371. Association for Computational Linguistics (2008)
5. Han, H., Giles, C.L., Manavoglu, E., Zha, H., Zhang, Z., Fox, E.A.: Automatic document metadata extraction using support vector machines. In: Proceedings of 2003 Joint Conference on Digital Libraries, pp. 37–48. IEEE (2003)
6. Justeson, J.S., Katz, S.M.: Technical terminology: some linguistic properties and an algorithm for identification in text. Nat. Lang. Eng. **1**(1), 9–27 (1995)
7. Krishnan, A., Sankar, A., Zhi, S., Han, J.: Unsupervised concept categorization and extraction from scientific document titles. CoRR abs/1710.02271 (2017)
8. Li, N., Zhu, L., Mitra, P., Mueller, K., Poweleit, E., Giles, C.L.: oreChem ChemXSeer: a semantic digital library for chemistry. In: Proceedings of the 10th Annual Joint Conference on Digital Libraries, pp. 245–254. ACM (2010)
9. Liu, J., Shang, J., Wang, C., Ren, X., Han, J.: Mining quality phrases from massive text corpora. In: Proceedings of the 2015 ACM SIGMOD International Conference on Management of Data, pp. 1729–1744. ACM (2015)
10. McKeown, K., et al.: Predicting the impact of scientific concepts using full-text features. J. Assoc. Inf. Sci. Technol. **67**(11), 2684–2696 (2016)
11. Pan, L., Wang, X., Li, C., Li, J., Tang, J.: Course concept extraction in MOOCs via embedding-based graph propagation. In: Proceedings of the Eighth International Joint Conference on Natural Language Processing (vol. 1: Long Papers), pp. 875–884 (2017)
12. Pedregosa, F., et al.: Scikit-learn: machine learning in python. J. Mach. Learn. Res. **12**(Oct), 2825–2830 (2011)
13. Prabhakaran, V., Hamilton, W.L., McFarland, D., Jurafsky, D.: Predicting the rise and fall of scientific topics from trends in their rhetorical framing. In: Proceedings of the 54th Annual Meeting of the Association for Computational Linguistics (vol. 1: Long Papers), vol. 1, pp. 1170–1180 (2016)
14. Teufel, S., Carletta, J., Moens, M.: An annotation scheme for discourse-level argumentation in research articles. In: Proceedings of the Ninth Conference on European chapter of the Association for Computational Linguistics, pp. 110–117. Association for Computational Linguistics (1999)
15. Teufel, S., Siddharthan, A., Batchelor, C.: Towards discipline-independent argumentative zoning: evidence from chemistry and computational linguistics. In: Proceedings of the 2009 Conference on Empirical Methods in Natural Language Processing: vol. 3, pp. 1493–1502. Association for Computational Linguistics (2009)
16. Tsai, C.T., Kundu, G., Roth, D.: Concept-based analysis of scientific literature. In: Proceedings of the 22nd ACM International Conference on Conference on Information & Knowledge Management, pp. 1733–1738. ACM (2013)

Knowledge Graph Embedding with Logical Consistency

Jianfeng Du$^{(\boxtimes)}$, Kunxun Qi, and Yuming Shen

Guangdong University of Foreign Studies, Guangzhou 510006, China
jfdu@gdufs.edu.cn

Abstract. Existing methods for knowledge graph embedding do not ensure the high-rank triples predicted by themselves to be as consistent as possible with the logical background which is made up of a knowledge graph and a logical theory. Users must take great effort to filter consistent triples before adding new triples to the knowledge graph. To alleviate users' burden, we propose an approach to enhancing existing embedding-based methods to encode logical consistency into the learnt distributed representation for the knowledge graph, enforcing high-rank new triples as consistent as possible. To evaluate this approach, four knowledge graphs with logical theories are constructed from the four great classical masterpieces of Chinese literature. Experimental results on these datasets show that our approach is able to guarantee high-rank triples as consistent as possible while preserving a comparable performance as baseline methods in link prediction and triple classification.

1 Introduction

Knowledge graph has become popular in knowledge representation nowadays. A knowledge graph is a directed graph with vertices labeled by entities and edges labeled by relations. It is represented as a set of triples of the form $\langle h, r, t \rangle$, where h is the *head entity* (simply *head*), r the *relation* and t the *tail entity* (simply *tail*). Although the current knowledge graphs have many triples, they are far from completeness. Traditional logic-based methods for knowledge graph completion become intractable as knowledge graphs evolve into a large scale.

To handle large knowledge graphs, embedding-based methods have emerged and become prevalent due to their high efficiency and scalability. These methods introduce the distributed representation for entities and relations and encode them into a continuous vector space, then the truth degrees of any new triples can be estimated by simple numerical calculation. In the scenario of knowledge graph completion, all new triples are ranked in the descending order of truth degrees computed by an embedding-based method, therein high-rank triples having passed the verification of domain experts are added to the knowledge graph.

Logical consistency is a prerequisite for new triples to be usable in knowledge graph completion. A new triple cannot be added to a knowledge graph if it is

© Springer Nature Switzerland AG 2018
M. Sun et al. (Eds.): CCL 2018/NLP-NABD 2018, LNAI 11221, pp. 123–135, 2018.
https://doi.org/10.1007/978-3-030-01716-3_11

inconsistent with the logical background which consists of the knowledge graph and a logical theory formalizing the domain knowledge. The logical theory can often be expressed by both *datalog rules* and *constraints*, where datalog rules are used to infer implicit triples and constraints are used to prevent conflicting triples from being present. For example, suppose $\langle$John, father, Tom$\rangle$ (meaning *John is the farther of Tom*) is an existing triple, $\langle x, \text{father}, y \rangle \rightarrow \langle x, \text{type}, \text{Man} \rangle$ (meaning *x is a man if x is the father of someone y*) and $\langle x, \text{mother}, y \rangle \rightarrow \langle x, \text{type}, \text{Woman} \rangle$ (meaning *x is a woman if x is the mother of someone y*) are two datalog rules, and $\langle x, \text{type}, \text{Man} \rangle \wedge \langle x, \text{type}, \text{Woman} \rangle \rightarrow \bot$ (meaning *anyone x cannot be both a man and a woman*) is a constraint, respectively, in the logical background. Then the new triple $\langle$John, mother, Tom$\rangle$ (meaning *John is the mother of Tom*) is inconsistent with the logical background and should not be added to the knowledge graph that underlies the logical background.

Although logical consistency is important, existing embedding-based methods for knowledge graph completion hardly address logical consistency in learning the distributed representation. Some methods such as the translational distance methods [2,3,9,11–15,24,26] and the semantic matching methods [1,10, 16,17,20,21,27] completely ignore the logical theory. Others [4,5,7,8,19,23,25] consider datalog rules but ignore constraints in the logical theory. None of them makes high-rank new triples as consistent as possible with the logical background. Users must take great effort to throw away inconsistent triples before adding new triples to the knowledge graph. Although automatic tools for checking logical consistency can be applied to filter high-rank triples, this postprocessing is still time consuming. It is wiser to encode logical consistency into the distributed representation to make high-rank triples as consistent as possible.

In this paper, we propose an approach to enhancing existing embedding-based methods to encode logical consistency into the learnt distributed representation. The key idea is to adapt knowledge graph embedding into an optimization problem that minimizes the sum of the global margin-based loss function in translational distance methods and a loss function on a portion of negative triples that are inconsistent with the logical background, where inconsistent triples are computed from the logical background and selected by adaptive relation-specific thresholds. We do not use the complete set of inconsistent triples in the optimal objective function for consideration of efficiency and scalability.

Since there is no benchmark knowledge graph coming with a logical theory including constraints, we construct new knowledge graphs and the corresponding logical theories from a domain that we are familiar with, which is about character relationships in the four great classical masterpieces of Chinese literature, to evaluate our approach. Experimental results demonstrate that our approach is able to guarantee the high-rank new triples as consistent as possible while it also preserves a comparable performance as baseline methods in the two traditional prediction tasks namely link prediction and triple classification.

The main contributions of this work are two-fold:

(1) We propose knowledge graphs and logical theories from the four great classical masterpieces of Chinese literature. These datasets give complex theories

on the domain of character relationships and are suitable for evaluating hybrid approaches that combine distributed representation learning with logic inference.

(2) We propose an effective and efficient approach to enhancing embedding-based methods to guarantee high-rank new triples as consistent as possible with the logical background while keeping comparable scores on traditional metrics.

2   Preliminaries

2.1   Knowledge Graph Embedding

Knowledge graph embedding (KGE) encodes a knowledge graph into a continuous vector space to support a variety of prediction tasks. There are roughly two main categories for KGE methods, namely the translational distance methods and the semantic matching methods.

A translational distance method measures the truth degree of a triple by the distance between the head and the tail, usually after a translation carried out by the relation. TransE [2] is a pioneer translational distance method which defines translation directly on entity vectors. Most subsequent methods define translation on projection of entity vectors. To name a few, TransH [24] defines projection on relation-specific hyperplanes, whereas TransR [15] defines projection by relation-specific matrices. All these methods uniformly learn vectors by minimizing a global margin-based loss function via a loss function for triples.

The loss function for a triple $\langle h, r, t \rangle$, written $\mathsf{loss}_r(h, t)$, can be defined as $||f_r(l_h) + l_r - f_r(l_t)||_{L_1}$ or $||f_r(l_h) + l_r - f_r(l_t)||_{L_2}$, where l_x denotes $(x_1, \ldots, x_n)$ which is the vector representation for x, $||l_x||_{L_1}$ is the L1-norm of l_x defined as $\sum_{i=1}^{n} |x_i|$, $||l_x||_{L_2}$ is the L2-norm of l_x defined as $\sqrt{\sum_{i=1}^{n} x_i^2}$, and $f_r(\cdot)$ is a relation-specific projection function which maps an entity vector into another one. For example, $f_r(l_e)$ is defined as l_e in TransE, whereas it is defined as $l_e - w_r^T l_e w_r$ for w_r an r-specific normal vector in TransH.

The global margin-based loss function to be minimized is defined over the set $\mathcal{G}$ of training triples (namely positive triples) and a set $\overline{\mathcal{G}}$ of negative triples which is disjoint with $\mathcal{G}$ and constructed from $\mathcal{G}$ by randomly corrupting triples in either heads or tails. By introducing the margin γ as a hyper-parameter, the global margin-based loss function to be minimized is given by

$$\sum_{\langle h,r,t \rangle \in \mathcal{G}} \sum_{\langle h',r,t' \rangle \in \overline{\mathcal{G}}} \max(0, \gamma + \mathsf{loss}_r(h, t) - \mathsf{loss}_r(h', t')). \tag{1}$$

A semantic matching method measures truth degrees of triples by matching the latent semantics for entities and relations in their vector space. It often employs a neural network to learn the distributed representation. Due to the diversity of network structures, semantic matching methods often have not a uniform optimal objective function. Nevertheless, a state-of-the-art method Analogy [16] has been proved to be generalized from several semantic matching methods including DistMult [27], HolE [17] and ComplEx [21]. Simply speaking, Analogy minimizes the energy function $\mathbb{E}_{h,r,t,y\in\mathcal{D}} - \log_2 \sigma(y \cdot l_h^T B_r l_t)$ on the data distribution $\mathcal{D}$ constructed from the training triples $\langle h, r, t \rangle$, where $y = +1$ for training triples (namely positive triples) and $y = -1$ for negative triples, l_h and l_t are respectively the vector representations for h and t, B_r is an $m \times m$ almost-diagonal matrix with $n < m$ real scalars on the diagonal, and $\sigma(\cdot)$ denotes the sigmoid function. DistMult (resp. ComplEx) can be treated as a special case of Analogy such that $n = m$ (resp. $n = 0$). Moreover, Analogy can be reduced to HolE by setting B_r as a certain circulant matrix without real scalars.

High-rank triples that have the top-k highest truth degrees (i.e. lowest loss values or energies) are often used as candidates in knowledge graph completion.

2.2   Logical Consistency

First-order logic is a traditional and popular approach to knowledge representation. A logical theory expressed by first-order logic is a set of rules R of the form $\forall x \ (\phi(x) \rightarrow \exists y \ \varphi(x, y))$, where $\phi(x)$ is a conjunction of atoms on the universally quantified variables x, and $\varphi(x, y)$ is a disjunction of atoms on both universally quantified variables x and existentially quantified variables y. The part of R at the left (resp. right) of $\rightarrow$ is called the *body* (resp. *head*) of R. By body(R) (resp. head(R)) we denote the set of atoms in the body (resp. head) of R. If the head of R has no atoms, R is also called a *constraint* and the empty head is written as $\perp$. If the head of R has a single atom without existentially quantified variables, R is also called a *datalog* rule. We omit $\forall x$ and simply write constraints as $\phi(x) \rightarrow \perp$ and datalog rules as $\phi(x) \rightarrow \varphi(x)$.

Existing KGE methods that take logical theories into account [4,5,7,8,19, 23,25] focus on datalog rules but ignore constraints. In this work we study KGE under a *Horn theory* which is a logical theory consisting of both datalog rules and constraints. A *model* of a Horn theory can be defined as a set S of *ground atoms* (which can be treated as a set of triples) such that (1) body(R) $\theta \subseteq S$ implies head(R) $\theta \cap S \neq \emptyset$ for any datalog rule R and any ground substitution θ for var(R) in the theory, and (2) body(R) $\theta \not\subseteq S$ for any constraint R and any ground substitution θ for var(R) in the theory, where var(R) denotes the set of variables in R. A constraint R is said to be *entailed* by a Horn theory if every model of the theory is also a model of $\{R\}$. A *logical background* is made up of a knowledge graph (namely a set of triples) and a Horn theory. A model of a logical background $\mathcal{B}$ that is made up of a knowledge graph $\mathcal{G}$ and a Horn theory $\mathcal{T}$ is a model of $\mathcal{T}$ that is also a superset of $\mathcal{G}$. A logical background *is* said to be *consistent* if it has a model. A triple is said to be *logically consistent* (simply *consistent*) with a consistent logical background $\mathcal{B}$ if adding the triple to $\mathcal{B}$ keeps $\mathcal{B}$ consistent. A triple is said to be *entailed* by a consistent logical background $\mathcal{B}$ if the triple is in the unique least model of $\mathcal{B}$.

3   Knowledge Graph Embedding Under a Horn Theory

Given a logical background with a Horn theory, new triples that can be added to the knowledge graph underlying the background should be consistent with the logical background. We focus on the task of finding consistent high-rank new triples. It is unwise to filter consistent triples after applying an existing KGE method to compute the truth degrees of new triples, since it is rather time-consuming—it needs to perform consistency checking for every new triple separately, where consistency checking in Horn theories is polynomial-time complete. It is wiser to encode logical consistency into the distributed representation to guarantee consistency for new triples as possible, so that the costly consistency checking need not to be performed after new triples are computed.

We propose an approach to enhancing translational distance KGE methods under a logical background. It is also applied to any semantic matching KGE method after the energy function is treated as a loss function and the optimal objective function is rewritten to a global margin-based loss function. The enhancements are two-fold. On one hand, to take datalog rules into account, we treat the set of entailed triples (which is actually the least model of the logical background) as the set of positive triples. On the other hand, to take constraints into account, we add to the optimal objective function a loss function on certain negative triples that are inconsistent with the logical background, where inconsistent triples are computed from the logical background and selected by adaptive relation-specific thresholds. While the former enhancement has been proposed in existing work [5,7], the latter enhancement is novel. These two enhancements can be formalized to the following optimal objective function:

$$\sum_{\langle h,r,t\rangle \in \mathcal{G}^+} \sum_{\langle h',r,t'\rangle \in \overline{\mathcal{G}^+}} \max(0, \gamma + \mathsf{loss}_r(h,t) - \mathsf{loss}_r(h',t'))$$

$$+ \sum_{\langle h,r,t\rangle \in \mathcal{G}^-} \max(0, \sigma_r - \mathsf{loss}_r(h,t)), \tag{2}$$

where $\mathcal{G}^+$ is the unique least model of the logical background, $\overline{\mathcal{G}^+}$ is constructed from $\mathcal{G}^+$ by randomly corrupting triples on either heads or tails, $\mathcal{G}^-$ is the set of negative triples inconsistent with the logical background, and σ_r is the r-specific threshold that classifies positive or negative triples on the relation r; i.e., $\langle h,r,t\rangle$ is a positive triple if $\mathsf{loss}_r(h,t) < \sigma_r$, or a negative triple otherwise.

More details are provided as follows. Let $\mathcal{B}$ denote logical background with a knowledge graph $\mathcal{G}$ and a Horn theory $\mathcal{T}$, where the set of datalog rules in $\mathcal{T}$ is $\mathcal{T}_1$ and the set of constraints in $\mathcal{T}$ is $\mathcal{T}_2$. Following [5] $\mathcal{G}^+$ is computed as the least fixpoint of $\mathcal{G}^{(t)}$, where $\mathcal{G}^{(t)} = \mathcal{G}^{(t-1)} \cup \bigcup \{\mathsf{head}(R)\,\theta \mid R \in \mathcal{T}_1, \mathsf{body}(R)\,\theta \subseteq \mathcal{G}^{(t-1)}\}$ for $t > 0$, and $\mathcal{G}^{(0)} = \mathcal{G}$. Following [24] $\overline{\mathcal{G}^+}$ is constructed from $\mathcal{G}^+$ by the *bern* strategy e.g. by randomly corrupting triples on either heads or tails with different probabilities proportional to the relation-specific frequency of heads or tails. Let R_{-1} denote the set of constraints modified from the constraint R by deleting one body atom. For efficiency, $\mathcal{G}^-$ can be approximated as $\bigcup_{R \in \mathcal{T}_2} \bigcup_{R' \in R_{-1}} \{\langle h,r,t\rangle \mid$

body$(R')\,\theta \in \mathcal{G}^+$, body$(R)\,\theta = \{\langle h, r, t\rangle\} \cup$ body$(R')\,\theta\}$. Such a $\mathcal{G}^-$ is a subset of the complete set of inconsistent triples, since for every triple $\langle h, r, t\rangle \in \mathcal{G}^-$, adding $\langle h, r, t\rangle$ to $\mathcal{G}^+$ will instantiate the body of some constraint in $\mathcal{T}_2$ to a subset of $\mathcal{G}^+$, which means that it will render $\mathcal{B}$ inconsistent and thus $\langle h, r, t\rangle$ is an inconsistent triple. We can make $\mathcal{G}^-$ closer to the complete set of inconsistent triples by adding to $\mathcal{T}_2$ more constraints that are entailed by $\mathcal{T}$. Finally, with the learnt vectors for entities and relations up to now, following [20] σ_r is set as the value that maximizes $\Sigma_{\langle h,r,t\rangle \in \mathcal{G}^+} I(\mathsf{loss}_r(h,t) < \sigma_r) + \Sigma_{\langle h',r,t'\rangle \in \overline{\mathcal{G}^+}} I(\mathsf{loss}_r(h',t') \geq \sigma_r)$, namely the number of correctly classified triples on r, where $I(C)$ is the indicator function such that $I(C) = 1$ if C is true or $I(C) = 0$ otherwise.

Following existing translational distance methods, the above optimal objective function is minimized by stochastic gradient descent. To guarantee efficiency and convergence, the set of relation-specific thresholds $\{\sigma_r\}$ is updated for every k training rounds, where k can be set from tens to hundreds. For the first k rounds, the loss values of inconsistent triples are ignored; i.e., the optimal objective function is reduced to the standard global margin-based loss function. Afterwards, the set of relation-specific thresholds is computed for the first time. In other iterations, since there have been relation-specific thresholds, the optimal objective function is recovered as given in Formula (2).

4   Experimental Evaluation

4.1   Implementation

We enhanced two translational distance methods (TransE [2] and TransH [24]) and three semantic matching methods (DistMult [27], ComplEx [21] and Analogy [16]) by our proposed approach. We chose these methods because they are the most efficient methods in their respective categories. To uniformly use a global margin-based loss function as the optimal objective function for semantic matching methods, the loss function for triples $\langle h, r, t\rangle$ in Analogy is defined as $\mathsf{loss}_r(h,t) = -l_h^\mathsf{T} B_r l_t$. This loss function is restricted by setting n real scalars (resp. 0 real scalar) on the diagonal of B_r in DistMult (resp. ComplEx), where n is the dimension of entity vectors. By DistMult^{tr}, ComplEx^{tr} and Analogy^{tr} we denote the variants of DistMult, ComplEx and Analogy that use Formula (1) as the optimal objective function, respectively. Moreover, by X-lc we denote the method enhanced from the baseline method X by our proposed approach.

We implemented all the above methods with multi-threads in Java, using stochastic gradient descent with fixed mini-batch size 1, and evaluated them in the RapidMiner platform[1] to ensure all methods to be compared in the same environment. For the baseline methods we replaced the input training set $\mathcal{G}$ in Formula (1) with the triple set $\mathcal{G}^+$ entailed by the union of the logical thoery and the training set. This implementation of baseline methods has been shown to be effective in improving the predictive performance [5,7]. For the implementation

---

[1] https://www.rapidminer.com/.

of Analogy^{tr} and Analogy^{tr}-lc, we fixed the number of real scalars on the diagonal of B_r as $\frac{n}{2}$ for n the dimension of entity vectors.

To tune the hyper-parameters in our implemented methods, we focus on three prediction tasks, namely link prediction, triple classification and consistency checking for high-rank triples, where the last task is studied for the first time.

Link prediction, originated from [2], is a benchmark task in the field of KGE which aims to predict the missing head or the missing tail in a triple. We used the conventional metrics Mean Reciprocal Rank (MRR) and Hits@k (where $k = 1, 10$) to evaluate the performance in the test set. Since a corrupted triple in the evaluation is actually correct if it is entailed by the logical background, we computed the ranks of triples without counting any entailed triples.

Triple classification, originated from [20], is another benchmark task aiming to judge whether a triple is correct or not. We used the conventional metric Classification Accuracy (CA) to evaluate the performance in the test set. This metric is computed by first determining the relation-specific thresholds on loss values of positive triples in the training set and then calculating the ratio of correctly classified positive triples and negative triples in the test set, where one negative triple is generated from every positive triple.

Consistency checking for high-rank triples aims to judge whether a high-rank triple is consistent with the logical background or not. We introduced the metrics Prec@k (where $k = 10, 100, 1000$) to evaluate the performance. Prec@k is defined as the proportion of triples that are consistent with the logical background among the top-k *new* triples that have the smallest loss values, where the set of new triples is defined as the set of triples which are constructed from existing entities and relations but are not entailed by the logical background.

In our experiments, we fixed the number of training rounds to 2000 in all the above methods and the interval for updating the relation-specific thresholds to 100 rounds in all enhanced methods. Moreover, we selected the learning rate among $\{0.001, 0.005, 0.01, 0.05, 0.1\}$, the margin among $\{0.1, 0.15, 0.2, 0.5, 1, 2\}$, the dimension of entity vectors among $\{20, 50, 100, 200\}$, and the dissimilarity measure as either L1-norm or L2-norm. Since our goal is to learn an embedding model that works well for all the aforementioned tasks, we determined the optimal hyper-parameters that achieve the highest geometric mean of three metrics Hits@1, CA and Prec@10 by five-fold cross-validation on the training set.

4.2   Dataset Construction

Existing benchmark datasets for KGE do not have corresponding logical theories with constraints. It is hard to add adequate constraints to these datasets since they belong to general domains using our unfamiliar relations. Thus we constructed new knowledge graphs and the corresponding logical theories from a domain that we are familiar with. The domain is about character relationships in the four great classical masterpieces of Chinese literature, namely Dream of the Red Chamber (DRC), Journey to the West (JW), Outlaws of the Marsh (OM), and Romance of the Three Kingdoms (RTK). We manually built the

Table 1. Knowledge graphs and their corresponding logical theories

KG	#rel	#ent	#all	#train	#ent-train	#test	#datalog	#const	#ent-const
DRC	48	392	382	344	5,377	323	110	107	737
JW	31	104	109	99	1,136	154	94	30	154
OM	49	156	201	181	1,975	339	111	47	282
RTK	51	123	153	138	3,123	342	142	63	389

Note: #rel/#ent/#all are respectively the number of relations/entities/triples in the knowledge graph, #train/#test are respectively the cardinality of the training set/test set, #ent-train is the number of triples entailed by the union of the logical theory and the training set, #datalog/#const/#ent-const are respectively the number of datalog rules/constraints/entailed constraints in the logical theory.

logical theories for these four masterpieces separately in Protege[2], a well-known ontology editor. These theories model character relationships which are originally expressed in OWL 2 [6], a W3C recommended language for modeling ontologies, and then translated to Horn theories by standard syntactic transformation. We collected triples on character relationships from e-books, yielding four knowledge graphs each of which corresponds to one of these masterpieces.

We split a knowledge graph into a training set and a seed set by 9:1. Initially the training set is the complete knowledge graph and the seed set is empty. Then the seed set is enlarged by randomly picking out triples from the training set one by one, until the number of triples in the seed set reaches one-tenth of the cardinality of the knowledge graph, where every triple picked out is composed by entities in the remainder of the training set but is not entailed by the union of the logical theory and the remainder of the training set. The test set is defined as the set of triples entailed by the union of the logical theory and the seed set.

Recall that we approximated the set $\mathcal{G}^-$ of negative triples in Formula (2). When more entailed constraints are used, the approximation will be closer to the exact result. To find more entailed constraints, we treated constraints in the logical theory as conjunctive queries and applied the REQUIEM tool[3] [18] to compute all constraints entailed by the $\mathcal{ELHIO}$ fragment of the logical theory, where $\mathcal{ELHIO}$ is a description logic underlying OWL 2 [6].

Table 1 reports the statistics about our constructed datasets. These datasets and our implemented systems are available at OpenKG.CN[4].

4.3   Results on Link Prediction and Triple Classification

Table 2 reports the evaluation results on link prediction and triple classification for all baseline methods X and their enhanced methods X-lc, under their corresponding optimal hyper-parameters determined by five-fold cross-validation on the training set. It can be seen that, for all the datasets the metric score of

---

[2] https://protege.stanford.edu/.

[3] http://www.cs.ox.ac.uk/projects/requiem/.

[4] http://www.openkg.cn/tool/kge-wlc.

Table 2. Evaluation results on link prediction and triple classification

KG	DRC				JW			
Metric	MRR	Hits@1	Hits@10	CA	MRR	Hits@1	Hits@10	CA
TransE	0.453	0.432	0.483	0.698	0.811	0.786	0.838	**0.873**
TransE-lc	0.442	0.429	0.457	0.700	0.758	0.714	0.831	0.860
TransH	0.453	0.430	0.488	0.697	0.815	0.792	0.844	0.870
TransH-lc	0.440	0.430	0.454	**0.704**	0.812	0.773	**0.899**	0.864
DistMult^{tr}	0.452	0.432	**0.489**	0.676	0.817	0.795	0.844	0.870
DistMult^{tr}-lc	**0.461**	**0.441**	0.486	0.683	0.800	0.763	0.838	0.821
ComplEx^{tr}	0.453	**0.441**	0.474	0.693	**0.821**	**0.805**	0.851	0.864
ComplEx^{tr}-lc	0.432	0.415	0.450	0.684	0.812	0.799	0.831	0.870
Analogy^{tr}	0.441	0.427	0.457	0.697	0.790	0.773	0.812	0.864
Analogy^{tr}-lc	0.430	0.416	0.440	**0.704**	0.814	0.799	0.831	0.860
KG	OM				RTK			
Metric	MRR	Hits@1	Hits@10	CA	MRR	Hits@1	Hits@10	CA
TransE	0.842	0.811	**0.912**	0.844	0.869	0.849	**0.905**	**0.930**
TransE-lc	0.845	0.814	**0.912**	0.844	0.868	0.848	0.901	0.925
TransH	0.848	0.813	0.909	0.845	0.863	0.842	0.893	0.920
TransH-lc	0.847	0.813	0.910	0.892	0.863	0.842	0.898	0.923
DistMult^{tr}	**0.857**	**0.835**	0.886	0.886	0.855	0.830	0.893	0.918
DistMult^{tr}-lc	0.828	0.788	0.883	**0.907**	0.858	0.842	0.895	0.909
ComplEx^{tr}	0.827	0.799	0.882	0.897	0.860	0.845	0.886	0.923
ComplEx^{tr}-lc	0.837	0.811	0.878	0.875	**0.874**	**0.860**	0.890	0.928
Analogy^{tr}	0.820	0.802	0.845	0.894	0.855	0.841	0.870	0.918
Analogy^{tr}-lc	0.815	0.798	0.850	0.883	0.860	0.845	0.890	0.917

Table 3. The mean±std metric score and the p-value in t-test for all datasets

Metric	MRR	Hits@1	Hits@10	CA
Baseline methods	0.740 ± 0.173	0.719 ± 0.171	0.773 ± 0.177	0.839 ± 0.090
Enhanced methods	0.735 ± 0.176	0.712 ± 0.172	0.771 ± 0.188	0.838 ± 0.089
p-value	0.92484	0.89759	0.97512	0.97042

an enhanced method can be higher or lower than the metric score of the corresponding baseline method, while the difference between them is rather small.

Table 3 further summarizes the comparison results between baseline and enhanced methods for all datasets. It can be seen that the mean metric score of the baseline methods is slightly higher than that of their enhanced methods, but the difference is trivial as the p-value in t-test for accepting the null hypothesis that the based methods have the same mean metric score as their enhanced

Table 4. Evaluation results on consistency checking for high-rank triples

KG	DRC			JW			OM			RTK		
Prec@k	@10	@100	@1000	@10	@100	@1000	@10	@100	@1000	@10	@100	@1000
TransE	0.4	0.48	0.438	0.9	0.90	0.491	**1.0**	0.95	0.842	**1.0**	0.99	0.981
TransE-lc	**1.0**	**1.00**	**1.000**	**1.0**	0.96	0.927	**1.0**	**1.00**	0.984	**1.0**	**1.00**	0.985
TransH	0.5	0.47	0.434	0.8	0.88	0.586	**1.0**	0.98	0.969	**1.0**	0.98	0.981
TransH-lc	**1.0**	**1.00**	0.968	**1.0**	**0.99**	0.913	**1.0**	**1.00**	0.979	**1.0**	0.99	0.985
DistMult^{tr}	0.5	0.37	0.670	0.1	0.80	0.953	0.7	0.92	0.979	0.6	0.60	0.549
DistMult^{tr}-lc	**1.0**	**1.00**	**1.000**	0.9	0.91	0.941	**1.0**	**1.00**	**1.000**	**1.0**	**1.00**	**1.000**
ComplEx^{tr}	0.7	0.65	0.735	0.6	0.65	0.898	0.8	0.83	0.972	0.9	0.72	0.645
ComplEx^{tr}-lc	**1.0**	**1.00**	0.999	0.9	0.90	**0.966**	**1.0**	0.97	0.982	**1.0**	**1.00**	0.949
Analogy^{tr}	0.8	0.77	0.802	0.5	0.82	0.947	0.9	0.92	0.981	0.8	0.64	0.620
Analogy^{tr}-lc	**1.0**	**1.00**	**1.000**	**1.0**	0.95	0.963	**1.0**	**1.00**	**1.000**	**1.0**	**1.00**	**1.000**

Table 5. The mean±std metric score and the p-value in t-test for all datasets

Metric	Prec@10	Prec@100	Prec@1000
Baseline methods	0.725 ± 0.240	0.766 ± 0.187	0.774 ± 0.202
Enhanced methods	0.990 ± 0.031	0.984 ± 0.031	0.977 ± 0.026
p-value	0.00002	0.00001	0.00007

methods have is close to 1. This implies that the enhanced methods achieve comparable performance as baseline methods in traditional prediction tasks.

4.4 Results on Consistency Checking for High-Rank Triples

Table 4 reports the evaluation results on consistency checking for high-rank triples for all baseline methods X and their enhanced methods X-lc, under their corresponding optimal hyper-parameters. It can be seen that, for all the datasets the metric score of an enhanced method is very high (always ≥ 0.9) and is consistently not lower than the metric score of the corresponding baseline method.

Table 5 further summarizes the comparison results between baseline and enhanced methods for all datasets. It can be seen that the mean metric score of the baseline methods is clearly lower than that of their enhanced methods, while this difference is significant since the p-value in t-test for accepting the null hypothesis that the baseline methods have the same mean metric score as their enhanced methods have is nearly 0.

5   Related Work

As mentioned in Subsect. 2.1, translational distance methods and semantic matching methods constitute the two main categories for KGE methods. The well-known translational distance methods include Structured Embedding (SE) [3], TransE [2], TransH [24], TransR [15], TransD [11], PTransE [14], TranSparse [12], TransA [13], KG2E [9], TransG [26], etc. Among the above methods, except for KG2E and TransG that use distribution functions to represent entities and relations, others represent entities and relations by vectors and employ Formula (1) as the optimal objective function. On the other hand, the well-known semantic matching methods include Latent Factor Model (LFM) [10], Single Layer Model (SLM) [20], Neural Tensor Network (NTN) [20], Semantic Matching Energy (SME) [1], DistMult [27], ComplEx [21], HolE [17], Analogy [16], etc. These methods exploit neural networks with various structures to learn a distributed representation. Among these methods Analogy can be treated as a generalized model for DistMult, ComplEx and HolE [16]. We refer the interested readers to [22] for a thorough review of existing KGE methods.

The most related work to us is about combining KGE with logic inference. Wang et al. [23] and Wei et al. [25] tried to combine embedding models with datalog rules for KGE. But in their work, rules are modeled separately from embedding models and would not help to learn a more predictive distributed representation. Rocktäschel et al. [19], Guo et al. [7] and Du et al. [5] proposed several joint learning paradigms that encode the inference of implicit triples into KGE, where the inference is guided by datalog rules. To avoid the costly propositionalization of datalog rules, Demeester et al. [4] further proposed a method for injecting lifted datalog rules to KGE. The above join models make a one-time injection of hard rules, taking them as either additional training instances [5,7,19] or regularization terms [4]. To make the best of background information, Guo et al. [8] proposed a paradigm for interactively injecting soft rules to KGE, which learns simultaneously from labeled triples, unlabeled triples and soft rules with various confidence levels in an iterative manner. However, the above work considers only datalog rules that are used to infer implicit triples, but ignores constraints that prevent new triples from being present.

6   Conclusions and Future Work

To ensure the high-rank triples predicted by an embedding-based method to be as consistent as possible with the logical background, which consists of a knowledge graph and a Horn theory, we have proposed an approach to enhancing the method to encode logical consistency into the learnt distributed representation. To evaluate this approach, we also constructed four knowledge graphs and their corresponding Horn theories from the four great classical masterpieces of Chinese literature. Experimental results on these datasets demonstrated the efficacy of our proposed approach. For future work, we plan to study improved methods to achieve higher metric scores in link prediction and triple classification, as well as to consider more expressive logical theories beyond Horn theories to guarantee logical consistency in knowledge graph embedding.

Acknowledgements. This work was partly supported by National Natural Science Foundation of China (61375056 and 61876204), Science and Technology Program of Guangzhou (201804010496), and Scientific Research Innovation Team in Department of Education of Guangdong Province (2017KCXTD013).

References

1. Bordes, A., Glorot, X., Weston, J., Bengio, Y.: A semantic matching energy function for learning with multi-relational data - application to word-sense disambiguation. Mach. Learn. **94**(2), 233–259 (2014)
2. Bordes, A., Usunier, N., García-Durán, A., Weston, J., Yakhnenko, O.: Translating embeddings for modeling multi-relational data. In: NIPS, pp. 2787–2795 (2013)
3. Bordes, A., Weston, J., Collobert, R., Bengio, Y.: Learning structured embeddings of knowledge bases. In: AAAI (2011)
4. Demeester, T., Rocktäschel, T., Riedel, S.: Lifted rule injection for relation embeddings. In: EMNLP, pp. 1389–1399 (2016)
5. Du, J., Qi, K., Wan, H., Peng, B., Lu, S., Shen, Y.: Enhancing knowledge graph embedding from a logical perspective. In: Wang, Z., Turhan, A.-Y., Wang, K., Zhang, X. (eds.) JIST 2017. LNCS, vol. 10675, pp. 232–247. Springer, Cham (2017). https://doi.org/10.1007/978-3-319-70682-5_15
6. Grau, B.C., Horrocks, I., Motik, B., Parsia, B., Patel-Schneider, P.F., Sattler, U.: OWL 2: the next step for OWL. J. Web Semant. **6**(4), 309–322 (2008)
7. Guo, S., Wang, Q., Wang, L., Wang, B., Guo, L.: Jointly embedding knowledge graphs and logical rules. In: EMNLP, pp. 192–202 (2016)
8. Guo, S., Wang, Q., Wang, L., Wang, B., Guo, L.: Knowledge graph embedding with iterative guidance from soft rules. In: AAAI, pp. 4816–4823 (2018)
9. He, S., Liu, K., Ji, G., Zhao, J.: Learning to represent knowledge graphs with Gaussian embedding. In: CIKM, pp. 623–632 (2015)
10. Jenatton, R., Roux, N.L., Bordes, A., Obozinski, G.: A latent factor model for highly multi-relational data. In: NIPS, pp. 3176–3184 (2012)
11. Ji, G., He, S., Xu, L., Liu, K., Zhao, J.: Knowledge graph embedding via dynamic mapping matrix. In: ACL, pp. 687–696 (2015)
12. Ji, G., Liu, K., He, S., Zhao, J.: Knowledge graph completion with adaptive sparse transfer matrix. In: AAAI, pp. 985–991 (2016)
13. Jia, Y., Wang, Y., Lin, H., Jin, X., Cheng, X.: Locally adaptive translation for knowledge graph embedding. In: AAAI, pp. 992–998 (2016)
14. Lin, Y., Liu, Z., Luan, H., Sun, M., Rao, S., Liu, S.: Modeling relation paths for representation learning of knowledge bases. In: EMNLP, pp. 705–714 (2015)
15. Lin, Y., Liu, Z., Sun, M., Liu, Y., Zhu, X.: Learning entity and relation embeddings for knowledge graph completion. In: AAAI, pp. 2181–2187 (2015)
16. Liu, H., Wu, Y., Yang, Y.: Analogical inference for multi-relational embeddings. In: ICML, pp. 2168–2178 (2017)
17. Nickel, M., Rosasco, L., Poggio, T.A.: Holographic embeddings of knowledge graphs. In: AAAI, pp. 1955–1961 (2016)
18. Pérez-Urbina, H., Motik, B., Horrocks, I.: Tractable query answering and rewriting under description logic constraints. J. Appl. Logic **8**(2), 186–209 (2010)
19. Rocktäschel, T., Singh, S., Riedel, S.: Injecting logical background knowledge into embeddings for relation extraction. In: NAACL, pp. 1119–1129 (2015)
20. Socher, R., Chen, D., Manning, C.D., Ng, A.Y.: Reasoning with neural tensor networks for knowledge base completion. In: NIPS, pp. 926–934 (2013)

21. Trouillon, T., Welbl, J., Riedel, S., Gaussier, É., Bouchard, G.: Complex embeddings for simple link prediction. In: ICML, pp. 2071–2080 (2016)
22. Wang, Q., Mao, Z., Wang, B., Guo, L.: Knowledge graph embedding: a survey of approaches and applications. IEEE Trans. Knowl. Data Eng. **29**(12), 2724–2743 (2017)
23. Wang, Q., Wang, B., Guo, L.: Knowledge base completion using embeddings and rules. In: IJCAI, pp. 1859–1866 (2015)
24. Wang, Z., Zhang, J., Feng, J., Chen, Z.: Knowledge graph embedding by translating on hyperplanes. In: AAAI, pp. 1112–1119 (2014)
25. Wei, Z., Zhao, J., Liu, K., Qi, Z., Sun, Z., Tian, G.: Large-scale knowledge base completion: Inferring via grounding network sampling over selected instances. In: CIKM, pp. 1331–1340 (2015)
26. Xiao, H., Huang, M., Zhu, X.: TransG: a generative model for knowledge graph embedding. In: ACL, pp. 992–998 (2016)
27. Yang, B., Yih, W., He, X., Gao, J., Deng, L.: Embedding entities and relations for learning and inference in knowledge bases. CoRR abs/1412.6575 (2014). http://arxiv.org/abs/1412.6575

An End-to-End Entity and Relation Extraction Network with Multi-head Attention

Lishuang Li$^{(\boxtimes)}$, Yuankai Guo, Shuang Qian, and Anqiao Zhou

Dalian University of Technology, Dalian 116024, China
lilishuang314@163.com,
{guoyuankai,QShuang,ahashi_syuu}@mail.dlut.edu.cn

Abstract. Relation extraction is an important semantic processing task in natural language processing. The state-of-the-art systems usually rely on elaborately designed features, which are usually time-consuming and may lead to poor generalization. Besides, most existing systems adopt pipeline methods, which treat the task as two separated tasks, i.e., named entity recognition and relation extraction. However, the pipeline methods suffer two problems: (1) Pipeline model over-simplifies the task to two independent parts. (2) The errors will be accumulated from named entity recognition to relation extraction. Therefore, we present a novel joint model for entities and relations extraction based on multi-head attention, which avoids the problems in the pipeline methods and reduces the dependence on features engineering. The experimental results show that our model achieves good performance without extra features. Our model reaches an F-score of 85.7% on SemEval-2010 relation extraction task 8, which has competitive performance without extra feature compared with previous joint models. On publication, codes will be made publicly available.

Keywords: Relation extraction · End-to-End joint extraction
Named entity recognition

1 Introduction

Named entity recognition (NER) and relation extraction are important semantic processing tasks in natural language processing (NLP). Traditional pipeline methods divide the task into two parts: named entity recognition and relation extraction. Firstly, entities are recognized in the sentences. Then, the identified entities are combined into entity pairs. Finally, the relations between entities pairs are extracted. The mainstream pipeline methods are based on neural network models, such as convolutional neural networks (CNN) and recurrent/recursive neural networks (RNNs). Zeng et al. [1] exploited a CNN to extract lexical and sentence level features for relation classification and achieved an F1-score of 82.7%. Xu et al. [2] presented a LSTM model with shortest dependency path (SDP) for relation extraction, which reached 83.7% F1-score.

The paper is supported by the National Natural Science Foundation of China under No. 61672126.

M. Sun et al. (Eds.): CCL 2018/NLP-NABD 2018, LNAI 11221, pp. 136–146, 2018.
https://doi.org/10.1007/978-3-030-01716-3_12

Zhou et al. [3] proposed a BLSTM with an attention model for relation extraction and reached 84.0% F1-score. The above experiments are based on the SemEval-2010 task 8 dataset. NER and relation extraction can be improved independently in the pipeline method. However, the pipeline methods ignore the interaction between NER and relation extraction. In addition, the errors in the upstream components are propagated to the downstream components without any feedback [4].

Recent studies show that end-to-end (joint) modeling of entity and relation is important for high performance [5]. The joint model processes NER and relation extraction simultaneously, which can alleviate the errors propagation. Furthermore, the NER and the relation extraction components share some parameters in joint model, which could capture the interaction between the sub-tasks. Li et al. [4] presented a model using structured perceptron with efficient beam-search on ACE04, which employed many features such as global entity mention features and local features. Miwa and Sasaki [6] proposed a history-based structured learning approach using lexical, contextual features and so on, which obtained 69.8% F1-score on CONLL 04 dataset. Although the above models adopted the joint method for entity and relation extraction, they are all based on shallow machine learning methods, which still rely on feature engineering such as dependency features and syntactic features. Miwa et al. [5] proposed a LSTM based sequence and tree-structure model and achieved 85.5% F1-score on SemEval-2010 Task 8. Katiyar et al. [7] presented attention-based pointer network for joint extraction of entity mentions and relations, reaching an F1-score of 53.6% on ACE05 dataset. Zheng et al. [8] proposed a joint extraction of entities and relations based on a tagging scheme using LSTM and achieved 49.5% F1-score on NYT dataset. The above models utilized the deep-learning models with simple features. Some model also applied attention mechanism such as Katiyar et al.'s model. The application of attention mechanism obtains a representation weighted by the importance of tokens.

In this paper, we propose a novel end-to-end attention-based BLSTM model for entities and relation extraction without any handcraft features and structure features. The contributions of this paper are as follows:

1. Our model enhances the interaction between entities and relations. The model can utilize the output of the hidden layers in NER to provide more information for relation extraction. The entity labels and the output of hidden layers are fed into the BLSTM with a multi-head attention layer to extract the relation between entity pairs.
2. Our multi-head attention can focus on the words which have decisive effect for relation extraction. Comparing with other attention mechanisms, our multi-head attention can capture different relevant features for relation extraction.

The experimental results show that our model achieves an F1-score of 85.66% on the SemEval-2010 task 8 [9], which has competitive performance without extra feature compared with previous joint models.

2   Proposed Method

In this section, we describe our end-to-end multi-head attention model in detail. The framework of the model is illustrated in Fig. 1. The model mainly contains three components: the word embedding layer, the name entity recognition layer based on BLSTM-CRF and the relation extraction layer based on BLSTM with multi-head attention.

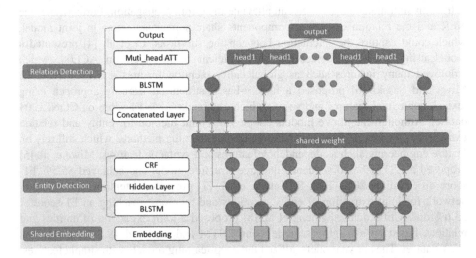

Fig. 1. End-to-End entity and relation extraction model with multi-head attention.

2.1   Entity Detection

LSTM units are firstly proposed by Hochreiter and Schmidhuber [10] to overcome gradient vanishing problem. LSTMs are more capable of capturing long-term dependencies between tokens, making it ideal for both entity mentions and relation extraction. We use the following implementation:

$$
\begin{aligned}
i_t &= \sigma(W_{xi}x_t + W_{hi}h_{t-1} + W_{ci}c_{t-1} + b_i) \\
f_t &= \sigma(W_{xf}x_t + W_{hf}h_{t-1} + W_{cf}c_{t-1} + b_f) \\
g_t &= \tanh(W_{xc}x_t + W_{hc}h_{t-1} + W_{cc}c_{t-1} + b_c) \\
c_t &= i_t g_t + f_t c_{t-1} \\
o_t &= \sigma(W_{xo}x_t + W_{ho}h_{t-1} + W_{co}c_t + b_o) \\
h_t &= o_t \tanh(c_t),
\end{aligned}
\tag{1}
$$

where i_t, f_t and o_t are the input gate, forget gate and output gate respectively. W_{xi}, W_{hi} and W_{ci} are the parameters of weights matrix of input gate. W_{xf}, W_{hf} and W_{cf} stand for the corresponding weights matrix of forget gate respectively. Similarly W_{xo}, W_{ho} and W_{co} are the weights matrix of output gate. b is the bias term and the hidden vectors are represented as $H = \{h_1, h_2, \ldots, h_n\}$, where n is the number of words in a sentence. All of those gates are set to generate some degrees, using the current input x_t, the state h_t that previous step generated, and the current state of this cell c_t. In this paper, we employ bidirectional LSTM (BLSTM) which contains two sub-networks for the left and right sequence context. For each sentence coming from the embedding layer, the forward LSTM encodes a sentence from left to right and the backward LSTM encodes a sentence from right to left. We concatenate the output of forward LSTM and backward LSTM as the following equation:

$$h_i = [\vec{h}_i \oplus \overleftarrow{h}_i]$$
(2)

2.2   CRF Layer

In some traditional methods, the output of BLSTM H is used to get independent tagging for the corresponding words in the sentence. However, NER tasks have strong dependencies across the output label. For example, it is illegal when the label "B" follows the label "I". Compared with the independent output layer, Conditional Random Fields (CRF) [11] can efficiently use the whole sentence tag from the output layer, which can obtain the best sequence of tags. The output H of BLSTM is marked as $X = \{x_1, x_2, \ldots, x_n\}$, which is the input of CRF layer. We consider P as the matrix of the output by BLSTM with size of $n*m$ (n is the length of the sentence, m is the number of tag's classes). $P_{i,j}$ denotes the score of the j-th tag of i-th word. For a sequence of predictions $y = \{y_1, y_2, \ldots, y_n\}$, the score is described as follows:

$$K(x, y) = \sum_{i=0}^{n} A_{y_i y_{(i+1)}} + \sum_{i=1}^{n} P_{i, y_i},$$
(3)

where A represents the matrix of transition scores. A softmax over all possible tag sequences yields a probability for the sequence y as Eq. (4).

$$p(y|X) = \frac{e^{s(X,y)}}{\sum_{\tilde{y} \in Y_X} e^{s(X,\tilde{y})}}.$$
(4)

We maximize the log-probability of the correct tag sequence as Eq. (5):

$$\log(p(y|S)) = K(X, y) - \log(\sum_{\tilde{y} \in Y_X} e^{K(X,\tilde{y})}),$$
(5)

where YX represents all possible tag sequences including the format which does not obey the "BIO" format constraints. We get the maximum score given by:

$$y^* = \arg\max_{\tilde{y} \in Y_X} K(X, \tilde{y})$$
(6)

The output of entity labels is transformed into one-hot vectors. The output of the hidden layer and the one-hot vectors are fed into the concatenated layer.

3   Relation Extraction Detection

3.1   Concatenated Layer

In the stage of relation detection, the input contains three parts: the hidden weights of NER $H = \{h_1, h_2, \ldots, h_n\}$. The label of tokens $L = \{l_1, l_2, \ldots, l_n\}$ from CRF layer, and the word embeddings shared with name entity recognition $embs = \{e_1, e_2, \ldots, e_n\}$. We concatenate the hidden weights of NER, token labels and word embeddings as the following format $d_i = \{h_i; l_i; e_i\}$. The concatenate vectors $D = \{d_1, d_2, \ldots, d_n\}$ are fed into the BLSTM with multi-head attention layer to extract the relations between entities pairs. The parameters of the entity layer are shared and they are jointly updated in entity recognition and relation extraction training. The shared weights enhance the relevance of entities and relations, and provide more features for relation extraction.

3.2   Multi-head Attention Layer on BLSTM

Some words play a key role in relation extraction. For example, in the sentence "The <e1> burst </e1> has been caused by water hammer <e2> pressure </e2> .", the word *'caused'* is critical to the relation extraction because the it may reflect relation type. These decisive words should get more attention than other words. Guided by this, we employ a multi-head attention mechanism to focus on the importance of different words for relation classification. The multi-head attention applies attention mechanism multiple times over the same inputs using separately attention heads and combines the results which could concentrate on the different relevant feature for relation extraction. The attention function can be described as Eqs. (7–8):

$$\alpha = \text{softmax}(\frac{QK^T}{\sqrt{d_w}})$$
$$head_i = \sum_n \alpha V \tag{7}$$
$$Q \in n * d_w, k \in n * d_w, V \in n * d_w, i \in H,$$

$$MultiHead(Q, K, V) = (head_1 \oplus \cdots \oplus head_H), \tag{8}$$

where Q, K and V represent the attention query matrix, key matrix and value matrix respectively. In our model, we use self-attention mechanism in each head of attention. Thus Q, K, and V represent the output sequences of BLSTM with size of $n * d_w$, where d_w is the dimension of BLSTM output. H represents the number of heads and i indicates the i-th head of attention. For each head of attention, we compute attention weights by the Eq. (7), which captures the different features from the sentences. We concatenate each head from the left to right and get the final results.

4   Relation Extraction Detection

4.1   Dataset and Tasks

We evaluate our model on SemEval-2010 task 8 dataset [9]. The dataset contains 8000 training instances and 2717 test instances annotated with 9 different relation types and an artificial relation "Other", which is used to indicate that the relation does not belong to any of the nine main relation types. Table 1 shows the details of the dataset.

Each instance contains a sentence marked with two nominals <e1> and <e2.>, and the task is used to predict the relation between the two nominals considering the directionality. It means that the relation Cause-Effect (e1, e2) is different from the relation Cause-Effect (e2, e1), as shown in the examples below:

"The current view is that the chronic <e1> inflammation </e1> in the distal part of the stomach caused by Helicobacter pylori <e2> infection </e2> results in an increased acid production from the non-infected upper corpus region of the stomach."

Cause-Effect (e2, e1)

"The <e1> singer </e1> , who performed three of the nominated songs, also caused a <e2> commotion </e2>   on the red carpet."

Cause-Effect (e1, e2)

The results are measured using official evaluation metric, which is based on macro-averaged F1-score for the nine proper relations and others. The definition of Precision (*P*), Recall (*R*) and *F1-score* are shown as Eq. (9):

$$P = \frac{TP}{TP+FP}, R = \frac{TP}{TP+FN}, F1-score = \frac{2*P*R}{P+R}, \tag{9}$$

where *TP* is short for true positives, *FP* represents false positives, and *FN* stands for false negatives.

Table 1. Annotation statistics of SemEval-2010 task 8 dataset.

Relation	Freq in train set	Freq in test set
Cause-Effect	(1003)12.53%	(328)12.07%
Instrument-Agency	(504)6.30%	(156)5.74%
Product-Producer	(717)8.96%	(231)8.50%
Content-Container	(540)6.75%	(192)7.06%
Entity-Origin	(716)8.95%	(258)9.49%
Entity-Destination	(845)10.56%	(292)10.47%
Component-Whole	(941)11.76%	(312)11.48%
Member-Collection	(690)8.62%	(233)8.57%
Message-Topic	(634)7.92%	(261)9.60%
Other	(1410)17.62%	(454)16.70%
Total	(8000)100%	(2717)100%

4.2   Hyper-parameters Settings

This subsection presents the hyper parameter tuning for our model. The model is implemented in Keras and trained on a single 1080Ti GPU. We employ Adam method [12] to optimize our model. The learning rate is set to 0.001 and batch size is 32. For multi-head attention we set the number of attention heads to 4. We use the publicly available GloVe [13] to train word embeddings. The dimensions of word vectors are set as 200. The dropout rate is set to be 0.4 to prevent overfitting [14]. L2-regularizations are also employed in training to prevent overfitting.

4.3   Overall Performance

Table 2 shows the experimental results of our method on SemEval-2010 task 8 dataset. The model achieves 84.21% F1-score only using the entity hidden weights, which is 1.31 percentage points higher than that of BLSTM pipeline model. The entity hidden weights can provide the information of entities and context in the process of NER. Experimental results demonstrate that the output of the hidden layer of NER is of benefit to the relation extraction and improves the performance of the model.

Table 2. Results on SemEval-2010 task 8 based on entity hidden weights and multi-head attention

Model	P(%)	R(%)	F(%)
BLSTM	82.76%	83.10%	82.93%
BLSTM + entity_hidden_weights	84.08%	84.35%	84.21%
BLSTM + multi-head attention	83.94%	84.74%	84.33%
BLSTM + entity_hidden_weights + multi-head attention	84.92%	86.41%	85.66%

In addition, we propose a multi-attention mechanism to capture the related semantic information of each word. The multi-head attention is able to focus on the words which are critical to the relation extraction. Comparing with other attention mechanisms, our multi-head attention can capture different relevant features for relation extraction. In Table 2, we show the results of models with multi-head attention mechanisms. From the results, we can observe that BLSTM with multi-head attention performs well, whose F-score is 1.4% percentage points higher than the BLSTM model, which proves the effectiveness of our attention mechanism.

Table 2 also shows the result of our model based on multi-head attention combined with entity hidden weights, which achieves 85.66% F1-score. The results indicate that our model with multi-head attention and entity hidden weights can promote the performance.

4.4   Comparison with Previous Models

The follow works are based on the SemEval-2010 task 8:

SDP-LSTM: Xu et al. [2] presented SDP-LSTM to classify the relation of two entities in a sentence. The model leveraged the SDP and multichannel RNNs with 8LSTM units picking up heterogeneous information along the SDP including word representations, part-of-speech tags (POS), grammatical relations and WordNet hypernyms. The model achieved an F1-score of 83.7%.

RNN: Zhang et al. [15] utilized a framework based on RNN and several modifications to enhance the model, including a max-pooling approach and a bi-directional architecture. This model used POS and position features, which achieved an F1-score of 80.0%.

LSTM-RNN: Miwa et al. [5] built an end-to-end LSTM based sequence and tree structured model. They extracted entities via a sequence layer and relations between the entities via the shortest path dependency tree network, which achieved an F1-score of 85%.

CR-CNN: Dos santos et al. [16] tackled the relation classification task using a CNN that performed classification by ranking and proposed a pairwise ranking loss function that made it easy to reduce the impact of artificial classes. CR-CNN achieved an F1-score of 84.1% with using position features.

depLCNN + NS: Xu et al. [17] employed CNN to learn more robust relation representations from the shortest dependency path. Furthermore, they proposed a straightforward negative sampling strategy to improve the assignment of subjects and objects. Their model reached an F1-score of 85.6%.

Att-BLSTM: Zhou et al. [3] proposed attention-based BLSTM to capture the most important semantic information in a sentence. We also use attention mechanism called multi-head attention, which can capture the words that play an import role in the sentence. Comparing with Zhou et al.'s pipeline model only using attention and word vectors, we leverage the end-to-end structure and a multi-head attention mechanism to enhance the effect of emphasis. Most of the above models used handcrafted features (in Table 3), while our model achieves competitive results only using the word vectors and output of the hidden layer of the NER. Our model leverages multi-head attention and entity hidden weights, which achieves an F1-score of 85.7%, without using lexical resources such as WordNet or NLP systems like dependency parser and NER to get high-level features.

4.5 Discussion

From the above section, we know that our method achieves the competitive results on the SemEval-2010 task 8 without feature engineering. The specific results of the analysis are as follows:

1. *Shared the output of hidden layers.* The output of the hidden layers in NER is shared with the relation extraction, which contains contextual information. Therefore, our model achieves competitive results without utilizing any handcrafted features and syntax features.
2. *Multi-head attention.* The model can highlight effective information by multi-head attention. Our multi-head attention can obtain the information from multiple respects compared with other attention mechanisms for relation extraction.

Table 3. Comparison with other models on SemEval-2010 task.

Model	Features	F1-score
Yan et al. (SDP-LSTM) [2]	Word vector + Grammer relation + POS + WordNet	83.7%
Zhang et al. (RNN) [15]	Word vector + POS + position feature	80.0%
Miwa et al. (LSTM-RNN) [5]	Word vector + SDPTree + WordNet	85.5%
Dos santos et al. (CR-CNN) [16]	Word vector + Position feature	84.1%
Xu et al. (depLCNN + NS) [17]	Word vector + shortest dependency paths	85.6%
Zhou et al. (Att-BLSTM) [3]	Word vector	84.0%
Our model	Word vector	85.7%

The experimental results show that our model achieves an F1-score of 85.7% on the SemEval-2010 task 8, which has competitive performance without extra feature compared with previous joint models.

4.6   Error Analysis

We visualize the model's predicted results to analyze the errors of our approach as shown in Fig. 2. The diagonal region indicates the correct prediction results and the other regions reflect the distribution of error samples. The highlighted region means the higher F-score. In Fig. 2 we can find that the F1-scores of relation of "Instrument-Agency" and the "Component-Whole" are lower than the other relations. The analysis for the errors are as follows:

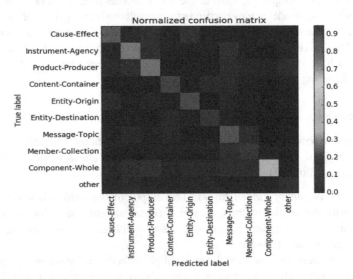

Fig. 2. The distribution of the predicted results for each relation class. The horizontal X-axis represents the predict label and the vertical Y-axis represents the true label.

1. The class imbalanced problem is one of the critical factor affecting the results. As shown in Table 1, only 6.3% of the instances belong to the relation of 'Instrument-Agency' in the train set, which is the smallest of all instances. Undersampling and oversampling will be adopted to balance the number of each class in our future works.
2. The lack of information between entities is another problem which leads to bad performance on some instances. For example, in the sentence of "This <e1> bed </e1> <e2> pole </e2> has a hook type handle, and fits under the mattress and provides a firm handle to assist with moving and positioning in bed.", there is no words between entities 'bed' and 'pole'. Due to the deficiency of key words, our model cannot effectively learn useful features for classification, thus such instances are classified by mistake.

5   Conclusion and Future Work

In this paper, we propose a novel model based on multi-head attention for joint entities and relations extraction. Instead of employing feature engineering, we utilize the hidden layer weights of NER which can supply the information of entities and context in the process of named entity recognition. We also leverage the multi-head attention for relation extraction, which helps our model extract the significant information from different aspects. Our model extracts entities and relations in joint model and does not apply extra handcraft features or NLP tools. The model achieves the competitive results on SemEval-2010 task 8 dataset, which indicates that the effectiveness of the hidden layer weights and multi-head attention. In future works, we will explore the effective way to deal with the problems of class imbalanced and insufficient information.

References

1. Zeng, D., Liu, K., Lai, S., Zhou, G., Zhao, J.: Relation classification via convolutional deep neural network. In: Proceedings of COLING, pp. 2335–2344 (2014)
2. Xu, Y., Mou, L., Li, G., Chen, Y., Peng, H., Jin, Z.: Classifying relations via long short term memory networks along shortest dependency paths. In: Proceedings of the 2015 Conference on Empirical Methods in Natural Language Processing, pp. 1785–1794 (2015)
3. Zhou, P., Shi, W., Tian, J., et al.: Attention-based Bidirectional Long Short-Term Memory Networks for relation classification. In: Meeting of the Association for Computational Linguistics, pp. 207–212 (2016)
4. Li, Q., Ji, H.: Incremental joint extraction of entity mentions and relations. In: Proceedings of the 52nd Annual Meeting of the Association for Computational Linguistics, pp. 402–412 (2014)
5. Miwa, M., Bansal, M.: End-to-End relation extraction using LSTMs on sequences and tree structures. In: Meeting of the Association for Computational Linguistics, pp. 1105–1116 (2016)
6. Miwa, M., Sasaki, Y.: Modeling joint entity and relation extraction with table representation. In: Proceedings of the 2014 Conference on Empirical Methods in Natural Language Processing (EMNLP), pp. 1858–1869 (2014)

7. Katiyar, A., Cardie, C.: Going out on a limb: joint extraction of entity mentions and relations without dependency trees. In: Meeting of the Association for Computational Linguistics, pp. 917–928 (2017)

8. Zheng, S., Wang, F., Bao, H., Hao, Y., Zhou, P., Xu, B.: Joint extraction of entities and relations based on a novel tagging scheme. arXiv preprint arXiv:1706.05075 (2017)

9. Hendrickx, I., et al.: Semeval-2010 task 8: multi-way classification of semantic relations between pairs of nominals. In: Proceedings of the Workshop on Semantic Evaluations: Recent Achievements and Future Directions, pp. 94–99 (2009)

10. Hochreiter, S., Schmidhuber, J.: Long short-term memory. Neural Comput. 9(8), 1735–1780 (1997)

11. Lafferty, J., McCallum, A., Pereira, F.C.N.: Conditional random fields: probabilistic models for segmenting and labeling sequence data. In: Proceedings of ICML, vol. 3, pp. 282–289 (2001)

12. Kingma, D.P., Ba, J.: Adam: a method for stochastic optimization. arXiv preprint arXiv:1412.6980 (2014)

13. Pennington, J., Socher, R., Manning, C.: Glove: global vectors for word representation. In: Proceedings of the 2014 Conference on Empirical Methods in Natural Language Processing (EMNLP), pp. 1532–1543 (2014)

14. Srivastava, N., Hinton, G., Krizhevsky, A., Sutskever, I., Salakhutdinov, R.: Dropout: a simple way to prevent neural networks from overfitting. J. Mach. Learn. Res. 15(1), 1929–1958 (2014)

15. Zhang, D., Wang, D.: Relation classification via recurrent neural network. arXiv preprint arXiv:1508.01006 (2015)

16. Santos, C.N.D., Xiang, B., Zhou, B.: Classifying relations by ranking with convolutional neural networks. arXiv preprint arXiv:1504.06580 (2015)

17. Xu, K., Feng, Y., Huang, S., Zhao, D.: Semantic relation classification via convolutional neural networks with simple negative sampling. arXiv preprint arXiv:1506.07650 (2015)

Attention-Based Convolutional Neural Networks for Chinese Relation Extraction

Wenya Wu, Yufeng Chen[✉], Jinan Xu, and Yujie Zhang

School of Computer and Information Technology, Beijing Jiaotong University,
Beijing, China
{wuwy, chenyf, jaxu, yjzhang}@bjtu.edu.cn

Abstract. Relation extraction is an important part of many information extraction systems that mines structured facts from texts. Recently, deep learning has achieved good results in relation extraction. Attention mechanism is also gradually applied to networks, which improves the performance of the task. However, the current attention mechanism is mainly applied to the basic features on the lexical level rather than the higher overall features. In order to obtain more information of high-level features for relation predicting, we proposed attention-based piecewise convolutional neural networks (PCNN_ATT), which add an attention layer after the piecewise max pooling layer in order to get significant information of sentence global features. Furthermore, we put forward a data extension method by utilizing an external dictionary HIT IR-Lab Tongyici Cilin (Extended). Experiments results on ACE-2005 and COAE-2016 Chinese datasets both demonstrate that our approach outperforms most of the existing methods.

Keywords: Relation extraction · Convolutional neural networks
Attention mechanism

1 Introduction

Relation extraction is one of the basic tasks of Natural Language Processing (NLP), and it is used to identifying the semantic relation holding between two nominal entities in a sentence. It converts unstructured data into structured data. Moreover, relation extraction provides significant technical support for many NLP tasks, such as mass information processing, Chinese information retrieval, knowledge base automatic construction, Machine Translation and automatic abstracting. At present, the scale of Chinese information has increased rapidly, and the technical requirements for Chinese texts processing have been gradually improved. However, there are relatively scarce researches about neural networks in Chinese relation extraction. Therefore, this paper focuses on Chinese corpus.

In the past ten years, deep neural networks have achieved the best results in various fields. For example, models CNNs-based (Zeng et al. 2014; Xu et al. 2015; dos Santos et al. 2015) and RNNs-based (Hashimoto et al. 2013; Zhou et al. 2016) have got fairly competitive results in relation extraction task. Methods that combine CNNs with RNNs (Liu et al. 2015; Cai et al. 2016) have emerged. Attention mechanism (Zhou et al. 2016;

© Springer Nature Switzerland AG 2018
M. Sun et al. (Eds.): CCL 2018/NLP-NABD 2018, LNAI 11221, pp. 147–158, 2018.
https://doi.org/10.1007/978-3-030-01716-3_13

Wang et al. 2016) is also frequently used in network structure. Experiments show that attention mechanism often can improve performance. Regrettably, most attention mechanisms focus on the low-level features like lexical characteristics. Although they are well-interpreted and achieve better results, we should also pay attention to more high-level features, such as the overall advanced features obtained after convolution and max pooling.

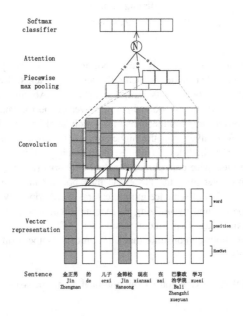

Fig. 1. The architecture of PCNN_ATT used for relation extraction, illustrating the procedure for handling one instance and predicting the relation between "金正男(Jin Zhengnan)" and "金韩松(Jin Hansong)".

Therefore, this paper proposes novel neural networks PCNN_ATT (Attention-based piecewise convolutional networks) for relation extraction. In this model, we choose a convolutional neural network (CNN) to extract features, since CNN is proficient in modeling flat structure and can generate a fixed-size vector with the most meaningful features (Liu et al. 2015). As illustrated in Fig. 1, we employ a convolution layer to extract the semantic features of sentences. Afterwards, to obtain more abundant information, we adopt a piecewise max pooling method presented by Zeng et al. (2015) to further extract high-level features. In the convolutional layer, we use multiple convolution kernels to obtain different types of text features. We believe that different features obtained by different kernels contribute differently to the final relation. Therefore, We add an attention layer to allocate weights effectively. Finally, we extract relation between two nominal entities with the relation vector produced by a softmax classifier.

The main contribution of this model is using PCNN with an attention mechanism, which could automatically focus on the high-level features obtained by max pooling that have decisive effects on relation predicting. In addition, we add a HowNet embedding of the entities in vector representation layer by using HowNet resource to express lexical features better. We evaluate the model on COAE2016 and ACE2005 datasets in relation extraction task. The experimental results show that this model is competitive.

Otherwise, recently most models about deep learning have achieved good results of relation extraction for English, while the effects on Chinese data sets are not so good because of the lack of Chinese annotation data (Sun et al. 2017). To tackle this problem, we use the HIT IR-Lab Tongyici Cilin to expand the existing corpus. The specific method and experiments will be introduced in Sect. 4.3.

The major contributions of this paper are as follows.

- A novel CNN architecture is proposed, which relies on a fancy attention mechanism to capture high-level features (obtained after pooling the results of different convolution kernels) attention. The experiments on COAE2016 and ACE2005 datasets show that our PCNN_ATT model on relation extraction task outperforms other state-of-the- art models.
- The shortest dependency path is utilized to increase the scale of Chinese dataset COAE2016, and we expand the corpus efficiently by taking advantage of external semantic dictionary Tongyici Cilin, which makes it more suitable for deep learning method.
- Hypernym features of nominals in the HowNet are integrated into our model. The experimental results on ACE2005 datasets demonstrate the validity of those features.

2   Related Work

It's well known that relation extraction is one of the important tasks of information extraction. In recent years, many scholars and experts in the Natural Language Processing field have devoted themselves to the construction of Knowledge Graph that makes search more depth and breadth. Information extraction is a key step in the construction processes of Knowledge Graph.

The methods of relation extraction include supervised method, unsupervised method, semi-supervised method and open domain relation extraction method. Compared with other methods, supervised method has more abundant researches. Generally speaking, relation extraction typically cast as a standard supervised multi-class or multi-label classification task. Depending on the representation of relation instances, these supervised means can be further divided into feature-based and kernel-based.

Feature-based methods usually need artificial design characteristics that typically express corpus property. For example, Huang et al. (2010) combined the basic features of words, entities and grammar. Unfortunately, it is not easy to design effective features artificially, which will cost lots of time and effort. Different from feature-based approach, kernel-based method (Bunescu and Mooney 2005) does not require the construction of feature vectors. It directly uses the original form of the string as the processing object, and then computes the similarity function between any two objects. However, the fatal disadvantage of this method is that the speeds of training and prediction are too slow.

Over the past decade, deep learning has shown its outstanding characteristics. Socher et al. (2012) proposed the model named Recursive Matrix-Vector Model (MV-RNN) that learns semantic compositionality from syntactic trees. To capture important information appeared at any position in the sentence, Zhou et al. (2016) put forward Attention-Based Bidirectional Long Short-Term Memory Networks (Att-BLSTM). Att-BLSTM can not only consider the past and future information of the sentence, but also determine the most pivotal information. In addition, CNN model also be used for relation extraction. Convolutional Deep Neural Network (CDNN) was used for relation classification by Zeng et al. (2014), and CDNN adopted position feature and WordNet hypernyms of nouns feature. In our work, position features also be utilized and we employ HowNet hypernyms of entities to make better performance. Zeng et al. (2015) proposed piecewise convolutional neural networks (PCNN) to alleviate the influence of noise caused by the feature extraction process, which are also adopted in this paper. Wang et al. (2016) employed a CNN architecture relying on a novel multi-level attention mechanism to capture both entity-specific attention and relation-specific pooling attention. Inspired by Wang et al. (2016), our work also adds an attention layer to capture more important information of high-level features.

3   Methodology

In this section, we will introduce the neural network architecture used in our paper. Figure 1 illustrates the procedure that handles one instance. This procedure includes five main parts: **Vector representation**, **Convolution**, **Piecewise max pooling**, **Attention** and **Softmax output**. We describe these parts below.

3.1   Vector Representation

The input embedding of the model consists of three parts: word embedding, position embedding and HowNet hypernym embedding.

Word Embedding. Word embeddings are distributed representations of words that map each word in texts to a low-dimensional vector. Given a sentence consisting of T words $S = [x_1, x_2, \cdots, x_T]$, every word x_i is converted into a real-valued vector e_i by looking up word embeddings. So $WE = [e_1, e_2, \cdots, e_T]$.

Position Embedding. To specify the target nouns in the sentence, Zeng et al. (2014) employed Position Feature(PF) that is the combination of the relative distances of the current word to ent_1 and ent_2. (ent_1 and ent_2 represents entity1 and entity2, respectively.) Figure 2 shows an example of the relative distance. The relative distance from word 儿子(erzi) to 金正男(Jin Zhengnan) and 金韩松(Jin Hansong) are 2 and −1. So the position embedding of the sentence S is $PE = [pe_1, pe_2, \cdots, pe_T]$, and $pe_i(i = 1, 2, \cdots T) = [d_{i1}, d_{i2}]$. ($d_{i1}$ and d_{i2} represent the distance vectors for the ith word of S to ent_1 and ent_2.)

Fig. 2. An example of relative distances.

HowNet Hypernym Embedding. HowNet is a network of semantic relationships among Chinese words. In HowNet, the formal description of words is organized in three layers: "word", "concept", and "sememe". Every word can be described by several concepts. And "concept" is described by a kind of knowledge expressing language, which is composed by sememes. "Sememe" is the basic semantic unit, all the sememes are organized into a hierarchical tree structure by the Hypernym-Hyponym relations. We use HowNet 2008 in our experiments, there are 1,700 sememes, 28,925 concepts, and 96,744 Chinese words.

The hypernym has been used to enhance the feature quality. For instance, WordNet usually be employed to perform better results (Rink and Harabagiu 2010; Zeng et al. 2014). HowNet contains multiple relations between Chinese words, including hypernyms. So for the entities in a sentence, we add its HowNet hypernym as a feature. The HowNet Hypernym Embeddings of S is $HHE = [hhe_1, hhe_2, \cdots, hhe_T]$.

$$hhe_i = \begin{cases} e_{x_i\text{HowNetH}}, & \text{if } x_i \text{ is entity,} \\ 0, & \text{if not.} \end{cases} \quad (1)$$

Where $e_{x_i\text{HowNetH}}$ is the word embedding of entity x_i hypernym in HowNet. 0 is a vector that has the same dimension as $e_{x_i\text{HowNetH}}$.

We concatenate the word representation, position representation and HowNet Hypernym representation as the input Emb_S of the network, and $Emb_S = [WE, PE, HHE]^T$.

3.2 Convolution

Collobert et al. (2011) considered convolution approach can merge all local features to perform prediction globally. Convolution is an operation between a vector of weights w and a vector of inputs that is treated as a sequence. The weights matrix w is regarded as the filter for the convolution. The ability to capture different features typically requires the use of multiple filters in the convolution. Under the assumption that we use n filters $(W = \{w_1, w_2, \cdots, w_n\})$. The convolution result is a matrix $C = \{c_1, c_2, \cdots, c_n\}$.

3.3 Piecewise Max Pooling

To capture the structural information between two entities better, we employ piecewise max pooling proposed by Zeng et al. (2015) instead of single max pooling. As shown in Fig. 1, the output of each convolutional filter c_i is divided into three segments

$\{c_{i1}, c_{i2}, c_{i3}\}$ by 金正男(Jin Zhengnan) and 金韩松(Jin Hansong). The piecewise max pooling procedure can be expressed as follows:

$$p_{ij} = \max(c_{ij}) \quad 1 \le i \le n, \ 1 \le j \le 3 \tag{2}$$

So, $p_i = [p_{i1}, p_{i2}, p_{i3}]$. We concatenate all vectors that obtained by piecewise max pooling on the outputs of convolutional filters as $P = [p_1, p_2, \cdots, p_n]$.

3.4   Attention

Attentive neural networks have recently achieved good results in relation extraction. For example, Zhou et al. (2016) added an attention layer after Bidirectional LSTM layer. And Wang et al. (2016) proposed a novel multi-level attention mechanism to capture both entity-specific attention and relation-specific pooling attention.

What is different from them is that our attention mechanism automatically focus on high-level global features obtained by convolution and pooling multiple filters that have decisive influence on prediction relations. Every convolution filter will extract a kind of high-level feature about sentence globally, and different features have different contributions to predict entity relation. Therefore, we add an attention layer to find the most advantageous features for relation predicting. The representation r of the sentence S is produced by the following formula:

$$M = \tanh(P) \tag{3}$$

$$\alpha = \text{softmax}\left(w^T M\right) \tag{4}$$

$$r = P\alpha^T \tag{5}$$

where $P \in R^{3 \times n}$, n is the number of filters, w is a trained parameter vector and w^T is a transpose.

We obtain the final sentence representation used for predicting relation from.

$$h^* = \tanh(r) \tag{6}$$

3.5   Softmax Output

In this part, we employ a softmax classifier to predict label $\hat{y}$ from a discrete set of classes Y for a sentence S. And h^* is involved in the following formula as input:

$$\hat{p}(y|S) = \text{softmax}(Wh^* + b) \tag{7}$$

$$\hat{y} = \arg\max_y \hat{p}(y|S) \tag{8}$$

We combine a dropout with L2 regularization to alleviate overfitting in our paper.

In this section, we have introduced each part of the model in detail, especially for our innovations, attention layer and HowNet embedding in vector representation layer.

4 Experiments

4.1 Datasets and Evaluation Metrics

COAE2016. The 8th Chinese Orientation Analysis and Evaluation (COAE2016) increased the task of relation classification for knowledge extraction. Our experiments are conducted on the dataset provided in this task, and the dataset consists of a training set of 988 sentences and a test set of 483 sentences. In this task, there are ten types of relation, as shown in Table 1.

ACE2005. ACE2005 Chinese corpus is divided into three categories: Newswire (NW), Broadcast News (BN), and Weblog (WL). Table 2 shows the distribution of the data. The corpus consists of data of various types annotated for entities and relations. The relation types are divided into 6 categories and 18 subcategories. Table 3 shows the distribution of the number for 6 categories. We randomly select 2/3 as training set and 1/3 as test set.

We use the macro-averaged Precision, Recall and F1-score to evaluate our systems.

Table 1. Relation type table.

Relation name	Relation symbol
Date of birth of PER	Cr2
Birthplace of PER	Cr4
Graduate institution of PER	Cr16
Spouse of PER	Cr20
Children of PER	Cr21
Senior executive of ORG	Cr28
Employees' number of ORG	Cr29
Founder of ORG	Cr34
Founding time of ORG	Cr35
Headquarters of ORG	Cr37

Table 2. Data source distribution table.

	Chars	Files	Rate (%)
NW	121797	238	40
BN	120513	298	40
WL	65681	97	20
Total	307991	633	100

Table 3. Data relation type distribution table.

Category	ART	PART-WHOLE	PHYS	ORG-AFF	PER-SOC	GEN-AFF	Total
Number	507	1862	1360	1874	486	1660	7749

4.2   Parameter Settings

Collobert et al. (2011) proved that word embedding learned from a large number of unlabeled data are far more satisfactory than the randomly initialized embeddings. Therefore, we use word embeddings pre-trained in 13 million words People's Daily corpus in our experiments. We tune all of the models using three-fold validation on the training set. We select the dimension of word embedding d_w among {50, 100, 200}, the dimension of position embedding d_p among {5, 10, 20}, the dimension of HowNet hypernym embedding d_h among {5, 10, 20}, the windows size w among {3, 5, 7}, the number of filters n among {100, 150, 200, 230, 250}, the learning rate λ among {0.001, 0.01, 0.1, 1.1}, while the batch size k among {50, 100, 150, 200}. The best configurations   are: $d_w = 100, d_p = 5, d_h = 10, w = 3, n = 230, \lambda = 0.01, and\ k = 50$. According to experience, the dropout rate is fixed to 0.5.

4.3   Corpus Expansion for COAE2016

As is well-know that large-scale labeled data is essential for deep learning methods. However, manually tagging large amounts of data is time consumed and vigour cost, and the existing Chinese labeled corpus is relatively scarce. Therefore, the effective way of data expansion brings new gospel. In this section, we will introduce how we use HIT IR-Lab Tongyici Cilin (Extended) and SDP to expand limited Chinese labeled corpus like COAE2016.

HIT IR-Lab Tongyici Cilin (Extended). We use the HIT IR-Lba Tongyic Cilin (Hereinafter referred to as Cilin), which was expanded by the Information Retrieval Laboratory of Harbin Institute of Technology. Cilin has increased the number of entries from 39,099 to 77,343 and it was organized according to the tree hierarchical structure.

Cilin encodes the words by category. Table 4 shows the method of word encoding. The fifth-level classification results can be divided into three specific situations, e.g., some lines are synonyms, some lines are related words, and some lines have only one word. That is to say, there are three markers in eighth places, namely "=", "#" and "@". "=" represents equality and synonym. "#" represents unequal and similar. "@" indicates self-sealing and independence, and it has neither synonyms nor related words in the dictionary. For example, Cb02A01 = 东南西北 四方. "东南西北(dongnanxibei)" and "四方(sifang)" have the same meaning. In our method, we mainly use synonyms.

Shortest Dependency Path. The dependency grammar reveals sentence syntactic structure by parsing the dependencies among the components in the language units. Intuitively speaking, grammatical components such as subject-predicate-object and attribute-adverbial-complement are identified in dependency syntactic analysis, and the relationships among the components are analyzed. As shown in Fig. 3, we perform dependency parsing on the sentence.

Table 4. Cilin words coding table.

Coding position	1	2	3	4	5	6	7	8
Symbol example	D	a	1	5	B	0	2	=\#\@
Symbolic nature	Big class	Middle class	Small class		Word group	Atomic word group		
Level	Level 1	Level 2	Level 3		Level 4	Level 5		

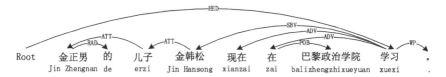

Fig. 3. Dependency parsing result

In succession, we find the shortest path between two entities in the result of dependency parsing. According to the algorithm, the shortest path between "金正男 (Jin Zhengnan)" and "金韩松(Jin Hansong)" is as follows:

金正男
Jin Zhengnan ←—ATT— 儿子
erzi ←—ATT— 金韩松
Jin Hansong

Corpus Expansion. In this part, we will describe the data expansion method in details with give examples. First, we preprocess the data, such as word segmentation. Then, we perform dependency parsing on sentences[1] and find the shortest path of two entities. For words that appear in the shortest path (exclude entities), we find their synonyms in Cilin and replace them in turns to get new data. Table 5 shows some examples.

Since Chinese word polysemy often occurs, we calculate the similarity between the word replaced and its synonyms. If the similarity exceeds the threshold, we replace it. We use Similarity Model proposed by Li et al. (2017) to compute word similarity.

4.4   Experimental Results and Analysis

Table 6 shows the experimental results on the COAE2016 dataset. We compare with previous approaches. In the case of a small data set (988 sentences for training), SVM method is superior to CNN. Our model PCNN_ATT is obviously better than baselines. It is not only outperforming an SVM-based approach, but also superior the CNN model with a relative improvement of 11.80%, which demonstrate PCNN_ATT have advantages in relation classification task. Our model is superior to PCNN (Zeng et al. 2015) with a relative improvement of 2.62%. Entity types, such as *person, location* and

---

[1] We use Stanford Parser to perform dependency parsing on sentences.

organization, are helpful to distinguish some relations. In this dataset, relation types are extremely specific. The addition of entity types can largely improve the classification results and solve some misclassification problems. After adding entity type (ET) information, the result is increased by 8.29%. Finally, the experiment was performed on the expanded dataset (11,328 sentences) with the result of 78.41%, increased 1.83%. The results of each model on the extended dataset have been improved, which prove that the data expanded method is effective.

Table 5. New data generation sample table

Original sentence	SDP	New data
金正男的儿子金韩松现在在巴黎政治学院学习。(Jin Zhengnan de erzi Jin Hansong xianzai zai bali zhengzhi xueyuan xuexi .)	金正男 <- 儿子 <- 金韩松	1.金正男 的 子嗣(zisi) 金韩松 现在 在 巴黎政治学院 学习 。 2.金正男 的 幼子(youzi) 金韩松 现在 在 巴黎政治学院 学习 。 3.金正男 的 犬子(quanzi) 金韩松 现在 在 巴黎政治学院 学习 。
松田耕平代替父亲松田恒次的位置，成为马自达的会长。(Songtiangengping daiti fuqin songtianhengci de weizhi , chengwei mazida de huizhang .)	松田耕平 <- 代替 -> 成为 -> 会长 -> 马自达	1.松田耕平 顶替(dingti) 父亲 松田恒次 的 位置 , 成为 马自达 的 会长 。 2.松田耕平 取代(qudai) 父亲 松田恒次 的 位置 , 成为 马自达 的 会长 。 3.松田耕平 代替 父亲 松田恒次 的 位置 , 成了(chengle) 马自达 的 会长 。 4.松田耕平 代替 父亲 松田恒次 的 位置 , 变成(biancheng) 马自达 的 会长 。

Table 6. Comparison of different models on COAE2016 dataset. WV, PF ET stand for word vectors, position features and entity types; *(Ext) stands for results on the extended dataset.

Methods	Features	Macro		
		P	R	F1
SVM	POS, entity order, entity distance, entity context	76.78	64.70	66.29
SVM(Ext)		77.23	65.72	67.96
CNN	WV, PF	60.83	55.60	56.69
CNN(Ext)	WV, PF	62.23	57.75	58.61
PCNN	WV, PF	69.95	61.88	65.67
PCNN(Ext)	WV, PF	71.63	62.31	66.65
PCNN_ATT	WV, PF	75.38	66.89	68.29
PCNN_ATT	WV, PF, ET	77.64	76.23	76.58
PCNN_ATT(Ext)	WV, PF, ET	**79.26**	**78.47**	**78.41**

Table 7 shows the experimental results on the ACE2005 dataset. We don't apply corpus extension method on ACE2005, because the scale of this dataset is adequate for model training. As we see that SVM method is much better than CNN method in this task. When we have the same features, our model PCNN_ATT performances well with a relative improvement of 13.95% and it is better than PCNN (Zeng et al. 2015) with a relative improvement of 5.63%. After adding entity type (ET) and entity subtype (ES), the result is raised by 0.5%. Finally, with the addition of HowNet hypernyms, the result is nearly 0.6% higher.

Table 7. Comparison of different models on ACE2005 dataset. WV, PF, ET, ES, HowNet stand for word vectors, position features, entity types, entity subtypes and HowNet hypernyms.

Methods	Features	Macro		
		P	R	F1
SVM	POS, entity order, entity type and subtype, entity context	76.13	70.18	73.27
CNN	WV, PF	59.65	58.26	58.95
PCNN	WV, PF	68.25	66.31	67.27
PCNN_ATT	WV, PF	73.11	73.01	72.90
PCNN_ATT	WV, PF, ET, ES	**73.96**	73.28	73.37
PCNN_ATT	WV, PF, ET, ES, HowNet	73.77	**74.12**	**73.94**

Based on the above experiments, our model PCNN_ATT shows competitive results. Adding an attention mechanism after the piecewise max pooling layer can improve performance. The attention mechanism can assign weights to the high-level features obtained by each convolution kernel, and gives greater weights to the features that contribute to the prediction of relation.

5   Conclusion

In this paper, we propose a novel neural network PCNN_ATT, to improve the performance of relation extraction. The PCNN_ATT model added an attention layer after piecewise max pooling layer, which pays more attention to high-level global features. And we put forward an effective corpus expansion method by utilizing external dictionary HIT IR-Lab Tongyici Cilin to make up insufficient data problem. We demonstrate the effectiveness of our method by evaluating the model on COAE-2016 and ACE-2005 datasets. PCNN_ATT achieves a better performance at capturing more important features, compared with some common neural networks. A significant improvement is observed when PCNN_ATT is used, outperforming most of existing methods.

Acknowledgements. The authors are supported by National Nature Science Foundation of China (Contract 61370130 and 61473294), and the Fundamental Research Funds for the Central Universities (2015JBM033), and International Science and Technology Cooperation Program of China under grant No. 2014DFA11350.

References

Zeng, D., Liu, K., Lai, S., Zhou, G., Zhao, J.: Relation classification via convolutional deep neural network. In: Proceedings of COLING, pp. 2335–2344 (2014)

Zeng, D., Liu, K., Chen, Y., Zhao, J.: Distant supervision for relation extraction via piecewise convolutional neural networks. In: Proceedings of EMNLP, pp. 17–21. Association for Computational Linguistics, Stroudsburg (2015)

Lin, Y., Shen, S., Liu, Z., Luan, H., Sun, M.: Neural relation extraction with selective attention over instances. In: Proceedings of ACL, pp. 2124–2133. Association for Computational Linguistics, Berlin (2016)

Jiang, X., Wang, Q., Li, P., Wang, B.: Relation extraction with multi-instance multi-label convolutional neural networks. In: Proceedings of COLING, pp. 1471–1480 (2016)

dos Santos, C.N., Xiang, B., Zhou, B.: Classifying relations by ranking with convolutional neural networks. In: Proceedings of ACL (2015)

Liu, Y., Wei, F., Li, S., Ji, H., Zhou, M., Wang, H.: A dependency-based neural network for relation classification. Comput. Sci. (2015)

Bunescu, R.C., Mooney, R.J.: A shortest path dependency kernel for relation extraction. In: Proceedings of HLT/EMNLP, pp. 724–731. Association for Computational Linguistics, Vancouver (2005)

Wang, L., Cao, Z., Melo, G.D., Liu, Z.: Relation classification via multi-level attention CNNs. In: Proceedings of ACL, pp. 1298–1307. Association for Computational Linguistics, Berlin (2016)

Li, S., Xu, J., Zhang, Y., Chen, Y.: A method of unknown words processing for neural machine translation using HowNet. In: Wong, D.F., Xiong, D. (eds.) CWMT 2017. CCIS, vol. 787, pp. 20–29. Springer, Singapore (2017). https://doi.org/10.1007/978-981-10-7134-8_3

Sun, J., Gu, X., Li, Y., Xu, W.: Chinese entity relation extraction algorithms based on COAE2016 datasets. J. Shandong Univ. (Nat. Sci.) 52(9), 7–12 (2017)

Liu, D., Peng, C., Qian, L., Zhou, G.: The effect of Tongyici Cilin in Chinese entity relation extraction. J. Chin. Inf. Process. 28(2), 91–99 (2014)

Cai, R., Zhang, X., Wang, H.: Bidirectional recurrent convolutional neural network for relation classification. In: Proceedings of ACL, pp. 756–765. Association for Computational Linguistics, Berlin (2016)

Xu, K., Feng, Y., Huang, S., Zhao, D.: Semantic relation classification via convolutional neural networks with simple negative sampling. Comput. Sci. 71(7), 941–949 (2015)

Zhou, P., et al.: Attention-based bidirectional long short-term memory networks for relation classification. In: Proceedings of ACL, pp. 207–212. Association for Computational Linguistics, Berlin (2016)

Hashimoto, K., Miwa, M., Tsuruoka, Y., Chikayama, T.: Simple customization of recursive neural networks for semantic relation classification. In: Proceedings of EMNLP, pp. 1372–1376. Association for Computational Linguistics, Seattle (2013)

Collobert, R., Weston, J., Bottou, L., Karlen, M., Kavukcuoglu, K., Kuksa, P.: Natural language processing (almost) from scratch. JMLR 12, 2493–2537 (2011)

Socher, R., Huval, B., Manning, C.D., Ng, A.Y.: Semantic compositionality through recursive matrix-vector spaces. In: Joint Conference on Empirical Methods in Natural Language Processing and Computational Natural Language Learning, pp. 1201–1211 (2012)

Rink, B., Harabagiu, S.: UTD: classifying semantic relations by combining lexical and semantic resources. In: Proceedings of the 5th International Workshop on Semantic Evaluation, pp. 256–259. Association for Computational Linguistics (2010)

A Study on Improving End-to-End Neural Coreference Resolution

Jia-Chen Gu, Zhen-Hua Ling$^{(\boxtimes)}$, and Nitin Indurkhya

National Engineering Laboratory for Speech and Language Information Processing,
University of Science and Technology of China, Hefei, China
gujc@mail.ustc.edu.cn, {zhling,nitin}@ustc.edu.cn

Abstract. This paper studies the methods to improve end-to-end neural coreference resolution. First, we introduce a coreference cluster modification algorithm, which can help modify the coreference cluster to rule out the dissimilar mention in the cluster and reduce errors caused by the global inconsistence of coreference clusters. Additionally, we tune the model from two aspects to get more accurate coreference resolution results. On one hand, the simple scoring function is replaced with a feedforward neural network when computing the head word scores for later attention mechanism which can help pick out the most important word. On the other hand, the maximum width of a mention is tuned. Our experimental results show that above methods improve the performance of coreference resolution effectively.

Keywords: Coreference resolution · End-to-end · Neural network

1 Introduction

Coreference resolution, the task of finding all expressions that refer to the same real-world entity in a text or dialogue, has become the core tasks of natural language processing (NLP) since the 1960s. An example of demonstrating the task of coreference resolution [1] that we need to resolve *I*, *my* and *she* as coreferential, *Nader* and *he* as coreferential respectively is showed in following text.

"***I*** *voted for **Nader** because **he** was most aligned with **my** values,*" ***she*** *said.*

Nowadays, people are paying more and more attention to applying neural network to coreference resolution because neural-network-based models [2–5] have achieved impressive coreference resolution performance, especially the end-to-end neural model [5], which does not rely on syntactic parsers and many hand-engineered features. This end-to-end model makes independent decisions about whether two mentions are coreferential and then establish a coreference cluster through this kind of coreference relation. For example, if we make decisions that {*President of the People Republic of China, Xi Jinping*} and {*Xi Jinping, Mr.Xi*} are coreferential respectively, then we can get a cluster that {*President of the People Republic of China, Xi Jinping, Mr.Xi*} are coreferential naturally.

© Springer Nature Switzerland AG 2018
M. Sun et al. (Eds.): CCL 2018/NLP-NABD 2018, LNAI 11221, pp. 159–169, 2018.
https://doi.org/10.1007/978-3-030-01716-3_14

However, this model sometimes makes globally inconsistent decisions, and gets an incompletely correct cluster because of the independence between these decisions. To avoid this kind of error, a coreference cluster modification algorithm is proposed in this paper which can help rule out the mentions which are not globally coreferential within each cluster on the basis of the span-ranking architecture. After getting a coreference cluster through locally coreferential decisions, we use a scoring function to measure the extent of coreference relation between every mention pair. Then we establish a standard to decide whether to rule out the dissimilar mention in a coreference cluster.

Furthermore, we tune the hyperparameters from two aspects to get more accurate coreference resolution results. On one hand, to get more accurate scoring function to help measure the extent of coreference relation, we replace the scoring function with a feed-forward neural network when applying an attention mechanism [6] to compute the head word score. This modification enables the system to pick out the most important word more accurately to help express the representation of a mention which can help incorporate more information over words in a span. On the other hand, our experiments and analysis show that the model is susceptible to the maximum width of a mention, i.e. the number of words a mention can comprise of most. Therefore, we tune the maximum width of a mention in experiments.

Our experimental results show that the proposed coreference cluster modification algorithm can improve the performance of coreference resolution on the English OntoNotes benchmark. Our approach outperforms the baseline single model with an F1 improvement of 0.3. Additionally, we can also obtain an F1 improvement of 1.2 when tuning the hyperparameters of the model.

2   Related Work

Machine-learning-based methods for coreference resolution have developed for a long time since the first paper on machine-learning-based coreference resolution [7] was published. Hand-engineered systems built on top of automatically produced parse trees [8,9] have achieved significant performance. Recently proposed neural-network-based models [2–4] outperformed all previous learning approaches. The more recent end-to-end neural model [5] has achieved further performance gains meanwhile it does not rely on syntactic parsers and hand-engineered features.

From a higher view of these approaches, all of the above models can be categorized as (1) mention-pair classifiers [10,11], (2) entity-level models [4,12,13], (3) latent-tree models [14–16], (4) mention-ranking models [3,9,17], (5) span-ranking models [5]. Our proposed methods are based on the span-ranking model [5], which relies on scoring span pairs and then uses the scores to make coreference decisions. However, the end-to-end span-ranking model only concentrates on the direct link between span pairs while neglects the indirect link between the interval spans, which is the motivation of our proposed coreference cluster modification algorithm.

3   Baseline Method

3.1   Task Definition

The end-to-end neural model [5] formulates the coreference resolution task as a set of antecedent assignments y_i for each of span i in the given document and our model follows the task formulation. The set of possible assignments for each y_i is $\mathcal{Y}(i) = \{\epsilon, 1, 2, ..., i - 1\}$ which consists of a dummy antecedent ϵ and all preceding spans. Non-dummy antecedents represent coreference links between i and y_i. The dummy antecedent ϵ represents two possible scenarios: (1) the span is not an entity mention or (2) the span is an entity mention but it is not coreferential with any previous span. We can get a final clustering through these decisions, which may lead to the problem of global inconsistence of coreference cluster we have just mentioned above.

3.2   Baseline Model

The aim of the end-to-end baseline model [5] is to learn a distribution $P(y_i)$ over antecedents for each span i as

$$P(y_i) = \frac{e^{s(i,y_i)}}{\sum_{y' \in \mathcal{Y}(i)} e^{s(i,y')}} \tag{1}$$

where $s(i, j)$ is a pairwise score for a coreference link between span i and span j. The baseline model includes three factors for this pairwise coreference score: (1) $s_m(i)$, whether span i is a mention, (2) $s_m(j)$, whether span j is a mention, and (3) $s_a(i, j)$, whether span j is an antecedent of span i. $s(i, j)$ is calculated as

$$s(i, j) = s_m(i) + s_m(j) + s_a(i, j) \tag{2}$$

$s_m(i)$ and $s_a(i, j)$ are both functions of the span representation vector $\boldsymbol{g}_i$, which is computed via bidirectional LSTMs [18] and attention mechanism [6]. The detailed calculation of $s_m(i)$ and $s_a(i, j)$ are as follows

$$s_m(i) = \boldsymbol{w}_m^\top FFNN_m(\boldsymbol{g}_i) \tag{3}$$

$$s_a(i, j) = \boldsymbol{w}_a^\top FFNN_a([\boldsymbol{g}_i, \boldsymbol{g}_j, \boldsymbol{g}_i \circ \boldsymbol{g}_j, \phi(i, j)]) \tag{4}$$

where $\circ$ denotes element-wise multiplication, FFNN denotes a feed-forward neural network, and the antecedent scoring function $s_a(i, j)$ includes explicit element-wise similarity of each span $\boldsymbol{g}_i \circ \boldsymbol{g}_j$ and a feature vector $\phi(i, j)$ encoding speaker and genre information from the metadata and the distance between the two spans.

The span representation $\boldsymbol{g}_i$ is composed of boundary representation, head word vector and feature vector. We will restrict our discussion to the head word vector. The baseline model learns a task-specific notion of headness using an attention mechanism [6] over words in each span:

$$\alpha_t = \boldsymbol{w}_\alpha \cdot Projection(\boldsymbol{x}_t^*) \tag{5}$$

$$a_{i,t} = \frac{exp(\alpha_t)}{\sum_{k=START(i)}^{END(i)} exp(\alpha_k)} \tag{6}$$

$$\hat{x}_i = \sum_{t=START(i)}^{END(i)} a_{i,t} \cdot x_t \tag{7}$$

where $\hat{x}_i$ is a weighted sum of word vectors in span i.

Given supervision of gold coreference clusters, the model is learned by optimizing the marginal log-likelihood of the possibly correct antecedents [5]. This marginalization is required since the best antecedent for each span is a latent variable.

3.3 Clustering Rules

The baseline model makes decisions about whether span i and span j are coreferential while these decisions are independent between each other. The baseline model obeys the following rules to make clustering decisions as follows.

- Span i has a set of scores, i.e. $s_a(i, j)$ with its every candidate antecedent in its antecedents set $\mathcal{Y}(i) = \{\epsilon, 1, 2, ..., i-1\}$ which can measure the extent of coreference relation.
- The span i picks out the one which has the highest score to be its antecedent and establishes a coreference link with its antecedent span.
- If span pairs {span i, span j} and {span j, span k} are both linked, span pair {span i, span k} will be linked naturally.

It is noticeable that span pair {span i, span k} will be linked naturally without confirming whether these two spans are truly coreferential, which incurs the problem of global inconsistence of coreference cluster. An example which demonstrates the problem of global inconsistence of coreference cluster is as follows.

__Chaoyang Road__ is a very important artery in the east-west direction. When people living in the west want to cross over from __the city__, they have to go via this road. Hence, if a traffic accident occurs at __this place__, we can indeed imagine how widespread the extent of the impact will be.

In above paragraph, mention pairs {*Chaoyang Road, this place*} and {*this place, the city*} are both locally coreferential, but the cluster of {*Chaoyang Road, this place, the city*} is not globally coreferential.

4   Proposed Methods

As we mentioned above, the baseline model always makes globally inconsistent decisions, and sometimes gets an incompletely correct cluster because of the independence between these decisions.

Therefore, we propose a coreference cluster modification algorithm to modify the clusters built by the baseline model in this paper. This method confirms the coreference relation between intra-cluster spans which can help rule out the dissimilar span after we get a coreference cluster. Through the modification procedure we can increase the possibility that these spans in a coreference cluster truly refer to the same entity.

We show how to conduct the modification procedure in the situation of n spans in a coreference cluster to demonstrate our algorithm. We first define some variables as follows.

- There are n spans in a coreference cluster and we name them as $\{1, 2, ..., n\}$ in order in a text.
- Span k has a direct link with span i which means span i, as one candidate antecedent of span k, has the highest score among all candidate antecedents.
- We set spans before span k except span i, i.e. $\mathcal{P}(i, k) = \{1, 2, ..., i - 1, i + 1, ..., k - 1\}$ as span k's indirect antecedents. Span k also has indirect links with each span in $\mathcal{P}(i, k)$ which mean extent of compatibility with span k.
- When an incompatible link $s_a(j, k)$ appears, spans before span k except span j form a set $\mathcal{Q}(j, k) = \{1, 2, ..., j - 1, j + 1, ..., k - 1\}$ which is used to consider which span to drop afterwards.

For span k, it originally takes only a direct link $s_a(i, k)$ into consideration while neglects the indirect links with spans in the set $\mathcal{P}(i, k)$. In our method the indirect links within the coreference clusters are labelled explicitly so that we can finally get an enriched coreference cluster full of links between every two spans in the cluster no matter direct or indirect (see Fig. 1).

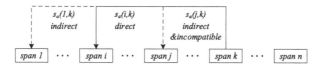

Fig. 1. Enriched coreference cluster after labeling indirect coreference links explicitly.

Value $s_a(i, j)$ can be positive and negative. The greater the abstract value of $s_a(i, j)$ is, the stronger the coreference relation of compatibility (positive) or incompatibility (negative) will be. Furthermore, we design some rules as demonstrated in Algorithm 1. The algorithm, from a high view, can be interpreted from the point of confidence degree because the abstract value of $s_a(i, j)$ represents the extent of coreference relation.

It takes two steps to conduct the algorithm:

- **First step: check.** We need to check whether there is the problem of global inconsistence of coreference cluster. However it is unsafe to judge the relation with the method of directly taking the indirect link into account because the model has the limitation to represent the coreference relation. To tolerate this kind of mistakes and increase robustness of the model, we introduce the inequity rules that taking direct link and average of all indirect links in $\mathcal{P}(i, k)$ into account to further confirm the coreference relation.

Algorithm 1. Coreference cluster modification

---

for $k = 3, 4, ..., n$ **do**

    if $s_a(i, k) + \frac{1}{k-2} \sum_{p \in \mathcal{P}(i,k)} s_a(p, k) < margin$ **then**

    $j = \arg\min_{p \in \mathcal{P}(i,k)} s_a(p, k)$

    if $\sum_{q \in \mathcal{Q}(j,k)} s_a(q, k) < \sum_{q \in \mathcal{Q}(j,k)} s_a(q, j)$ **then**

        drop span k

    else

        drop span j

    end if

    else

        drop none of these spans in a cluster

    end if

end for

- **Second step: drop.** If the problem of global inconsistence of coreference cluster truly happen, we need to consider which span to drop furthermore. It must be that some indirect link $s_a(j, k)$ is incompatible which means span k and span j are incompatible. We make a comparision between the sum of span j 's links with spans in $\mathcal{Q}(j, k)$ and the sum of span k 's links with spans in $\mathcal{Q}(j, k)$, then we make the modification decision that drop span j or span k.

5   Experiments

5.1   Experimental Setup

We used the English coreference resolution data from the CoNLL-2012 shared task [19] in our experiments. This dataset contains 2802 training documents, 343 development documents, and 348 test documents.

Our model reused the hyperparameters from Lee et al. [5] so that we can make comparisons with the baseline model. Some parameters of the baseline model are mentioned below.

- **Word representations.** The word embeddings were fixed concatenations of 300-dimensional GloVe embeddings [20] and 50-dimensional embeddings from Turian et al. [21]. In the character CNN, characters were represented as learned 8-dimensional embeddings. The convolutions had window sizes of 3, 4, and 5 characters, each consisting of 50 filters.
- **Hidden dimensions.** The hidden states in the LSTMs had 200 dimensions.
- **Feature encoding.** All features including speaker, genre, span distance and mention width were represented as learned 20-dimensional embeddings.
- **Pruning.** The baseline model pruned the spans such that the maximum span width $L = 10$, the number of spans per word $\lambda = 0.4$, and the maximum number of antecedents $K = 250$.

– **Learning.** The baseline model used ADAM [22] for learning with a minibatch size of 1. 0.5 dropout [23] was applied to the word embeddings and character CNN outputs and 0.2 dropout was applied to all hidden layers and feature embeddings.

5.2   Coreference Cluster Modification

The only hyperparameter in this method is *margin* in the inequilties, which is used to measure the possibility of global inconsistance of coreference cluster. Moreover, some other factors may also affect the performance of our proposed algorithm. We tuned these factors across experiments about different combinations of them on the development dataset as showed in Table 1.

Table 1. Some factors were tuned by experiments on the development dataset, where **number** means the number of spans in a coreference cluster and **function** means the function involved in the first check step.

Number	Function	Margin	Avg. F1
<5	Mean	0	67.4
<5	Min	0	67.3
<5	Mean	−2	67.6
<7	Mean	−2	**67.7**
<10	Mean	−2	67.6
All	Mean	−2	67.3

From this table, we can see that our proposed method still didn't work well for post-processing the clusters with more than 10 spans. The distribution of the size of coreference clusters on the development set given by the baseline model is showed in Fig. 2. We can see that the coreference clusters with less than 10 spans accounted for about 93% of all coreference clusters. Besides, the *mean* function worked slightly better than the *min* function during the check step. One possible reason was that the *mean* function took the information of all indirect links within the cluster into account. Finally, the last row was chosen as the configuration of our proposed method.

5.3   Parameter Tuning

The baseline model simply projects the outputs from the bidirectional LSTMs [18] to a scalar score as we decribe in Equity [5] When computing the weight of each word. We replace the simple function with a feed-forward neural network which can help incorporate more information about words in a span to get more accurate attention weights to pick out the head word. The feed-forward neural

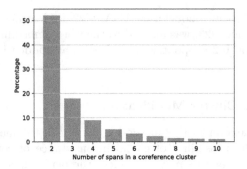

Fig. 2. Distribution of coreference clusters according to number of spans in a coreference cluster.

network in this method consists of two hidden layers with 150 dimensions and rectified linear units [24].

We analysed the error examples of the baseline model on the test dataset. We found that 3934 mentions were not detected, in which 576 mentions had more than 10 words in a span that exceeded the maximum span width. This implied that the model was susceptible to the maximum span width and an accuracy improvement may be achieved by increasing the maximum span width. Therefore, we increased the maximum span width from 10 to 30 words by experiments and obtained the gain of average F1 as shown in Fig. 3.

5.4   Results

We report the precision, recall, and F1 of the MUC, B^3, and $CEAF_{\phi 4}$ metrics using the official CoNLL-2012 evaluation scripts. The final measurement is the average F1 of the three metrics.

Results on the test set are shown in Table 2. The performances of the systems proposed in the last three years were included for comparison. The baseline model of our methods was the span-ranking model from Lee et al. [5] which achieved

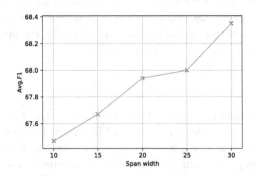

Fig. 3. Average F1 on the test dataset with different maximum width of spans.

an F1 score of 67.2. Our method achieved an F1 score of 67.5, improving the performance for coreference resolution. Furthermore, we can achieve a higher F1 score of 68.4 after parameter tuning.

Table 2. Results on the test set on the English CoNLL-2012 shared task. The final column (Avg. F1) is the main evaluation metric, computed by averaging the F1 of MUC, B^3, and $CEAF_{\phi4}$.

	MUC			B^3			$CEAF_{\phi4}$			
	Prec.	Rec.	F1	Prec.	Rec.	F1	Prec.	Rec.	F1	Avg. F1
Martschat and Strube [16]	76.7	68.1	72.2	66.1	54.2	59.6	59.5	52.3	55.7	62.5
Clark and Manning [13]	76.1	69.4	72.6	65.6	56.0	60.4	59.4	53.0	56.0	63.0
Wiseman et al. [17]	76.2	69.3	72.6	66.2	55.8	60.5	59.4	54.9	57.1	63.4
Wiseman et al. [2]	77.5	69.8	73.4	66.8	57.0	61.5	62.1	53.9	57.7	64.2
Clark and Manning [4]	79.9	69.3	74.2	71.0	56.5	63.0	63.8	54.3	58.7	65.3
Clark and Manning [3]	79.2	70.4	74.6	69.9	58.0	63.4	63.5	55.5	59.2	65.7
Lee et al. [5]	78.4	73.4	75.8	68.6	61.8	65.0	62.7	59.0	60.8	67.2
Lee et al. [25]	81.4	79.5	80.4	72.2	69.5	70.8	68.2	67.1	67.6	73.0
Our proposed	78.3	73.8	76.0	68.3	62.4	65.2	62.8	59.7	61.2	67.5
Our proposed + paramter tuning	79.3	73.9	76.5	70.2	62.7	66.2	63.5	61.2	62.3	68.4

Recently, Lee et al. [25] has just improved the baseline model by proposing a high-order inference model and tuning some model hyperparameters. The results of this work were also listed in Table 2 for comparison. Although our results were not as good as the ones of Lee et al. [25], our method has the advantage of simplicity and it can be considered as a rule-based post-processing of the output given by the baseline model.

6   Conclusion

We presented an improved neural coreference resolution method through a cluster modification algorithm which can help modify the coreference cluster to reduce errors caused by global inconsistence of coreference clusters. Additionally, we replace the scoring function with a feed-forward neural network when computing the head word score which can help pick out the most important word. The maximum mention width is also tuned because our experiments showed that the model is susceptible to the maximum width of mentions. Our experimental results demonstrated that these above procedures helped to increase the accuracy of coreference resolution. To improve the performance of the proposed cluster modification algorithm for clusters with large sizes will be a task of our future work.

Acknowledgements. This work was funded in part by Chinese Academy of Sciences President's International Fellowship Initiative (Grant No. 2018VTA0008).

References

1. Lee, H., Chang, A., Peirsman, Y., Chambers, N., Surdeanu, M., Jurafsky, D.: Deterministic coreference resolution based on entity-centric, precision-ranked rules. Comput. Linguist. **39**(4), 885–916 (2013)
2. Wiseman, S., Rush, A.M., Shieber, S.M.: Learning global features for coreference resolution, arXiv preprint arXiv:1604.03035 (2016)
3. Clark, K., Manning, C.D.: Deep reinforcement learning for mention-ranking coreference models, arXiv preprint arXiv:1609.08667 (2016)
4. Clark, K., Manning, C.D.: Improving coreference resolution by learning entity-level distributed representations, arXiv preprint arXiv:1606.01323 (2016)
5. Lee, K., He, L., Lewis, M., Zettlemoyer, L.: End-to-end neural coreference resolution, arXiv preprint arXiv:1707.07045 (2017)
6. Bahdanau, D., Cho, K., Bengio, Y.: Neural machine translation by jointly learning to align and translate, arXiv preprint arXiv:1409.0473 (2014)
7. Connolly, D., Burger, J.D., Day, D.S.: A machine learning approach to anaphoric reference. In: New Methods in Language Processing, pp. 133–144 (1997)
8. Raghunathan, K., et al.: A multi-pass sieve for coreference resolution. In: Proceedings of the 2010 Conference on Empirical Methods in Natural Language Processing, pp. 492–501. Association for Computational Linguistics (2010)
9. Durrett, G., Klein, D.: Easy victories and uphill battles in coreference resolution. In: Proceedings of the 2013 Conference on Empirical Methods in Natural Language Processing, pp. 1971–1982 (2013)
10. Ng, V., Cardie, C.: Identifying anaphoric and non-anaphoric noun phrases to improve coreference resolution. In: Proceedings of the 19th International Conference on Computational Linguistics, vol. 1, pp. 1–7. Association for Computational Linguistics (2002)
11. Bengtson, E., Roth, D.: Understanding the value of features for coreference resolution. In: Proceedings of the Conference on Empirical Methods in Natural Language Processing, pp. 294–303. Association for Computational Linguistics (2008)
12. Haghighi, A., Klein, D.: Coreference resolution in a modular, entity-centered model. In: Human Language Technologies: The 2010 Annual Conference of the North American Chapter of the Association for Computational Linguistics, pp. 385–393. Association for Computational Linguistics (2010)
13. Clark, K., Manning, C.D.: Entity-centric coreference resolution with model stacking. In: Proceedings of the 53rd Annual Meeting of the Association for Computational Linguistics and the 7th International Joint Conference on Natural Language Processing (vol. 1: Long Papers), pp. 1405–1415 (2015)
14. Fernandes, E.R., Dos Santos, C.N., Milidiú, R.L.: Latent structure perceptron with feature induction for unrestricted coreference resolution. In: Joint Conference on EMNLP and CoNLL-Shared Task, pp. 41–48. Association for Computational Linguistics (2012)
15. Björkelund, A., Kuhn, J.: Learning structured perceptrons for coreference resolution with latent antecedents and non-local features. In: Proceedings of the 52nd Annual Meeting of the Association for Computational Linguistics (vol. 1: Long Papers), pp. 47–57 (2014)
16. Martschat, S., Strube, M.: Latent structures for coreference resolution. Trans. Assoc. Comput. Linguist. **3**(1), 405–418 (2015)
17. Wiseman, S.J., Rush, A.M., Shieber, S.M., Weston, J.: Learning anaphoricity and antecedent ranking features for coreference resolution. Association for Computational Linguistics (2015)

18. Hochreiter, S., Schmidhuber, J.: Long short-term memory. Neural Comput. **9**(8), 1735–1780 (1997)
19. Pradhan, S., Moschitti, A., Xue, N., Uryupina, O., Zhang, Y.: CoNLL-2012 shared task: modeling multilingual unrestricted coreference in ontonotes. In: Joint Conference on EMNLP and CoNLL-Shared Task, pp. 1–40. Association for Computational Linguistics (2012)
20. Pennington, J., Socher, R., Manning, C.: Glove: global vectors for word representation. In: Proceedings of the 2014 Conference on Empirical Methods in Natural Language Processing (EMNLP), pp. 1532–1543 (2014)
21. Turian, J., Ratinov, L., Bengio, Y.: Word representations: a simple and general method for semi-supervised learning. In: Proceedings of the 48th Annual Meeting of the Association for Computational Linguistics, pp. 384–394. Association for Computational Linguistics (2010)
22. Kingma, D.P., Ba, J.: Adam: a method for stochastic optimization. arXiv preprint arXiv:1412.6980 (2014)
23. Srivastava, N., Hinton, G., Krizhevsky, A., Sutskever, I., Salakhutdinov, R.: Dropout: a simple way to prevent neural networks from overfitting. J. Mach. Learn. Res. **15**(1), 1929–1958 (2014)
24. Nair, V., Hinton, G.E.: Rectified linear units improve restricted Boltzmann machines. In: Proceedings of the 27th International Conference on Machine Learning (ICML 2010), pp. 807–814 (2010)
25. Lee, K., He, L., Zettlemoyer, L.: Higher-order coreference resolution with coarse-to-fine inference, arXiv preprint arXiv:1804.05392 (2018)

Type Hierarchy Enhanced Heterogeneous Network Embedding for Fine-Grained Entity Typing in Knowledge Bases

Hailong Jin, Lei Hou$^{(\boxtimes)}$, and Juanzi Li

Department of Computer Science and Technology,
Tsinghua University, Beijing 100084, China
jinh15@mails.tsinghua.edu.cn, {houlei,lijuanzi}@tsinghua.edu.cn

Abstract. Type information is very important in knowledge bases, but some large knowledge bases are lack of type information due to the incompleteness of knowledge bases. In this paper, we propose to use a well-defined taxonomy to help complete the type information in some knowledge bases. Particularly, we present a novel embedding based hierarchical entity typing framework which uses *learning to rank* algorithm to enhance the performance of word-entity-type network embedding. In this way, we can take full advantage of labeled and unlabeled data. Extensive experiments on two real-world datasets of DBpedia show that our proposed method significantly outperforms 4 state-of-the-art methods, with 2.8% and 4.2% improvement in Mi-F1 and Ma-F1 respectively.

Keywords: Entity typing · Knowledge base completion
Heterogeneous network embedding

1 Introduction

Types (also called concept or class) and entities are two fundamental elements in knowledge bases ($\mathcal{KB}$). Types in $\mathcal{KB}$ are usually organized as a hierarchical structure, namely *type hierarchy*. Type information is very valuable in $\mathcal{KB}$ because it is the glue that hold our mental world together [9]. **Entity typing**, assigning types (e.g., *Person, Athlete, BasketballPlayer*) to an entity (e.g., *Yao Ming*) in $\mathcal{KB}$, is a basic task in knowledge base construction.

Traditional entity typing focuses on a small set of types, (e.g., *Person, Location* and *Organization* [10,14]) which are too coarse-grained to be helpful for other NLP applications. While **fine-grained entity typing** assigns more specific types to an entity, which normally forms a *type-path* in the type hierarchy [15]. As shown in Fig. 1, *Yao Ming* is associated with a type-path */Thing/Agent/Person/Athlete/BasketballPlayer*. Characterizing an entity with fine-grained types (type-paths) benefits many real applications, such as *knowledge base completion* [2,12,17,22,23], *relation extraction* [26,27], *emerging entity typing* [1,6,11], and *heterogeneous knowledge base integration* [4,20].

© Springer Nature Switzerland AG 2018
M. Sun et al. (Eds.): CCL 2018/NLP-NABD 2018, LNAI 11221, pp. 170–182, 2018.
https://doi.org/10.1007/978-3-030-01716-3_15

Related researches on **fine-grained** entity typing mainly follows two lines. The first is **entity typing in text** by utilizing **contextual information**, which classifies an *entity mention* in the text to a broad set of types. It exploits the well-defined linguistic features based on the mention itself and its textual context, and then jointly learns feature and type representations to make type inference [8,15,24]. These methods require extensive human labor in the feature design and extraction (such as POS tags, dependency parsing results) and rich background knowledge (such as WordNet). The second is **entity typing in knowledge bases** by utilizing **structured information**, which tries to enrich the type information for entities in $\mathcal{KB}$. It usually regards the type inference as a hierarchical multi-label classification problem. But the model training heavily depends on the structured information within $\mathcal{KB}$, such as categories and infoboxes of entities [21].

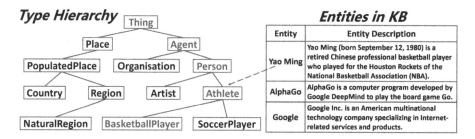

Fig. 1. An example of entity typing in $\mathcal{KB}$. The red ones are assigned types. (Color figure online)

To address the above issues, we propose a novel **Embedding-based Fine-grained Hierarchical Entity Typing method (EFHET)**. Specifically, we construct four heterogeneous networks to exploit different kinds of information effectively. We use a network embedding method [18] to learn low-dimensional representations for each entity and type based on the above networks. The learned entity and type embeddings not only preserve their semantic closeness, but also have a strong predictive power for the entity typing task. Finally, we use the learned embeddings to make type inference for each unlabeled entity in $\mathcal{KB}$. In summary, the contributions of this paper are as follows:

- Instead of manually defining complicated features, we present a simple and effective way to learn representations for each entity and type in $\mathcal{KB}$.
- To fully consider the structure of type hierarchy, we propose to use *learning to rank* algorithm to improve the quality of entity and type embeddings.
- Extensive experiments show that our proposed method significantly outperforms 4 state-of-the-art methods, with 2.8% and 4.2% improvement in Mi-F1 and Ma-F1 respectively.

The rest of this paper is organized as follows. In Sect. 2, we formally define the problem of fine-grained hierarchical entity typing in knowledge bases. Section 3 demonstrates the proposed approach in detail. The dataset description and evaluation are presented in Sect. 4. Section 5 outlines some related works and finally Sect. 6 concludes our work.

2 Problem Formulation

In this section, we formally define the problem of Fine-Grained Hierarchical Entity Typing in $\mathcal{KB}$. Before that, we first introduce the related definitions.

Definition 1. *Entity:* *Each entity e in the knowledge base describes a specific thing in the real world, and it can be represented as $e = (label(e), text(e))$, where $label(e)$ is the name and $text(e)$ is the text description of e respectively. Typically, the textual information $text(e)$ is described by N_e words $\{w_i | w_i \in V\}_{i=1}^{N_e}$, where V is the vocabulary. In other words, each entity is regarded as a document and it is the most generic representation.*

Most large scale knowledge bases, such as WordNet, YAGO and DBpedia, use hierarchical taxonomy to organize entities, and we define the taxonomy as type hierarchy.

Definition 2. *Type Hierarchy:* *Type hierarchy $\mathcal{H}$ is represented as a tree or a directed acyclic graph (DAG). It categorizes and organizes knowledge in large knowledge bases. It is defined as $\mathcal{H} = (\mathcal{T}, \mathcal{R})$, where $\mathcal{T}$ is the type set and $\mathcal{R} = \{(t_i, t_j) | t_i, t_j \in \mathcal{T} \text{ and } i \neq j\}$ is the relation set in which (t_i, t_j) means that t_i is subtype of t_j.*

Normally, an entity is associated with a type-path from the root (general type) to one internal/leaf node (specific type) in the type hierarchy. In Fig. 1, *Yao Ming* is associated with */Thing/Agent/Person/Athlete/BasketballPlayer*, where *Thing* is the root and *BasketballPlayer* is a specific type in the type hierarchy. We formally define type-path as follow:

Definition 3. *Type Path:* *A type-path p is denoted as root $\rightarrow t_1 \rightarrow t_2 \rightarrow \cdots \rightarrow t_n$, where $(t_{i+1}, t_i) \in \mathcal{R}$, namely t_{i+1} is subtype of t_i. A type-path in the type hierarchy $\mathcal{H}$ starts from the root and ends at one specific type. Note that it is not necessary to end at the leaf type.*

Large knowledge bases often contain millions of entities, which constitute a huge entity set. Normally, only a small number of entities are assigned with accurate type-paths, while the rest is either missing type-paths or associated with imprecise types. Take DBpedia as an example, only 3,048,928 entities are assigned type-paths. There are more than 1,500,000 unlabeled entities which need to deal with. Our goal is to solve the problem of incomplete and imprecise entity types in $\mathcal{KB}$, which can be formally defined as:

Definition 4. *Fine-Grained Hierarchical Entity Typing:* *Given a type hierarchy $\mathcal{H}$ and a partially-labeled entity set $\mathcal{E} = \mathcal{E}^l \cup \mathcal{E}^u$, where each labeled* *entity e^l in the labeled set $\mathcal{E}^l$ is associated with a type-path p^l, fine-grained hierarchical entity typing aims to learn a typing model from entity textual description and the partially-labeled information, and predict the type-path p_i for each unlabeled entity $e_i \in \mathcal{E}^u$.*

Different from most existing tasks, we are expected to only use the most common entity textual description to find its fine-grained type information (i.e., a type path in the given hierarchy). Therefore, we need to exploit and integrate all the available information to characterize entities, types and their relationship.

3 The Proposed Approach

To utilize different kinds of information (e.g., textual description, type hierarchy and labeled entity type-paths) without complicated feature engineering, we propose a network embedding based framework. It learns low-dimensional representations for entities and types from heterogenous network, and makes type-path inference for unlabeled entities based on the learned embeddings.

Figure 2 presents the framework of our proposed method, which consists of three key components:

- **Heterogeneous network construction:** given the entity collection, we first construct four networks which capture essential information in our task.
- **Joint entity and type representation:** based on the networks, we adapt a heterogeneous network embedding method to learn the low-dimensional representations for each entity and type.
- **Type-path Inference:** finally we employ the learned embeddings to infer a type-path for each unlabeled entity.

In the following parts, we will describe each component in detail.

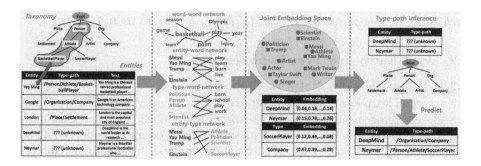

Fig. 2. Framework of the proposed EFHET (*In the second component, the blue nodes represent words, purple nodes represent entities, and green nodes represent types*) (Color figure online)

3.1 Heterogeneous Network Construction

In order to encode different kinds of information in knowledge bases, we construct four heterogeneous networks.

Definition 5. *Word-word (Co-occurrence) Network,* *denoted as* $G_{ww} = (V, E_{ww})$, *captures the word co-occurrence information in the entity description, which is the essential information for embedding method to preserve the word semantics. The weight* ω_{ij} *of the edge between word* w_i *and* w_j *is defined as the number of times that the two words co-occur in the context windows of a given window size.*

Entities with similar word descriptions are similar to each other, so we need to represent the entity-word relation, i.e., entity-word network.

Definition 6. *Entity-word Network,* *denoted as* $G_{ew} = (\mathcal{E} \cup V, E_{ew})$, *is a bipartite network between entities and words. The weight* ω_{ij} *between entity* e_i *and word* w_j *is defined as the number of times* w_j *appears in the description of* e_i.

The above two networks encode the unlabeled information in entity description. To fully use the partially labeled entity type-paths in $\mathcal{E}^l$, we introduce two networks which encode labeled information as below:

Definition 7. *Entity-type Network,* *denoted as* $G_{et} = (\mathcal{E}^l \cup \mathcal{T}, E_{et})$, *is a bipartite network that captures the labeled information. There is an edge between entity* e_i^l *and type* t_j *(i.e.,* $\omega_{ij} = 1$) *iff.* t_j *belongs to the type-path* p_i^l *of labeled entity* e_i^l, *namely there will be* n *edges between* e_i^l *and* n *types in* p_i^l.

Definition 8. *Type-word Network,* *denoted as* $G_{tw} = (\mathcal{T} \cup V, E_{tw})$, *is a bipartite network that captures the labeled information. The weight* ω_{ij} *between type* t_i *and word* w_j *is defined as:* $\omega_{ij} = \sum_{e_k \in S} n_{kj}$, *where* n_{kj} *is the term frequency of word* w_j *in the description of entity* e_k, *and* $S = \{e_k | t_i \in p_k^l\}$ *is an entity set, each entity in* S *is assigned with type* t_i *(i.e.,* t_i *belongs to its type-path).*

3.2 Joint Entity and Type Representation

3.2.1 Heterogeneous Network Embedding

We expect to learn low-dimensional representation for each entity and type from the above heterogeneous networks. We expect in the embedding space, close entities (types) should also have similar representations. A desirable model should be able to deal with arbitrary kinds of information networks. Inspired by **P**redictive **T**ext **E**mbedding (**PTE**) [18,19] which can handle both unsupervised and supervised information, we learn entity and type embeddings which have strong predictive power for classification, since entity typing in $\mathcal{KB}$ is a classification task essentially.

The entity-word, type-word and entity-type networks are bipartite networks, and the word-word network is essentially bipartite network by treating each undirected edge as two directed edges. Therefore, we first present the embedding method for individual bipartite network. The basic idea of bipartite network embedding is to preserve the second-order proximity, which can be interpreted as vertices with similar neighbors are similar to each other and thus should be represented closely in a low-dimensional space.

Formally, given a bipartite network $G = (V_A \cup V_B, E)$, where V_A and V_B are two disjoint vertex sets, and E is the set of edges between them. The conditional probability of vertex $v_i \in V_A$ generated by vertex $v_j \in V_B$ is defined as:

$$p(v_i|v_j) = \frac{\exp(\boldsymbol{u}_i \cdot \boldsymbol{u}_j)}{\sum_{i' \in A} \exp(\boldsymbol{u}_{i'} \cdot \boldsymbol{u}_j)} \tag{1}$$

where $\boldsymbol{u}_i$, $\boldsymbol{u}_j$ are the embedding vectors of vertex v_i and v_j respectively. For each vertex $v_j \in V_B$, Eq. (1) defines a conditional distribution $p(\cdot|v_j)$ over all the vertices in the set V_A. To preserve the second-order proximity, the conditional distribution $p(\cdot|v_j)$ should be close to its empirical distribution $\hat{p}(\cdot|v_j)$, which can be achieved by minimizing their KL-divergence, the objective function can be calculated as:

$$O = - \sum_{(i,j) \in E} w_{ij} \log p(v_j|v_i) \tag{2}$$

Now we can define the conditional probabilities $p(w_j|w_i)$, $p(w_j|e_i)$, $p(w_j|t_i)$ and $p(t_j|e_i)$ respectively, and reach the objective functions O_{ww}, O_{ew}, O_{et} and O_{tw} accordingly. The four objective functions are corresponding to four networks before (e.g., O_{ww} to G_{ww}). Intuitively, the heterogeneous network embedding could be learned via collectively embedding the four bipartite networks above, i.e., minimizing the joint objective function:

$$O_n = O_{ww} + O_{ew} + O_{et} + O_{tw} \tag{3}$$

The objective function can be optimized in a joint way, which means to train the model with the labeled data (the entity-type and type-word) and the unlabeled data (other networks) simultaneously. Interested readers can refer to the PTE paper [18,19] for the detailed training process. After this component, we achieve entity and type embeddings, $E_{emb} = \{\mathbf{e}_i\}$ and $T_{emb} = \{\mathbf{t}_i\}$.

3.2.2 Modeling Type Correlation

The most intuitive idea is using the embeddings $\mathbf{e}_i, \mathbf{t}_j$ learned by network embedding method (PTE) to make type inference directly. However, such straightforward inference suffers from two main issues: First, network embedding preserves second-order proximity in bipartite network which can make related entities (types) close to each other, but the relatedness between types and entities cannot be ensured. So we need to build a bridge between entities and types. Second, the network embedding method uses supervised information in an indirect way, which might lose some useful information. The type granularity is the most valuable information that is lost, because an edge in the entity-type network only indicates the entity belongs to the linked type but cannot tell the link strength. So, we need to take the structure of type hierarchy into account.

To solve the first issue, we learn two projection matrices ($\mathbf{U}$ and $\mathbf{V}$) for entity and type respectively to strengthen the relatedness between types and entities:

$$\Phi_e(\mathbf{e}_i) = \mathbf{U}\mathbf{e}_i \qquad \Phi_t(\mathbf{t}_j) = \mathbf{V}\mathbf{t}_j \qquad (4)$$

The projection matrices build a bridge between entities and types and map them into a closer semantic space.

For the second issue, we incorporate the idea of *learning to rank* into the projection matrices (i.e., $\mathbf{U}$ and $\mathbf{V}$) learning inspired by [8,15,24]. To better utilize the granularity, we use *type order* to model different relatedness between an entity and its related types, which reflects the structure information of type hierarchy to a certain degree. In particular, we define two kinds of *type order*:

- **Ancestor Order:** In the type-path of an entity, specific types should be closer to an entity than its ancestor. For example, *Yao Ming* should be closer to *BasketballPlayer* than *Athlete*.
- **Sibling Order:** In the type hierarchy, the correct type should be closer to an entity than its siblings. For example, *Yao Ming* should be closer to *BasketballPlayer* than *SoccerPlayer*.

Accordingly, we design two different loss functions in the learning to rank algorithm, such that the learned projection matrices can meet these two type orders.

For the **ancestor order**, we define adaptive margin $\gamma_{ij} = 1 + \frac{l(t_i,t_j)}{l(root,t_j)}$, where $l(t_i, t_j)$ denotes the number of steps going down from type t_i to type t_j.

$$L_a(e) = \sum_{t_k \in p(e)} \sum_{t_{k'} \in A(t_k)} R(r_a(e, t_k)) \Theta^a_{i,k,k'} \qquad (5)$$

$$\Theta^a_{i,k,k'} = max\{0, \gamma_{kk'} - s(e, t_k) + s(e, t_{k'})\} \qquad (6)$$

$$r_a(e, t_k) = \sum_{t_{k'} \in A(t_k)} I(\gamma_{kk'} + s(e, t_{k'}) > s(e, t_k)) \qquad (7)$$

Here, $p(e)$ is the labeled type-path for the entity e, $A(t_k)$ represents the set of ancestor types of type t_k, $R(x) = \sum_{i=1}^{x} \frac{1}{i}$ maps the ranking to a floating-point weight and $s(\mathbf{e}, \mathbf{t}_k)$ is the inner product of $\Phi_e(\mathbf{e})$ and $\Phi_t(\mathbf{t_k})$.

For the **sibling order**, we use the fixed margin. The loss function $L_s(e)$ can be analogously defined with $S(t_k)$ representing the set of sibling types of type t_k. The objective function for *learning to rank* model could be achieved by aggregating over all the entities with labeled type-path, namely:

$$O_l = \sum_{e_i \in \mathcal{E}^l} L_a(e_i) + L_s(e_i) \qquad (8)$$

Then we can use stochastic (sub) gradient descent to learn the parameters (two projection matrices). We achieve two projection matrices $\mathbf{U}$ and $\mathbf{V}$.

3.3   Type-Path Inference

With the learned embeddings $\{\mathbf{e}_i\}$ and $\{\mathbf{t}_i\}$, projection matrices $\mathbf{U}$ and $\mathbf{V}$, we perform top-down search in the given type hierarchy $\mathcal{H}$ to estimate the correct type-path $\mathcal{P}$. Starting from the tree's root, we recursively find the best type among the children types by measuring the dot product of the entity and type embeddings, i.e., $s(\mathbf{e}_i, \mathbf{t}_j) = \Phi_e(\mathbf{e}_i) \cdot \Phi_t(\mathbf{t}_j)$. The search process stops when we reach a leaf type, or the similarity score is below a pre-defined threshold $\eta > 0$.

4   Experiments and Analysis

In this section, we evaluate the proposed method using real world dataset collected from Wikipedia and DBpedia. We will introduce the dataset and experiment settings in Sect. 4.1, present the comparison results in Sect. 4.2 and investigate some method details in Sect. 4.3. Our source code is available[1] for reference.

4.1   Datasets and Experimental Setup

Datasets: To the best of our knowledge, there is no large-scale available dataset which is suitable for fine-grained entity typing in knowledge bases. Therefore, we use type hierarchy and entity description from DBpedia[2]. For each entity, we use the single type-path assigned in DBpedia as the ground truth. According to the text information used, we construct 2 datasets. **Long abstract:** Full abstract of each entity, contain 74.65 words on average. **Short abstract:** Short abstract of each entity, contain 53.5 words on average.

For each dataset, we remove those entities which are only labeled as *Thing*. Because they don't provide any semantic information. We follow a 80/20 ratio to partition it into training/test sets. Table 1 shows the detailed statistics. The average type-path length of each entity is 3.27.

Table 1. Statistics of the dataset

Datasets	#Types	#Entities	#Words
Long abstract	451	3,048,928	31,752
Short abstract	451	3,048,928	25,430

Metrics: To evaluate the performance of our proposed method, we use Accuracy (Strict-F1), Micro-averaged F1 (Mi-F1) and Macro-averaged F1 (Ma-F1), which have been used in many fine-grained typing systems [7,15,24].

Baseline: We denote our proposed **E**mbedding-based **F**ine-grained **H**ierarchical **E**ntity **T**yping method as **EFHET**, and compare it with 4 state-of-the-art entity

---

[1] https://github.com/Tsinghua-PhD/EFHET.
[2] http://downloads.dbpedia.org/2016-10/.

typing methods and its variants as follow: **FIGMENT** [22]: They designed two scoring models for pairs of entities and types: a global model and a context model. In the experiment, we use the released implementation[3]. **MuLR** [23]: They introduced representations of entities on different levels: character, word and entity. We also use the released implementation[4] in our experiment. **Global** [12]: It proposed an evaluation framework comprising of methods for dataset construction and evaluation metrics to evaluate KBC approaches for inferring missing entity type instances. **CE/HCE** [5]: They proposed a framework to learn entity and type embeddings to capture semantic relatedness between entities and types. **EFHET-3**: We don't use the word-word(co-occurrence) network in heterogeneous network embedding component.

Parameter: In heterogeneous network construction module, we remove stopwords and low frequency words (i.e., <1500 in long and short abstract). In joint entity and type representation module, we use the default parameter setting same with PTE [18,19] in heterogenous network embedding model. We use adaptive and fixed margin for ancestor order and sibling order respectively in learning to rank model (readers can refer to relevant section). In type-path inference module, the threshold η is set as 0.25. For baseline methods, we adopt the same parameter setting as those reported in the original papers.

4.2   Overall Comparison Results

Table 2 shows the overall performance on the entity typing task, and we compare **EFHET** with two sets of methods:

Entity Typing Methods. Our **EFHET** outperforms the state-of-the-art entity typing systems (i.e., **FIGMENT**, **MuLR** and **Global**), and achieves 2.8% and 4.2% improvement in Mi-F1 and Ma-F1 respectively, because it utilizes more structure information: (1) *shallow structure information*. Heterogeneous network embedding makes related entities (or types) embed close to each other if they share similar types and words (entities). (2) *deeper structure information*. In learning to rank component, we define two kinds of type order over the type hierarchy: ancestor order makes entities closer to specific types than general types, and sibling order makes the learned entity embeddings have strong discriminative power between sibling types.

Embedding Methods. Our model also performs better than the embedding method **CE/HCE**, achieves 20.9%, 19.0% and 21.7% improvement in accuracy, Ma-F1 and Mi-F1 respectively. **CE/HCE** uses target entity e_t and its types to predict its contextual entity e_c (similar to Skip-gram model). Thus it needs a large set of entity pairs $\{(e_t, e_c)\}$, which requires recognizing all entity mentions from the description in advance. Meanwhile, the entity level co-occurrence relation is much sparser than word, so the learned embeddings are limited by lack

---

[3] https://github.com/yyaghoobzadeh/figment.
[4] https://github.com/yyaghoobzadeh/figment-multi.

of data. Compare with **CE/HCE**, our model only needs entity description itself which is easier to obtain.

Variants. **EFHET** outperforms **EFHET-3**, demonstrating the effectiveness of word-word (co-occurrence) network. Word-word network encodes textual information. Two words are embedded close to each other if they are semantically close. For example, "computer" and "PC" are similar words, "computer" and "PC" may often occur in entity e_i and e_j respectively, but e_i and e_j may be embedded close benefit from the closeness of "computer" and "PC". In other words, word-word networks may deal with synonym problem in entity or type representation.

Table 2. Typing performance

Dataset	Long abstract			Short abstract		
	Acc	Ma-F1	Mi-F1	Acc	Ma-F1	Mi-F1
FIGMENT	0.305	0.441	0.457	0.304	0.436	0.479
MuLR	0.335	0.479	0.503	0.327	0.481	0.497
Global	0.297	0.425	0.438	0.292	0.411	0.437
CE/HCE	0.164	0.331	0.314	0.179	0.336	0.344
EFHET-3	0.369	0.513	0.528	0.339	0.508	0.513
EFHET	0.373	0.521	0.531	0.344	0.510	0.519

4.3   Results Analysis

Results at Different Type Levels. Table 3 reports the accuracy of **EFHET**, **FIGMENT**, **MuLR** and **Global** at different levels of the type hierarchy. General types (e.g., *Person*) are at first/second level, while specific types (e.g., *BasketballPlayer*) are at deeper levels. From the results, we can see that **EFHET** consistently outperforms the other three methods, and achieves 12.6% improvement in accuracy than other methods on the fourth level types. The performance becomes worse as the level depth increases, because we use top-down search to make type inference. If we find the wrong type at level-1, we can not get the correct type at level-2. **EFHET** always outperform other methods, because in learning to rank component we use **sibling order** to make the learned embeddings have strong distinguishing power between sibling types in type inference.

Results for Infrequent Types. We also investigate the performance on infrequent types using the evaluation metric (Ma-F1) for a type proposed by [25] which is different from Table 2. The results for infrequent types (occur less than 3,000 times) and frequent types (occur more than 30,000 times) are shown in Table 4. Generally, the performance on infrequent types is worse than frequent ones. **EFHET** consistently outperforms the other methods on infrequent types, which demonstrates its ability on dealing with rare types.

Table 3. Performance at different type levels

	EFHET	FIGMENT	MuLR	Global
Level-1	0.762	0.445	0.579	0.518
Level-2	0.584	0.294	0.390	0.342
Level-3	0.391	0.173	0.242	0.287
Level-4	0.274	0.112	0.137	0.148

Table 4. Performance on infrequent types

Type	Infrequent		Frequent	
	Painter	Average	Company	Average
# training data	2,028	1,804	55,173	65,241
# test data	445	417	12,111	15,176
FIGMENT	0.294	0.287	0.507	0.529
MuLR	0.316	0.323	0.618	0.651
EFHET	0.490	0.447	0.783	0.775

5   Related Work

There are two branches of research that are related to our work in this paper, namely, entity typing in knowledge base and entity typing in text.

Entity Typing in Knowledge Base. *Tipalo* automatically assigns types for an entity in DBpedia by interpreting its natural language definition [3]. Paulheim and Bizer design a heuristic link-based type inference mechanism *SDType* [13], and *FIGMENT* [22] types entities in corpus-level using contextual information. Xu *et al.* employ a multi-label hierarchical classification method to assign Chinese entities with DBpedis types [21]. However, none of them integrates type hierarchy information properly.

Entity Typing in Text. Yogatama *et al.* first present a ranking lost based embedding model to learn embeddings for types and entity mentions simultaneously [24]. Ren *et al.* propose a general framework to jointly embed entity mentions, textual features and entity types [16], and further consider the type correlation and wrong labeling problem in [15]. Ma *et al.* present a prototype-driven label embedding method for fine-grained named entity typing which can predict both seen and unseen types [8]. Li *et al.* try to include structured knowledge from taxonomy hierarchy in their entities and categories joint embedding [5].

6   Conclusion

In this paper, we propose an embedding-based model to solve the problem of Fine-Grained Hierarchical Entity Typing in knowledge bases. To encapsulate various information, we construct four heterogeneous networks and employ the network embedding method to learn low-dimensional representations for entities and types simultaneously. To effectively capture the semantic relatedness between entities and types with different granularities, we propose two kinds of type order and incorporate them into the representation learning via learning to rank model. Finally, the learned embeddings are used to predict type-path for unlabeled entities. Experiments on real world datasets demonstrate the effectiveness of the proposed model.

Acknowledgment. The work is supported by the national key research and development program of China (No. 2017YFB1002101), NSFC key project (U1736204, 61661146007), Fund of Online Education Research Center, Ministry of Education (No. 2016ZD102), and THU-NUS NExT Co-Lab.

References

1. Brambilla, M., Ceri, S., Della Valle, E., Volonterio, R., Acero Salazar, F.X.: Extracting emerging knowledge from social media. In: Proceedings of WWW 2017, pp. 795–804. International World Wide Web Conferences Steering Committee (2017)
2. Dong, X., et al.: Knowledge vault: a web-scale approach to probabilistic knowledge fusion. In: Proceedings of SIGKDD 2014, pp. 601–610. ACM (2014)
3. Gangemi, A., Nuzzolese, A.G., Presutti, V., Draicchio, F., Musetti, A., Ciancarini, P.: Automatic typing of DBpedia entities. In: Cudré-Mauroux, P., et al. (eds.) ISWC 2012. LNCS, vol. 7649, pp. 65–81. Springer, Heidelberg (2012). https://doi.org/10.1007/978-3-642-35176-1_5
4. Li, J., Wang, Z., Zhang, X., Tang, J.: Large scale instance matching via multiple indexes and candidate selection. Knowl.-Based Syst. **50**(3), 112–120 (2013)
5. Li, Y., Zheng, R., Tian, T., Hu, Z., Iyer, R., Sycara, K.: Joint embedding of hierarchical categories and entities for concept categorization and dataless classification. In: Proceedings of COLING 2016, pp. 2678–2688. ACL (2016)
6. Lin, T., Etzioni, O., et al.: No noun phrase left behind: detecting and typing unlinkable entities. In: Proceedings of EMNLP 2012, pp. 893–903. ACL (2012)
7. Ling, X., Weld, D.S.: Fine-grained entity recognition. In: AAAI, pp. 94–100. AAAI (2012)
8. Ma, Y., Cambria, E., Gao, S.: Label embedding for zero-shot fine-grained named entity typing. In: Proceedings of COLING 2016, pp. 171–180. ACL (2016)
9. Murphy, G.: The Big Book of Concepts. MIT Press, Cambridge (2004)
10. Nadeau, D., Sekine, S.: A survey of named entity recognition and classification. Lingvist. Investig. **30**(1), 3–26 (2007)
11. Nakashole, N., Tylenda, T., Weikum, G.: Fine-grained semantic typing of emerging entities. In: Proceedings of ACL 2013, pp. 1488–1497. ACL (2013)
12. Neelakantan, A., Chang, M.W.: Inferring missing entity type instances for knowledge base completion: new dataset and methods. arXiv:1504.06658 (2015)
13. Paulheim, H., Bizer, C.: Type inference on noisy RDF data. ISWC 2013. LNCS, vol. 8218, pp. 510–525. Springer, Heidelberg (2013). https://doi.org/10.1007/978-3-642-41335-3_32
14. Ratinov, L., Roth, D.: Design challenges and misconceptions in named entity recognition. In: Proceedings of CoNLL 2009, pp. 147–155. ACL (2009)

15. Ren, X., He, W., Qu, M., Huang, L., Ji, H., Han, J.: AFET: automatic fine-grained entity typing by hierarchical partial-label embedding. In: Proceedings of EMNLP 2016, pp. 1369–1378. ACL (2016)
16. Ren, X., He, W., Qu, M., Voss, C.R., Ji, H., Han, J.: Label noise reduction in entity typing by heterogeneous partial-label embedding. In: Proceedings of SIGKDD 2016, pp. 1825–1834. ACM (2016)
17. Sterckx, L., Demeester, T., Deleu, J., Develder, C.: Knowledge base population using semantic label propagation. Knowl.-Based Syst. **108**, 79–91 (2015)
18. Tang, J., Qu, M., Mei, Q.: PTE: predictive text embedding through large-scale heterogeneous text networks. In: Proceedings of SIGKDD 2015, pp. 1165–1174. ACM (2015)
19. Tang, J., Qu, M., Wang, M., Zhang, M., Yan, J., Mei, Q.: LINE: large-scale information network embedding. In: Proceedings of WWW 2015, pp. 1067–1077. ACM (2015)
20. Wang, Z., Li, J., Wang, Z., Tang, J.: Cross-lingual knowledge linking across wiki knowledge bases. In: Proceedings of WWW 2012, pp. 459–468. ACM (2012)
21. Xu, B., Zhang, Y., Liang, J., Xiao, Y., Hwang, S., Wang, W.: Cross-lingual type inference. In: Navathe, S.B., Wu, W., Shekhar, S., Du, X., Wang, X.S., Xiong, H. (eds.) DASFAA 2016. LNCS, vol. 9642, pp. 447–462. Springer, Cham (2016). https://doi.org/10.1007/978-3-319-32025-0_28
22. Yaghoobzadeh, Y., Schütze, H.: Corpus-level fine-grained entity typing using contextual information. In: Proceedings of EMNLP 2015, pp. 715–725. ACL (2015)
23. Yaghoobzadeh, Y., Schütze, H.: Multi-level representations for fine-grained typing of knowledge base entities. arXiv:1701.02025 (2017)
24. Yogatama, D., Gillick, D., Lazic, N.: Embedding methods for fine grained entity type classification. In: Proceedings of ACL 2015, pp. 26–31. ACL (2015)
25. Yosef, M.A., Bauer, S., Hoffart, J., Spaniol, M., Weikum, G.: HYENA: hierarchical type classification for entity names. In: Proceedings of COLING 2012, pp. 1361–1370. ACL (2012)
26. Zhang, C., Xu, W., Ma, Z., Gao, S., Li, Q., Guo, J.: Construction of semantic bootstrapping models for relation extraction. Knowl.-Based Syst. **83**(C), 128–137 (2015)
27. Zheng, S., Xu, J., Zhou, P., Bao, H., Qi, Z., Xu, B.: A neural network framework for relation extraction: learning entity semantic and relation pattern. Knowl.-Based Syst. **114**(C), 12–23 (2016)

Scientific Keyphrase Extraction: Extracting Candidates with Semi-supervised Data Augmentation

Qianying Liu[1], Daisuke Kawahara[3], and Sujian Li[2(✉)]

[1] School of Mathematical Science, Peking University, Beijing, China
yingliu96@pku.edu.cn
[2] Key Laboratory of Computational Linguistics, MOE, Peking University, Beijing, China
lisujian@pku.edu.cn
[3] Graduate School of Informatics, Kyoto University, Kyoto, Japan
dk@i.kyoto-u.ac.jp

Abstract. Keyphrase extraction can provide effective ways of organizing scientific documents. For this task, neural-based methods usually suffer from performance unstability due to data scarcity. In this paper, we adopt the pipeline two-step method including candidate extraction and keyphrase ranking, where candidate extraction is a key to influence the whole performance. In the candidate extraction step, to overcome the low-recall problem of traditional rule-based method, we propose a novel semi-supervised data augmentation method, where a neural-based tagging model and a discriminative classifier boost each other and get more confident phrases as candidates. With more reasonable candidates, keyphrase are identified with recall promoted. Experiments on SemEval 2017 Task 10 show that our model can achieve competitive results.

Keywords: Keyphrase extraction · Neural networks
Semi-supervised learning

1 Introduction

With the number of scientific papers increasing dramatically, how to retrieve and manage them is a big problem. Keyphrases are usually used to organize scientific papers, since they carry the core information of an article in a concise way. With keyphrases, readers do not need to read the whole article and can have a general understanding of the content in a short time. Thus, automatic keyphrase extraction has drawn much attention recently and can benefit many downstream applications such as document summarization and question answering.

For keyphrase extraction there have been various approaches [8]. One research line casts it into a sequence tagging problem and builds an end-to-end neural model to extract keyphrases directly [1,5,10,16,17]. Neural models often suffer

© Springer Nature Switzerland AG 2018
M. Sun et al. (Eds.): CCL 2018/NLP-NABD 2018, LNAI 11221, pp. 183–194, 2018.
https://doi.org/10.1007/978-3-030-01716-3_16

from time-consuming parameter fine-tuning and require a large amount of expensive training data. Due to the limited size of the training set, it is hard for neural based models to get reliable results. The out-of-vocabulary (OOV) problem can also deteriorate the performance of neural models. Another research line is to use a multi-step pipeline to extract keyphases. Pipeline models are often composed of two main steps, which are candidate generation and keyphrase ranking. While only using a few hyper-parameters, pipeline models reported competitive performance over neural models with robust performance. However, the performance of candidate extraction is a key that determines the recall of the whole model and limits the overall performance. Wang et al. [27] used N-gram statistics for candidate extraction, although their methods can not extract those keyphrases that appear rarely in the text. Wang and Li [28] relied on rule-based candidate generation which mainly depended on part-of-speech (POS) patterns and performed especially poor on specific categories (Fig. 1).

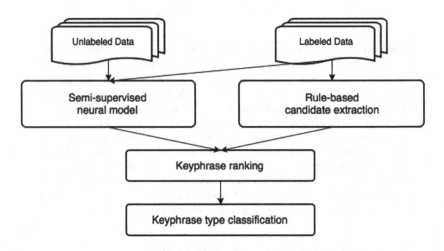

Fig. 1. A brief flow chart of the pipeline of our model.

To overcome the deficiencies of these candidate generation methods, we combine the two research lines by adding a neural model into the pipeline. We design a neural-based model to extract more keyphrase candidates. Following Wang and Li [28], we mix these candidates with the rule-based candidates and feed them into a multi-step pipeline. While extracting candidates with the neural model, we also cast this problem as a sequence tagging problem. As supplement of the rule-based candidates, the neural model can detect candidates of a wide variety of patterns and significantly improve the recall of candidates. Character level features and other handcrafted features are also used to relieve the OOV problem.

Due to the limited size of the training set, most studies utilize external knowledge or unlabeled data to improve the performance of keyphrase extraction [1,16]. However, a preliminary experiment shows that simply adding more candidates could make the keyphrase ranking unit suffer from imbalanced data and drag down its performance. Thus, we need to extract extra data from unlabeled data which has both quality and quantity. Here, we introduce a semi-supervised learning framework to get extra training data of high quality. This approach can both provide data augmentation for the neural model and balance the input data for the ranking unit. Experiments on the ScienceIE task[1] of SemEval 2017 show the effectiveness of our model.

2   Related Work

Early approaches of keyphrase extraction often build a pipeline model, which first selects candidate phrases and then casts this problem into a binary classification problem. Handcrafted rules are designed by experts for the candidate selection step. Phrases with certain part-of-speech tags or n-grams that fit in certain syntactic patterns are chosen as candidates. While these methods often reach a high recall, pruning methods are often added to remove the candidates that are unlikely to be keyphrases [13,14,27]. Many different algorithms have been used for the second classification stage. Researchers often use various handcrafted features including statistical features, syntactic features and structural features [15,26].

Wang and Li [28] proposed a traditional pipeline model on this task. A rule-based candidate generator feeds candidates to a linear model based keyphase ranking unit to identify keyphrases. Then random forest stacked upon a linear model is used for keyphrase classification. The performance of the model is limited because the rule-based candidate generator cannot deal with the wide range of keyphrase patterns, and simply adding new rules would make the keyphrase ranking unit suffer from serious imbalanced data.

Various neural models have been proposed in the sequence tagging problem recently. Chiu and Nichols [5] introduced a model for Name Entity Recognition (NER). Using the BIOES(Begin, Inside, Outside, End, Single) tagging scheme, they extracted character level features with a character level convolutional neural network (char-CNN) and used token level bi-directional Long Short-Term Memory Network (BiLSTM) for tagging. While using this tagging scheme, it is obvious that some label sequences (e.g., 'O I' or 'O E') are unreasonable, so adding a Conditional Random Field (CRF) layer which can label a sequence jointly is a natural choice to improve the performance. Huang et al. [10] compared the performance of LSTM , BiLSTM, LSTM-CRF and BiLSTM-CRF on POS tagging, chunking and NER datasets, and BiLSTM-CRF outperformed all the other models. Ma and Hovy [17] proposed an end-to-end sequence labeling model via BiLSTM-CNN-CRF without task-specific resources, feature engineering or data pre-processing beyond pretrained word embedding. But keyphrase extraction for scientific articles have two main differences from NER that makes a simple BiLSTM-CRF perform poor. Firstly, keyphrase extraction has much

---

[1] https://scienceie.github.io/.

smaller training data so that the neural network suffers from overfitting. Secondly, keyphrases extraction suffers from far more OOV words since most scientific keywords are rare words.

Our approach takes the advantage of these methods, proposing a model that combines neural models and traditional machine learning algorithms to improve the performance.

Current approaches of data augmentation in NLP include Dong et al. [7] which built a neural architecture based on NMT and PPDB for paraphrasing for question answering. Yasunaga et al. [29] generated adversarial examples for POS tagging by giving continuous perturbations directly to the embedding. There are also some approaches of synthesizing adversarial samples to examine the performance of models. Rule based approaches include Hossini et al. [9] and Samanta and Mehta [22]. Jia and Liang [11] created adversarial examples for question answering to evaluate the performance of QA systems under the perturbation of the question sentence. To the best of our knowledge, there is still no previous work which adopts the neural-based data augmentation strategy for keyphrase extraction.

3   Model

Our model is composed of two stages: candidate extraction in Sect. 3.1 and keyphrase ranking in Sect. 3.2.

3.1   Candidate Extraction

Our candidate extracting unit includes two parts: a neural model extractor and a rule based extractor.

Neural Model. We introduce a neural model of three layers: The embedding layer, the token-level BiLSTM layer and the CRF tagging layer. The embedding combines pretrained word embedding, character level features, POS features and other handcrafted features. The LSTM layer encodes the surrounding information of the tokens. The CRF tagging layer jointly models the tags considering the dependencies across the tags.

Feature Vector. For tokens $\{x_1, x_2, \ldots, x_n\}$ in a sentence, the embedding of each token is a concatenation of these four following parts: Word embedding $w(x_i)$, Character level embedding $c(x_i)$, Part-of-Speech tags $pos(x_i)$ and Handcrafted features $hc(x_i)$.

Word Embedding. $w(x_i)$ We initialized our model with the pretrained GloVe [21] 100d embedding. An unknown word token is added to represent the OOV words and its embedding is initialized as a zero vector.

Character Level Embedding. $c(x_i)$ All tokens are padded to a fix size and we use CNNs of the filter size 2 and 3 to capture character level features. The character lookup table is initialized randomly. We use an full-connected layer to map the embedding into a fixed size.

Part-of-Speech Tags. $pos(x_i)$ We map the POS tags of the tokens to a vector space and randomly initialize them. The POS tags are labeled by the NLTK toolkit [3].

Handcraft Features. $hc(x_i)$ We use one hot embedding to map these following features: (1) Rarity, which depends on the TF-IDF of the token, (2) Capitalization, which depends on whether the token is all capitalized and whether the tokens first character is capitalized, (3) Whether the token is an IEEE word or not. We also combine these features to the embedding: (1) The TF-IDF score of the token based on English Wikipedia, (2) The length of the token, (3) The frequency of the token in wikiwords.

The final feature vector of a token x_i is $v(x_i) = (w(x_i); c(x_i); pos(x_i); hc(x_i))$

LSTM. We concatenate all features and feed them into a token LSTM layer. The LSTM is bidirectional to capture the contextual information. We use dropout to avoid over-fitting. The LSTM hidden states are passed to a linear layer to map it into the size of the label space.

CRF. Because of the strong dependencies across tags, we propose a CRF layer to model the tags jointly. A 'Begin of Sentence (BOS)' tag is added to the start of a sentence and a 'End of Sentence (EOS)' tag is added to the end of a sentence. When testing we use Viterbi algorithm to decode the tags.

Rule-Based Candidates. We follow Wang and Li [28] and add rule-based candidates into the model. We use five rules to generate rule-based candidates. We extract phrases that match the POS pattern $NP = (NN * |JJ*) * (NN*)$ or consist of capital letters. Then we examine their length, whether they contain special digit or consists of only digits. If they pass all the check rules, then we add them as a candidate.

3.2 Keyphrase Ranking

For identifying the keywords, we use stacking to ensemble random forest models. Ensemble models allow to combine more features without damping the performance.

Linear models are used to apply the weights of the stacking [23]. We use Linear Regression for this layer. A hyper-parameter α is used here as a score bound to balance the precision and the recall. Candidates whose scores are higher than α would be chosen as keyphrases. The identifier can be represented as follows:

$$C(p) = \begin{cases} 1, & \phi(p) > \alpha \\ 0, & \phi(p) \le \alpha \end{cases} \tag{1}$$

where p stands for a candidate phrase, $C(p)$ stands for the whole identifier and $\phi(p)$ stands for the ranking model.

Instead of training the random forest and linear model jointly, we train them separately here. The low layer is fitted first and their results are fed into the linear model. The low layer actually works like a feature transformer.

Here we incorporate the linguistic features, context features and external knowledge features. We also use TextRank [18] and SGRank [6] as unsupervised features, examining whether the phrase is in the top n keyphrases according to these two algorithms.

4   Semi-supervised Data Augmentation

Semi-supervised data augmentation is mainly used to balance the data of the keyphrase ranking unit and augment the data for the neural model. In Algorithm 1 and Fig. 2, y stands for the labeled data and y^* stands for the unlabeled data. We call the original ranking unit C and the neural candidate extractor G. R stands for the rule-based candidate generation unit. An identifier is built for

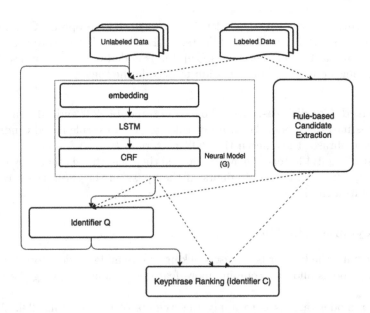

Fig. 2. A brief flow chart of our semi-supervised model. The solid line stands for the semi-supervised training process and the dotted line stands for the supervised training process.

this process and we call it Q. Q is similar to C except that we choose to use SGD Regression for Q and turn up the hyper-parameter α to ensure its precision remains high.

The training of G and Q includes two steps which are presented by the two kinds of lines in Fig. 2. The dotted line in Fig. 2 shows the process of pretraining the sequence tagging model G and the identifier Q on labeled data. The solid line shows the semi-supervised learning process. We use G to tag unlabeled data and feed the candidates into Q. In an ideal situation, the neural model should improve its generalization ability and the independent identifier should only allow high quality data to update the parameters of the neural model. We show the detailed architecture in Sect. 4.1 and how we use the simulated data to update our models in Sect. 4.2.

4.1   Architecture

We first train G and Q through supervised training with labeled data. Then we let G extract candidates from the unlabeled data and use Q to rank them. For tokens $\{y_1, y_2, \ldots, y_n\}$ in an unlabeled sentence, the neural model labels them with the BILOS tagging scheme as $\{l(y_1), l(y_2), \ldots, l(y_n)\}$. We use these labels to extract phrases $P = \{p_1, p_2, \ldots, p_m\}$. Q identifies them similar to the keyphrase ranking unit and we pick out the positive results $S = \{p_{k_1}, p_{k_2}, \ldots, p_{k_j}\}$. We call the candidates that Q predict as *simulated data* which is denoted as S. Because of the high precision of Q, we can assume that these simulated data are reliable.

Algorithm 1. Semi(Q, C, G, R, y, y^*)

1: **for** y_i in y **do**
2:      train G with $\{y_i, groundtruth\}$
3: **for** p_i in $G(y), R(y)$ **do**
4:      train Q with $\{p_i, groundtruth\}$
5: **for** p_i^* in $G(y^*)$ **do**
6:      $S = \{p_i^* \mid$ if $Q(p_i^*)$ is true$\}$
7:      train Q with $\{p_i^*, false\}$
8: **for** p_i^S in S **do**
9:      train G with $\{p_i^S, true\}$
10: **for** p_i in $S, G(y), R(y)$ **do**
11:      train C with $\{p_i, true\}$

During semi-supervised training, we develop G and Q jointly while C is trained afterwards. To update G, the simulated data is saved and then fed back to G as positive examples for data augmentation. Q also updates its parameters by fitting all the candidates that the neural model extracts as negative data. While updating the two models in the semi-supervised training process, we mix the simulated data and real data together. After all unlabeled data goes through this structure, we stop the process and use G and R to extract candidates on the training data to fit C. When we fit C, S is added to the training data as positive examples to balance the data.

This architecture is only activated when we train the model. While testing, the test data only goes through the neural model G for candidate extraction.

4.2 Model Updating Details

Weights. While this process can augment reliable new data, the simulated data is still treated differently when we update the parameters. When we update the neural model with the simulated data, we add weights to these data and the weights are determined by the score of the candidates ranked by Q. These weights are also used when we update the ranking unit C. When we update C, we add a hyper-parameter β as a weight for all the simulated data. This hyper-parameter is set to 0.5 in our experiment.

Updating Scheme. We update the parameters of G by batch, and Q is updated after every epoch. Only the linear stacking layer of Q is updated while the random trees are fixed.

The purpose of updating Q is to force the precision of Q to go higher and the recall to go lower. This will lead to a result that less and less simulated data is used to update G every epoch. In other words, during training G, the potential amount of noise in the mixture of real data and simulated data declines every epoch, which exactly meets the need of G's trimming.

We separate the simulated data into 3 epochs when the neural model is trained. Then we let real data go through the model again until the model reaches convergence.

5 Experiment

5.1 Dataset

We use the SemEval 2017 Task 10 ScienceIE dataset in this approach. In this dataset, 500 paragraphs among the domains Computer Science, Material Sciences and Physics were selected, providing both the full text and additional metadata. Only one paragraph of each article is manually labeled that the size of the training data is very limited. Meanwhile, because of the wide range of terminology vocabulary in those domains, the huge size of OOV words in the pretrained embedding influences the performance of the model. In the process of the unsupervised training, we use the unlabeled context of the articles to extract simulated data.

5.2 Tagging Scheme

In the process of extracting candidates with a neural model, we follow Luan et al. [16] and Ammer et al.[1] and cast the problem into a sequence tagging problem. We use the BIOES(Begin, Inside, Outside, End, Single) system to tag the context and assign three labels to each token. For example, the PRO-CESS keyword 'okta Markov Chain' is labeled as 'B-PROCESS I-PROCESS E-PROCESS'.

5.3   Hyperparameters and Details

All parameter are tuned on the development set. We select the hyperparameters for the neural model and the identifiers when we warm them up separately. We use two filter sizes for CNN which are 2 and 3. The character embedding dimension is 50. The word embedding dimension is 100. The handcraft features' dimension is 8. The Part of Speech feature dimension is 50. The token-level hidden dimension is 200. The token level LSTM layer is made up of two layers of BiLSTM. We use 0.18 and 0.5 for the hyperparameter α in the two identifiers.

While training, the batch size is 32 and dropout is set to 0.5 for the LSTM model and CNNs. We use the Adam [12] optimizer with a learning rate of 0.001 and a weight decay of 1e-8. We monitor the performance on the development set and use early stopping. The best parameters are fixed on the development set while we experiment supervised learning.

We use scikit-learn [20] to implement the linear models and random forest models. The neural model is implemented by Pytorch [19].

5.4   Performance Comparisons

Table 1 reports the results of our model and compares them with other SemEval Systems. We can see that the performance of our model is competitive against other models and out performs all the supervised single models, which proves the effectiveness of our model.

Table 1. Overall F1 scores for identification and classification(SemEval Subtask A and Subtask B). SM stand for single model.

Span level	Identification
Wang and Li [28]	0.510
Ammer et al. [1] (SM)	0.499
Ammer et al. [1] (Semi SM)	0.541
LuanOH17 [16]	0.521
LuanOH17 [16] (Semi)	0.576
Ours	0.543

Table 2 compares our model with Wang and Li [28] whose candidates are rule-based. It shows the significant boost of candidate recall of our system and the improvement in the **TASK** category. This shows that our model can capture candidates of a wider variety while balancing between recall and precision.

We also examine how the semi-supervised architecture affects overall performance in Table 2. We report results of simply adding more candidates to show the importance of our semi-supervised architecture. It shows that simply adding more candidates does not work because the imbalanced data harms the performance of the ranking unit. Adding candidates without consideration of data balance seriously drags down the performance.

Table 2. Comparisons of Wang and Li [28] and our model.

Span level	Candidate recall (train)				Results				
	M	P	T	Overall	K	M	P	T	Overall
Wang	0.715	0.608	0.334	0.683	0.510	**0.460**	0.399	0.072	0.409
Wang + IEEE	0.715	0.608	0.334	0.683	0.509	0.421	0.372	0.066	0.377
Wang + Unsupervised	0.715	0.608	0.334	0.683	0.501	0.418	0.340	0.059	0.361
Wang + Ngram	0.776	0.646	0.340	0.649	0.499	0.398	0.361	0.056	0.360
Ours without semi	0.814	0.707	0.511	0.717	0.494	0.413	0.357	0.084	0.366
Ours	**0.820**	**0.721**	**0.531**	**0.728**	**0.543**	0.450	**0.417**	**0.104**	**0.414**

Table 3. Comparisons of our model and other data balancing methods

Span level		Results				
		K	M	P	T	Overall
Ours without semi		0.494	0.413	0.357	0.084	0.366
+ Undersampling	AllKNN [24]	0.507	0.422	0.357	0.067	0.371
	TomekLinks [25]	0.485	0.408	0.348	0.101	0.359
+ Oversampling	SMOTE [4]	0.520	0.420	0.364	**0.115**	0.373
+ Combined Resampling	SMOTEENN [2]	0.510	0.414	0.366	0.065	0.371
Ours		**0.543**	**0.450**	**0.417**	0.104	**0.414**

Table 3 shows the results of applying other data balancing methods before the ranking unit. Here we remove the semi-supervised architecture and use other methods to balance data before the random forest model for keyphrase ranking. We can see that our semi-supervised architecture works better than other data balancing algorithms that are based on resampling in the feature space.

6   Conclusion

In this paper we propose a system for the task of scientific keyphrase extraction which has a wide range of OOV words and keyphrase patterns. To improve the pipeline of candidate extraction and keyphrase ranking, we design a semi-supervised data augmentation method to identify more reliable keyphrase candidates to ensure the final extraction performance. We also use abundant features including character level features, linguistic features, context features and external knowledge features to relieve the OOV problem.

Acknowledgement. We thank the anonymous reviewers for their insightful comments on this paper. This work was partially supported by National Natural Science Foundation of China (61572049 and 61273278).

References

1. Ammar, W., Peters, M.E., Bhagavatula, C., Power, R.: The AI2 system at SemEval-2017 Task 10 (ScienceIE): semi-supervised end-to-end entity and relation extraction. In: Proceedings of the 11th International Workshop on Semantic Evaluation, SemEval@ACL 2017, 3–4 August 2017, Vancouver, Canada, pp. 592–596 (2017)
2. Batista, G.E., Bazzan, A.L., Monard, M.C.: Balancing training data for automated annotation of keywords: a case study. In: WOB, pp. 10–18 (2003)
3. Bird, S., Loper, E.: NLTK: the natural language toolkit. In: Proceedings of the ACL 2004 on Interactive Poster and Demonstration Sessions, p. 31. Association for Computational Linguistics (2004)
4. Chawla, N.V., Bowyer, K.W., Hall, L.O., Kegelmeyer, W.P.: Smote: synthetic minority over-sampling technique. J. Artif. Intell. Res. **16**, 321–357 (2002)
5. Chiu, J.P., Nichols, E.: Named entity recognition with bidirectional LSTM-CNNs. Trans. Assoc. Comput. Linguist. **4**, 357–370 (2016)
6. Danesh, S., Sumner, T., Martin, J.H.: Sgrank: combining statistical and graphical methods to improve the state of the art in unsupervised keyphrase extraction. In: Proceedings of the Fourth Joint Conference on Lexical and Computational Semantics, *SEM 2015, 4–5 June 2015, Denver, Colorado, USA, pp. 117–126 (2015)
7. Dong, L., Mallinson, J., Reddy, S., Lapata, M.: Learning to paraphrase for question answering. In: Proceedings of the 2017 Conference on Empirical Methods in Natural Language Processing, EMNLP 2017, 9–11 September 2017, Copenhagen, Denmark, pp. 875–886 (2017)
8. Hasan, K.S., Ng, V.: Automatic keyphrase extraction: a survey of the state of the art. In: Proceedings of the 52nd Annual Meeting of the Association for Computational Linguistics (vol. 1: Long Papers), pp. 1262–1273 (2014)
9. Hosseini, H., Kannan, S., Zhang, B., Poovendran, R.: Deceiving Google's perspective API built for detecting toxic comments. arXiv preprint arXiv:1702.08138 (2017)
10. Huang, Z., Xu, W., Yu, K.: Bidirectional LSTM-CRF models for sequence tagging. arXiv preprint arXiv:1508.01991 (2015)
11. Jia, R., Liang, P.: Adversarial examples for evaluating reading comprehension systems. In: Proceedings of the 2017 Conference on Empirical Methods in Natural Language Processing, EMNLP 2017, 9–11 September 2017, Copenhagen, Denmark, pp. 2021–2031 (2017)
12. Kingma, D.P., Ba, J.: Adam: a method for stochastic optimization. CoRR abs/1412.6980 (2014)
13. Liu, Z., Huang, W., Zheng, Y., Sun, M.: Automatic keyphrase extraction via topic decomposition. In: Conference on Empirical Methods in Natural Language Processing, pp. 366–376 (2010)

14. Liu, Z., Li, P., Zheng, Y., Sun, M.: Clustering to find exemplar terms for keyphrase extraction. In: Conference on Empirical Methods in Natural Language Processing, pp. 257–266 (2009)
15. Lopez, P., Romary, L.: HUMB: automatic key term extraction from scientific articles in grobid. In: Proceedings of the 5th International Workshop on Semantic Evaluation, SemEval 2010, pp. 248–251. Association for Computational Linguistics, Stroudsburg (2010)
16. Luan, Y., Ostendorf, M., Hajishirzi, H.: Scientific information extraction with semi-supervised neural tagging. In: Proceedings of the 2017 Conference on Empirical Methods in Natural Language Processing, EMNLP 2017, 9–11 September 2017, Copenhagen, Denmark, pp. 2641–2651 (2017)
17. Ma, X., Hovy, E.H.: End-to-end sequence labeling via bi-directional LSTM-CNNs-CRF. In: Proceedings of the 54th Annual Meeting of the Association for Computational Linguistics, ACL 2016, 7–12 August 2016, Berlin, Germany, vol. 1: Long Papers (2016)
18. Mihalcea, R., Tarau, P.: TextRank: bringing order into text. In: Proceedings of the 2004 Conference on Empirical Methods in Natural Language Processing, EMNLP 2004, A meeting of SIGDAT, a Special Interest Group of the ACL, held in conjunction with ACL 2004, 25–26 July 2004, Barcelona, Spain, pp. 404–411 (2004)
19. Paszke, A., et al.: Automatic differentiation in PyTorch. In: NIPS-W (2017)
20. Pedregosa, F., et al.: Scikit-learn: machine learning in python. J. Mach. Learn. Res. **12**(Oct), 2825–2830 (2011)
21. Pennington, J., Socher, R., Manning, C.D.: Glove: global vectors for word representation. In: Proceedings of the 2014 Conference on Empirical Methods in Natural Language Processing, EMNLP 2014, 25–29 October 2014, Doha, Qatar, A meeting of SIGDAT, a Special Interest Group of the ACL, pp. 1532–1543 (2014)
22. Samanta, S., Mehta, S.: Towards crafting text adversarial samples. arXiv preprint arXiv:1707.02812 (2017)
23. Schwenker, F.: Ensemble methods: foundations and algorithms [book review]. IEEE Comput. Int. Mag. **8**(1), 77–79 (2013)
24. Tomek, I.: An experiment with the edited nearest-neighbor rule. IEEE Trans. Syst. Man Cybern. **SMC–6**(6), 448–452 (1976)
25. Tomek, I.: Two modifications of CNN. IEEE Trans. Syst. Man Cybern. **SMC–6**(11), 769–772 (1976)
26. Wang, C., Li, S.: CoRankBayes: Bayesian learning to rank under the co-training framework and its application in keyphrase extraction. In: Proceedings of the 20th ACM International Conference on Information and Knowledge Management, CIKM 2011, pp. 2241–2244. ACM, New York (2011)
27. Wang, C., Li, S., Wang, W.: Experiment research on feature selection and learning method in keyphrase extraction. In: Li, W., Mollá-Aliod, D. (eds.) ICCPOL 2009. LNCS (LNAI), vol. 5459, pp. 305–312. Springer, Heidelberg (2009). https://doi.org/10.1007/978-3-642-00831-3_29
28. Wang, L., Li, S.: PKU_ICL at SemEval-2017 Task 10: Keyphrase extraction with model ensemble and external knowledge. In: Proceedings of the 11th International Workshop on Semantic Evaluation (SemEval-2017), pp. 934–937 (2017)
29. Yasunaga, M., Kasai, J., Radev, D.: Robust multilingual part-of-speech tagging via adversarial training. In: Proceedings of the 2018 Conference of the North American Chapter of the Association for Computational Linguistics: Human Language Technologies, vol. 1 (Long Papers), pp. 976–986. Association for Computational Linguistics (2018)

Linguistic Resource Annotation and Evaluation

Using a Chinese Lexicon to Learn Sense Embeddings and Measure Semantic Similarity

Zhuo Zhen and Yuquan Chen[✉]

Department of Computer Science and Engineering, Shanghai Jiao Tong University,
Shanghai 200240, People's Republic of China
HHee@sjtu.edu.cn, chen-yq@cs.sjtu.edu.cn

Abstract. Word embeddings have recently been widely used to model words in Natural Language Processing (NLP) tasks including semantic similarity measurement. However, word embeddings are not able to capture polysemy, because a polysemous word is represented by a single vector. To address this problem, learning multiple embedding vectors for different senses of a word is necessary and intuitive. We present a novel approach based on a Chinese lexicon to learn sense embeddings. Every sense is represented by a vector that consists of semantic contributions made by senses explaining it. To make full use of the lexicon's advantages and address its drawbacks, we perform representation expansion to make sparse embedding vectors dense and disambiguate in gloss polysemous words by semantic contribution allocation. Thanks to the use of an intuitive way of noise filtering, we achieve noticeable improvement both in dimensionality reduction and semantic similarity measurement. We perform experiments on a translated version of Miller-Charles dataset and report state-of-the-art performance on semantic similarity measurement. We also apply our approach to SemEval-2012 Task4: Evaluating Chinese Word Similarity, which uses a translated version of wordsim353 as the standard dataset, and our approach also noticeably outperforms conventional approaches.

1 Introduction

To date, word embeddings [1] are widely used in NLP tasks including semantic similarity measurement and are proved to be effective. Most word embeddings, such as word2vec [2] and GloVe [3], are based on the distributional model that leverages neural networks to model expected contexts of words. However, word embeddings suffer from two major limitations. The first is that word embeddings are unable to capture polysemy since every word is represented by a single vector. The second is that most word embeddings are based on distributional statistics of corpora, which have no connection to semantic inventories, making it hard to learn reliable and accurate representations for infrequent words.

© Springer Nature Switzerland AG 2018
M. Sun et al. (Eds.): CCL 2018/NLP-NABD 2018, LNAI 11221, pp. 197–208, 2018.
https://doi.org/10.1007/978-3-030-01716-3_17

There are many research works seeking to address these problems by learning sense embeddings. [4–6] decompose word embeddings into multiple prototypes corresponding to distinct meanings of words. The main drawback of these approaches is that they have no connection to sense inventories. Thus they do not know which words are polysemous, how many senses they have. Besides, the mapping from sense embeddings to any sense inventory has to be conducted manually. [7,8] construct synsets using semantic resources like WordNet [9] and Wikipedia first and then leverage them to decompose word embeddings. Word Sense Disambiguation(WSD) algorithms or sense clustering algorithms are adopted to construct synsets. So the unsatisfactory accuracy and coverage of WSD algorithms and sense clustering algorithms become the bottlenecks. Instead of decomposing word embeddings, [10,11] learn sense embeddings from sense annotated corpus. Because there is no proper dataset with manual sense annotation, they utilize WSD algorithms or sense clustering algorithms to generate sense annotated corpus first. So they suffer from the non-optimal WSD algorithms and sense clustering algorithms too. Bilingual resources are leveraged by [12,13] for sense embedding learning. They hold an assumption that the translation of words and their contexts can at least partially be helpful for polysemy resolving in the original language. Establishing a connection to some sense inventory is hard for them too.

With the shortage of proper datasets with manual sense annotation, we realize lexicons have several advantages for sense embedding learning. First, lexicons are redacted by linguistic experts and are semantically authoritative. Besides, we can directly specify polysemous words' senses. What's more, infrequent words and senses are also explained clearly, which would significantly help to improve coverage and accuracy for embedding learning of infrequent senses.

In this article, we propose a novel approach based on a Chinese lexicon to learn sense embeddings. In Chinese, the combination of words usually accompanies with semantic coupling. So we assume words in glosses have semantic contributions to the words explained. Thus a word can be represented by the semantic contributions made by words explaining it. Similarly, a sense can be represented by a vector that consists of semantic contributions made by senses explaining it too. But there are some problems to be addressed if we want to use a lexicon as the only semantic resource. The first is that some glosses are so short that the embedding vectors learned are very sparse. The second is that we can not directly distinguish what a polysemous word in some gloss actually means. The third is that the amount of senses is large making learned embedding vectors high dimensional.

To address the problem caused by short glosses, we perform representation expansion to make sparse embedding vectors dense. To guarantee and accelerate convergence, we use an attenuation parameter to control the process. Then, instead of telling what a polysemous word in some gloss actually means, we assign each of its senses a weight using softmax algorithm and allocate its semantic contributions to its senses, which we call semantic contribution allocation. Instead of matrix calculation, we perform dimensionality reduction in a more intuitive way. If a sense makes few contributions to other senses, we treat its contributions as noise and filter them out. So vector dimension is reduced during noise filtering. Experimental results report that executing representation expansion and semantic contribution allocation in sequence iteratively improves the performance of similarity measurement rapidly. And a proper threshold for noise filtering is significantly helpful to performance improvement and dimensionality reduction.

Our approach has following advantages. The first is that a lexicon is the only semantic resource. Thus no more manual annotation effort is needed. The second is that it can learn accurate representations for infrequent senses. The third is that there are few parameters to tune with low training cost. The final one is the intuitive way of noise filtering successfully balances dimensionality reduction and performance improvement.

2   Algorithm

As introduced in the previous section, we assume that a word explaining another word in gloss has a semantic contribution to the one explained. So a word can be represented by the semantic contributions made by words explaining it. Similarly, a sense can be represented by a vector that consists of semantic contributions made by senses in sense gloss. Here we use a matrix $M \in R^{N_{\text{sense}} \times N_{\text{sense}}}$ to denote all the semantic contributions, where N_{sense} is the number of sense gloss in the lexicon. And we want to learn an accurate estimation of all the entries in M. More specifically, we set $M_{i,j}$ as the semantic contribution to the i_{th} sense made by the j_{th} sense. Then M^i, the i_{th} row of the matrix, is a vector consists of semantic contributions to the i_{th} sense made by all senses, which is the representation of the i_{th} sense. If we observe the matrix in another perspective, we can find M_j, the j_{th} column of the matrix, consists of contributions made by the j_{th} sense to all senses.

A lexicon has several advantages for sense embedding learning but it also has some drawbacks. The first is that some glosses are so short that sense embedding vectors learned are very sparse. Besides, we can not directly tell what a polysemous word in some gloss actually means. What's more, the large amount of sense makes embedding vectors high dimensional.

To address the problem of sparse embeddings caused by short glosses, we do representation expansion to make sparse vectors dense. And instead of trying to figure out what a polysemous word actually means, we assign each of its senses a weight using softmax algorithm and allocate its semantic contributions to its senses, which we call semantic contribution allocation. It needs to be pointed out that every time M is modified by initialization or representation expansion, semantic contribution allocation should be carried out immediately. This is because changes in vectors induce changes in sense weight assignment. So we try to learn more accurate representations by executing representation expansion and semantic contribution allocation in sequence iteratively. Finally, dimensionality reduction is realized by filtering out the semantic contributions regarded as noise. The overview of the entire algorithm is shown in Algorithm 1.

Algorithm 1. Overview

1: Initialization
2: Semantic contribution allocation
3: **while** (consistency with standard dataset increases):
4:　　　Representation expansion
5:　　　Semantic contribution allocation
6: Dimensionality reduction

2.1 Initialization

We initialize M with zeroes at first. Then we scan sense glosses in the lexicon and compute initial semantic contributions made by words. Here we use $wordContribution_{i,k}$ to denote the semantic contribution to the i_{th} sense made by the k_{th} word. It is calculated like tf-idf [14]. Here $tf(sense_i, word_k)$ denotes the term frequency of the k_{th} word in the i_{th} sense's gloss and $ef(word_k)$ denotes the number of glosses that the k_{th} word appears in.

$$wordContribution_{i,k} = (1 + \log tf(sense_i, word_k)) \times \log \frac{N_{\text{sense}}}{ef(word_k)} \quad (1)$$

According to the principle of maximum entropy, we allocate a polysemous word's semantic contributions to its senses equally. Here N_{word_k} is the number of senses of the k_{th} word. For simplicity and without loss of generality, we suppose $sense_j \in word_k$, which means the j_{th} sense is from the k_{th} word.

$$M_{i,j} = \frac{wordContribution_{i,k}}{N_{word_k}} \quad (2)$$

2.2 Semantic Contribution Allocation

Since we can not directly tell what a polysemous word in some gloss actually means, we choose to allocate its semantic contributions to its senses. Then what we should consider is weight assignment. We assume that if one sense is semantically closer to the explained sense, it deserves higher weight. Here we use $weight(i,j)$ to denote the weight of the j_{th} sense explaining the i_{th} sense. We try to measure the semantic distance in two different ways. $distanceE(i,j)$ is the Euclidean distance between the i_{th} sense and j_{th} sense, i.e. M^i and M^j. And $distanceC(i,j)$ is the cosine of the angle of M^i and M^j. For simplicity and without loss of generality, we suppose $sense_j \in word_k$.

When using $distanceE(i,j)$ to measure the semantic distance, we assign weights in the following way. To make sure the weights lie in a reasonable interval, we use μ to control the process.

$$weight(i,j) = \frac{e^{-\frac{distanceE(i,j)^2}{\mu}}}{\sum_{sense_l \in word_k} e^{-\frac{distanceE(i,l)^2}{\mu}}} \quad (3)$$

When using $distanceC(i,j)$ to measure distance, the weights are assigned in a different way. And we use ϵ for smoothing.

$$weight(i,j) = \frac{e^{distanceC(i,j)} - 1 + \epsilon}{\sum_{sense_l \in word_k}(e^{distanceC(i,l)} - 1 + \epsilon)} \tag{4}$$

After the weights are determined, senses' semantic contributions are updated in the following way.

$$M_{i,j}^{\text{new}} = (\sum_{sense_l \in word_k} M_{i,l}) \times weight(i,j) \tag{5}$$

2.3 Representation Expansion

Suppose we have learned embedding vectors for all senses in the lexicon but they are sparse. How can we make them dense? Since every sense is represented by a vector that consists of semantic contributions, we can expand a sense's sparse embedding vector by adding a weighted sum of learned embedding vectors. An intuitive explanation is that if sense B has a direct semantic contribution to sense A, then senses that have direct semantic contributions to sense B have indirect contributions to sense A. And it can be implemented by adding embedding vector of sense B with a weight to the one of sense A.

Intuitively, a sense's semantic contribution plays a key role in representation expansion. For example, suppose the embedding vector of sense A is sparse, and the entries in it corresponding to sense B and sense C are positive, which means sense B and sense C have semantic contributions to sense A. If sense B has a larger semantic contribution than sense C, the vector of sense B deserves a larger weight than the one of sense C in the expansion.

Besides, we assume that a sense's embedding vector is not semantically equal to the sense itself. While we are expanding representations, we are losing semantic accuracy. And the more times we do representation expansion, the more accuracy we lose. Here n denotes how many times we do representation expansion and η is the attenuation parameter to denote accuracy loss. We express their relationship in the following way, while α is the initial value we need to figure out.

$$\eta(n) = \alpha^n \tag{6}$$

Then the expansion of a sparse embedding vector is realized by adding all senses' embedding vectors multiplied by corresponding semantic contributions and η. The process can be described as follows, where M_{new}^i is the updated representation of the i_{th} sense.

$$M_{\text{new}}^i = M^i + \eta \times \sum_{j=1}^{N_{\text{sense}}} M_{i,j} \cdot M^j \tag{7}$$

After integration, we can directly update M ,where M_{new} denotes the updated M.

$$M_{\text{new}} = M + \eta \times M^2 \tag{8}$$

2.4 Dimensionality Reduction

As illustrated above, M_j is the j_{th} column of M and consists of the semantic contributions to all senses made by the j_{th} sense. Here we use $contribution(i)$ to denote the sum of semantic contributions made by the i_{th} sense.

$$contribution(i) = \sum_{j=1}^{N_{sense}} M_{j,i} \qquad (9)$$

We assume if $contribution(i)$ is very small, it is very likely to be noise and has a negative effect on semantic similarity measurement. More specifically, if $contribution(i)$ is small, the i_{th} column would be removed from M. Thus we set a threshold to filter noise out, which at the same time reduces the dimension of embedding vectors.

3 Experiment and Evaluation

In this section, we first perform experiments on a translated version of Mill-Charles dataset [15]. With attenuation parameter set to 0.7, which is randomly chosen, we try to figure out whether a stop word list is helpful to semantic similarity measurement and using which distance metric is better for semantic contribution allocation. Then, we try to find out the relationship between the attenuation parameter and the correlation coefficient with the dataset. With an ideal attenuation parameter, we try noise filtering, which is also a way of dimensionality reduction, and we want to find a threshold that balances dimensionality reduction and semantic similarity measurement performance. After all these are finished, we can determine attenuation parameter for representation expansion, distance metric for semantic contribution allocation and the threshold for dimensionality reduction. Then, we compare with some conventional approaches. Finally, we apply our approach to SemEval-2012 Task4, Evaluating Chinese Word Similarity, for another evaluation.

3.1 Preparation

We use Modern Chinese Dictionary [16] as our corpus. Before applying it to sense embedding learning, some preprocessing is necessary. First, we filter out meaningless symbols like phonetic notations. Then we filter out some sentences that redirect pages. Finally, we filter out some descriptions about pragmatics.

The cosine of the angle of two embeddings is computed as semantic similarity. When we compute the semantic similarity between two words, especially polysemous words, we choose one sense from each word and try all possible combinations. For example, the similarity between the m_{th} word and the n_{th} word, $wordSimilarity(m, n)$, is computed as follows.

$$wordSimilarity(m, n) = \max_{sense_i \in word_m, sense_j \in word_n} \frac{M^i \cdot M^j}{||M^i|| \times ||M^j||} \qquad (10)$$

We first perform experiments on a dataset based on Miller and Charles dataset [17]. Because the original dataset is in English, we choose a translated version [15]. Some words that are not written as an entry in the lexicon are removed. The correlation between the dataset and our result is evaluated by Pearson correlation coefficient.

We then switch to the standard dataset of SemEval-2012 Task4: Evaluating Chinese Word Similarity, which is a translated version of wordsim353 [18]. And the task uses Kendall correlation coefficient as the evaluation metric.

3.2   Experiment and Result

Whether to Use a Stop Word List and Which Distance Metric Is Better. Setting attenuation parameter to 0.7, we perform semantic contribution allocation based on the two distance metrics mentioned above, with or without a stop word list. As is shown in Table 1, results show us that an improper stop word list really does harm to semantic similarity measurement. And using cosine to allocate semantic contribution is better. To achieve the best performance, we should allocate semantic contribution using cosine as distance metric without a stop word list.

Table 1. Correlation under different conditions

With Stop Word List or Not	Yes	Not
Euclidean	0.7690	0.8170
Cosine	0.8005	0.8227

Relationship Between Attenuation Parameter and Correlation. Without a stop word list, we try to figure out the relationship between the attenuation parameter and correlation with the dataset. Allocating semantic contribution based on cosine, we get the curve shown in Fig. 1. Bigger attenuation parameter brings better performance. But as the attenuation parameter increases, the curve tends to be flat. So if we go on increasing attenuation parameter, the correlation would only increase slightly. To achieve the best performance, we fix the attenuation parameter to 0.9 for following experiments and evaluations.

Dimensionality Reduction Threshold's Effect on Correlation. As illustrated in previous sections, we treat a column of M as the semantic contributions made by corresponding sense. And we assume if a sense makes very few semantic contributions, its contributions are very likely to be noise and should be removed. It may be helpful to both dimensionality reduction and performance improvement. Then we need to figure out a proper threshold. In this part, we try different thresholds and want to figure out the relationship between threshold and correlation. And before dimensionality reduction, the highest correlation we achieve is 82.55%.

First, we set the threshold small and increase it slowly. We find that correlation coefficient fluctuates in a very narrow interval and there is no significant change. The result is shown in Fig. 2. Then we keep on increasing the threshold. And we find that although the correlation coefficient is fluctuating, it is obviously higher than the former best result. As is shown in Fig. 3, when the threshold is set to 11000, the correlation coefficient reaches its peak to 90.52%. Keeping on increasing the threshold, we find that larger threshold makes larger information loss. And the correlation declines rapidly, as is shown in Fig. 4.

Setting the threshold to 11000, we want to check out how many dimensions are filtered out. Result reports that 89.66% of dimensions are ignored. So the intuitive noise filtering method filters out nearly 90% of features and brings nearly 8% performance improvement from 82.55% to 90.52%.

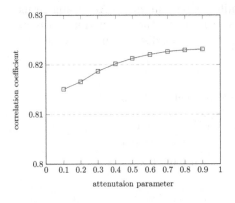 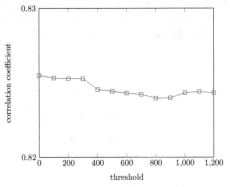

Fig. 1. When the attenuation parameter increases, the correlation coefficient increases but the curve tends to be flat.

Fig. 2. When a small threshold increases slowly, the correlation coefficient fluctuates in a very narrow interval. This means a small threshold has no significant effect on correlation coefficient.

3.3   Evaluation

Evaluation on Translated Miller-Charles Dataset. With the attenuation parameter set to 0.9, we use cosine to allocate semantic contribution and do not use a stop word list. After dimensionality reduction, the Person correlation coefficient we get is 0.9052. We compared this result with some conventional approaches. *How* leverages the taxonomy and attribute knowledge from a Chinese knowledge base, HowNet, which is similar to WordNet, to calculate semantic similarity [19]. *Dc* utilizes snippets of query results from Google for similarity measurement. If a word appears in the snippets of the query of another word and

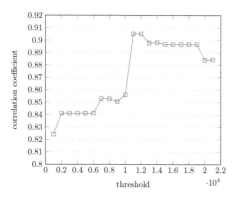

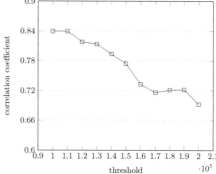

Fig. 3. Using a moderate threshold for dimensionality reduction, the correlation coefficient obviously exceeds the former best one.

Fig. 4. When a large threshold increases, the correlation coefficient declines rapidly. This is because the threshold is so large that it induces serious information loss.

vice versa, these two words are semantically similar [20]. *Attr* extracts attribute sets from the Internet and measures semantic similarity by measuring the overlapping extent of two words' attribute sets [15]. *word2vec* is the famous word embedding algorithm and we leverage its Skip-Gram model for this evaluation [2]. *Dic* learns word embeddings using Modern Chinese Standardized Dictionary as the corpus [21]. *Multi-sense* relies on both monolingual and bilingual information to learn sense embeddings. This model consists of an encoder to choose senses for given words, and a decoder to predict expected context words for chosen senses. When the autoencoder's training is finished, sense embeddings are learned [12]. *We* is our result. As is shown in Table 2, our approach outperforms others.

Table 2. Comparison with some conventional approaches on translated M&C dataset

Approach	Dc	How	Attr	word2vec	Dic	Multi-sense	We
Pearson	0.5027	0.6394	0.7169	0.7876	0.7922	0.8049	**0.9052**

Evaluation on Translated WordSim353 Dataset. After experiments on the M&C dataset, we carry out an evaluation on a translated version of wordsim353 [18], which is the standard for SemEval-2012 Task4: Evaluating Chinese Word Similarity [22]. *We* is our novel approach. *word2vec* represents word embeddings learned by word2vec with Skip-Gram model [2]. *Multi-sense* learns sense embeddings by an autoencoder based on both monolingual and bilingual information [12]. *MIXCC*, *MIXCD*, *GUO-ngram* and GUO-words are systems that participated in the task [22]. As is shown in Table 3, our novel approach outperforms other approaches.

Table 3. Comparison with some conventional approaches on translated wordsim353 dataset

Approach	Guo-words	Guo-ngram	MIXCD	MIXCC	Multi-sense	word2vec	We
Kendall	-0.011	0.007	0.040	0.050	0.130	0.295	**0.305**

4 Related Work

Decomposing word embeddings into multiple prototypes corresponding to distinct meanings of words is conducted by [4–6]. These methods have no connection to sense inventories. Thus they can not specify which words are polysemous and how many senses they have. And the mapping from sense embeddings to some sense inventory has to be carried out manually. [7] tries to combine word embeddings and synsets constructed from WordNet [9] to learn more accurate sense embeddings. [8] uses WordNet and Wikipedia as semantic resources. On the one hand, it extracts contextual information to learn word-based representations for words, on the other hand, it constructs synsets to learn synset-based representations for senses. Both these two kinds of representations are utilized to measure semantic similarity. [23] learns word embeddings first and initializes a polysemous word's sense embeddings by averaging semantically similar words in sense glosses. Then it uses two WSD algorithms to obtain senses' relevant occurrences that are used to learn more accurate sense embeddings. There are still some researchers using the sense-annotated corpus to exploit semantic knowledge. But they have to generate a sense-annotated corpus first. [10] uses a WSD system to annotate corpus while [11] chooses a sense clustering algorithm, i.e. k-means, for polysemous word annotation. However, the non-optimal WSD techniques and sense clustering algorithms seriously limit the annotation accuracy, which makes it hard to learn accurate and high-coverage sense embeddings. Bilingual resources are also leveraged with the assumption that "polysemy in one language can be at least partially resolved by looking at the translation of the word and its context in another language" [12,13]. [24] proposes a probabilistic model for sense embedding. The model takes the dependency between sense choices of neighboring words into account. Then it uses an algorithm similar to hard-EM to optimize a max-margin objective. It uses a knowledge base to help model training, but the knowledge base is only used to determine the numbers of polysemous words' senses, which makes it hard to map from sense embeddings to sense inventories.

5 Conclusion

In this article, we present a novel method that learns sense embeddings using a Chinese lexicon as the only corpus. Using a lexicon to learn sense embeddings has several advantages. A lexicon is redacted by linguistic experts and is semantically authoritative. It specifies polysemous words' senses and is significantly helpful to learn embeddings for infrequent senses. But it also has some drawbacks. Short glosses tend to make sense embeddings vectors sparse, polysemous words in sense glosses induce semantic ambiguity and the sense embedding vectors are high dimensional before dimensionality reduction. First, to address the problem caused by short glosses, we conduct representation expansion to make sparse vectors dense and use an attenuation parameter to accelerate convergence. Then, instead of determining what a polysemous word in some gloss actually means, we assign each of its senses a possibility by softmax algorithm. Finally, if a sense makes few semantic contributions to other senses, we treat its contributions as noise and filter them out. Filtering out noise by a proper threshold is significantly helpful to dimensionality reduction and semantic similarity measurement.

Our novel method has the following advantages. A Chinese lexicon is all we need and no annotation effort is needed. The coverage and accuracy of infrequent sense embeddings are guaranteed. Few parameters need to be tuned and the learning cost is low. The intuitive and efficient way of noise filtering is remarkably helpful to both dimensionality reduction and semantic similarity measurement.

We carry out an evaluation on a translated version of Miller-Charles dataset and the Pearson correlation coefficient is 0.9052, which is state-of-the-art. We also apply our method to SemEval-2012 Task 4: Evaluating Chinese Word Similarity, and report noticeably better results compared with conventional approaches.

References

1. Bengio, Y., Ducharme, R., Vincent, P., Jauvin, C.: A neural probabilistic language model. J. Mach. Learn. Res. **3**(Feb), 1137–1155 (2003)
2. Mikolov, T., Chen, K., Corrado, G., Dean, J.: Efficient estimation of word representations in vector space. arXiv preprint arXiv:1301.3781 (2013)
3. Pennington, J., Socher, R., Manning, C.: Glove: global vectors for word representation. In: Proceedings of the 2014 Conference on Empirical Methods in Natural Language Processing (EMNLP), pp. 1532–1543 (2014)
4. Reisinger, J., Mooney, R.J.: Multi-prototype vector-space models of word meaning. In: Human Language Technologies: The 2010 Annual Conference of the North American Chapter of the Association for Computational Linguistics, pp. 109–117. Association for Computational Linguistics (2010)
5. Huang, E.H., Socher, R., Manning, C.D., Ng, A.Y.: Improving word representations via global context and multiple word prototypes. In: Proceedings of the 50th Annual Meeting of the Association for Computational Linguistics: Long Papers, vol. 1, pp. 873–882. Association for Computational Linguistics (2012)
6. Tian, F., et al.: A probabilistic model for learning multi-prototype word embeddings. In: Proceedings of COLING 2014, the 25th International Conference on Computational Linguistics: Technical Papers, pp. 151–160 (2014)
7. Pilehvar, M.T., Collier, N.: De-conflated semantic representations. arXiv preprint arXiv:1608.01961 (2016)
8. Camacho-Collados, J., Pilehvar, M.T., Navigli, R.: Nasari: a novel approach to a semantically-aware representation of items. In: Proceedings of the 2015 Conference of the North American Chapter of the Association for Computational Linguistics: Human Language Technologies, pp. 567–577 (2015)

9. Miller, G.A.: Wordnet: a lexical database for English. Commun. ACM **38**(11), 39–41 (1995)
10. Iacobacci, I., Pilehvar, M.T., Navigli, R.: Sensembed: learning sense embeddings for word and relational similarity. In: Proceedings of the 53rd Annual Meeting of the Association for Computational Linguistics and the 7th International Joint Conference on Natural Language Processing (Volume 1: Long Papers), vol. 1, pp. 95–105 (2015)
11. Zhou, H., Jia, C., Yang, Y., Ning, S., Lin, Y., Huang, D.: Combining large-scale unlabeled corpus and lexicon for Chinese polysemous word similarity computation. In: Wen, J., Nie, J., Ruan, T., Liu, Y., Qian, T. (eds.) CCIR 2017. LNCS, vol. 10390, pp. 198–210. Springer, Cham (2017). https://doi.org/10.1007/978-3-319-68699-8_16
12. Šuster, S., Titov, I., van Noord, G.: Bilingual learning of multi-sense embeddings with discrete autoencoders. arXiv preprint arXiv:1603.09128 (2016)
13. Guo, J., Che, W., Wang, H., Liu, T.: Learning sense-specific word embeddings by exploiting bilingual resources. In: Proceedings of COLING 2014, the 25th International Conference on Computational Linguistics: Technical Papers, pp. 497–507 (2014)
14. Baeza-Yates, R., Ribeiro-Neto, B.: Modern Information Retrieval, vol. 463. ACM Press, New York (1999)
15. Zhao, J., Liu, H., Lu, R.: Attribute-base computing of word similarity. In: The 11th China Conference on Machine Learning (2008)
16. Lv, S., Ding, S.: Chinese Modern Dictionary. The Commercial Press, Beijing (2005)
17. Miller, G.A., Charles, W.G.: Contextual correlates of semantic similarity. Lang. Cogn. Process. **6**(1), 1–28 (1991)
18. Evgeniy, L.F., Finkelstein, L., Gabrilovich, E., Matias, Y., Rivlin, E., Solan, Z., Wolfman, G.: Placing search in context: the concept revisited. ACM Trans. Inf. Syst. **20**(1), 116–131 (2002)
19. Liu, Q.: Word similarity computing based on hownet. Comput. Linguist. Chin. Lang. Process. **7**(2), 59–76 (2002)
20. Chen, H.H., Lin, M.S., Wei, Y.C.: Novel association measures using web search with double checking. In: Proceedings of the 21st International Conference on Computational Linguistics and the 44th Annual Meeting of the Association for Computational Linguistics, pp. 1009–1016. Association for Computational Linguistics (2006)
21. Liu, H., Zhao, J., Lu, R.: Computing semantic similarities based on machine-readable dictionaries. In: IEEE International Workshop on Semantic Computing and Systems, WSCS 2008, pp. 8–14. IEEE (2008)
22. Jin, P., Wu, Y.: Semeval-2012 task 4: evaluating Chinese word similarity. In: Joint Conference on Lexical and Computational Semantics, pp. 374–377 (2012)
23. Chen, X., Liu, Z., Sun, M.: A unified model for word sense representation and disambiguation. In: Proceedings of the 2014 Conference on Empirical Methods in Natural Language Processing (EMNLP), pp. 1025–1035 (2014)
24. Qiu, L., Tu, K., Yu, Y.: Context-dependent sense embedding. In: Proceedings of the 2016 Conference on Empirical Methods in Natural Language Processing, pp. 183–191 (2016)

Revisiting Correlations between Intrinsic and Extrinsic Evaluations of Word Embeddings

Yuanyuan Qiu[1,2], Hongzheng Li[3], Shen Li[1,2], Yingdi Jiang[1,2], Renfen Hu[1,2(✉)], and Lijiao Yang[1,2]

[1] Institute of Chinese Information Processing, Beijing Normal University, Beijing, China
{reesechiu,shen,de,irishere}@mail.bnu.edu.cn, yanglijiao@bnu.edu.cn
[2] UltraPower-BNU Joint Laboratory for Artificial Intelligence, Beijing Normal University, Beijing, China
[3] School of Computer Science and Technology, Beijing Institute of Technology, Beijing, China
lihongzheng@mail.bnu.edu.cn

Abstract. The evaluation of word embeddings has received a considerable amount of attention in recent years, but there have been some debates about whether intrinsic measures can predict the performance of downstream tasks. To investigate this question, this paper presents the first study on the correlation between results of intrinsic evaluation and extrinsic evaluation with Chinese word embeddings. We use word similarity and word analogy as the intrinsic tasks, Named Entity Recognition and Sentiment Classification as the extrinsic tasks. A variety of Chinese word embeddings trained with different corpora and context features are used in the experiments. From the data analysis, we reach some interesting conclusions: there are strong correlations between intrinsic and extrinsic evaluations, and the performance of different tasks can be affected by training corpora and context features to varying degrees.

Keywords: Word embedding · Intrinsic evaluation Extrinsic evaluation

1 Introduction

Word embeddings are proved to be beneficial to various Natural Language Processing (NLP) tasks, such as part-of-speech tagging (POS), chunking, named entity recognition (NER), and syntactic parsing [2,8,25].

With the increasing usage of word embedding, the issue of evaluation becomes important. Current evaluation methods have two major categories: intrinsic and extrinsic. Intrinsic evaluations directly test for syntactic or semantic relationships between words through word similarity or analogical reasoning tasks [12,16,18]. While in extrinsic evaluation, embeddings are exploited as input features for

© Springer Nature Switzerland AG 2018
M. Sun et al. (Eds.): CCL 2018/NLP-NABD 2018, LNAI 11221, pp. 209–221, 2018.
https://doi.org/10.1007/978-3-030-01716-3_18

downstream NLP tasks, and their performance can indirectly reflect the effects of embeddings [1]. However, most intrinsic and extrinsic evaluations are conducted separately and few research studies the correlation between them. Chiu et al. [7] argue that most intrinsic evaluations are poor predictors of downstream tasks performance. In their experiment, they compare the embeddings trained with different window sizes on word similarity task and three extrinsic tasks, while another important intrinsic task word analogy is not considered. Moreover, some effective features such as training corpora and context features have not yet been explored either.

On the other hand, existing discussion of embedding evaluation is mostly about English word embeddings, and there are rich benchmarks in English for both intrinsic and extrinsic evaluation. Although Chinese NLP has grown rapidly in recent years, few attempts have been made in the evaluation of Chinese word embeddings [19].

Based on the above consideration, this paper studies the correlation between intrinsic evaluation and extrinsic evaluation by using 21 Chinese word embeddings trained with different settings. Specifically, we choose word similarity and word analogy as the intrinsic tasks, Named Entity Recognition and Sentiment Classification as the extrinsic tasks. 7 corpora of different sizes and domains are used for training. In addition to the corpus factors, we examine the effectiveness of two important context features during training, i.e. character features and ngram features.

The experimental results demonstrate that both intrinsic and extrinsic performance can be affected by the size and domain of a training corpus, as well as the context features, but influence degrees vary among different tasks. By analyzing the data, we find that there is a consistency between intrinsic and extrinsic evaluations to some extent. Effective features in intrinsic tasks can also improve the performance of extrinsic tasks, but each task may have a preference of specific features. For example, domain-specific corpora have a distinct advantage for extrinsic tasks, and character features are particularly favorable to intrinsic tasks, e.g. analogical reasoning on morphological relations. Thus, this study can not only offer greater and deeper insight on training and evaluating word embeddings, but also some practical suggestions on selecting the suitable word embeddings for NLP tasks.

The contributions of this paper can be summarized as follows: first of all, we present a comprehensive study on the correlation between intrinsic and extrinsic evaluation of word embeddings. We find that intrinsic evaluation can serve as a good predictor for downstream tasks, and different tasks may favor different features. Secondly, we build domain-specific NER and sentiment classification datasets, which could serve as extrinsic benchmarks for evaluation of Chinese word embeddings, as well as other NLP models.

The remaining parts of this paper are organized as follows: Sect. 2 discusses the related work. Sections 3 and 4 describe the intrinsic and extrinsic tasks respectively. Section 5 conducts experiments and gives analysis in detail. And we give conclusions in Sect. 6.

2   Related Work

There have been a lot of discussion of the evaluation of word embeddings in recent years. These works study either intrinsic evaluation approaches such as word similarity [4,11] and word analogy [21], or extrinsic tasks such as POS tagging and Name Entity Recognition. Schnabel et al. [23] present a comprehensive study of intrinsic and extrinsic evaluation of embeddings. Ghannay et al. [13] conduct a detailed comparison of different kinds of word embeddings on various NLP tasks. Relevant works can be found in [14,22,28]. As a typical shared task officially proposed in 2002 [24], NER is one of the downstream tasks commonly used to evaluate the embeddings in most works related to extrinsic evaluations.

However, there are still some challenges and debates in the field of embedding evaluation, for example, Schnabel et al. [23] argue that extrinsic evaluation only provides one way to specify the goodness of an embedding, and it is not clear how it connects to other measures. On the other hand, there are not so many works studying the correlation between intrinsic and extrinsic evaluations. One representative work is [7]. They state that most intrinsic evaluations are poor predictors of downstream tasks performance. However, the experiments consider only one factor in the training of word embeddings, i.e. the window size, a hyper-parameter. Moreover, the intrinsic evaluation only includes the word similarity task, which is insufficient because the effectiveness of word similarity task in evaluation has been questioned a lot, for example, human judgment of word similarity is subjective and similarity is often confused with relatedness [3,10].

As for evaluation of Chinese word embedding, related work and datasets are much less than that of English. In Chinese, a word is composed of one or more graphical characters, known as Hanzi, which could encode rich semantic and phonetic information. It has attracted considerable attention to use character relevant features to enhance the word representations [6,20,27]. To evaluate the newly proposed methods, Chen et al. [6] build a small analogy dataset covering 230 unique Chinese words by translating part of an English dataset. Chen and Ma [5] create several evaluation sets for Chinese word embeddings on both word similarity and analogical tasks. Li et al. [19] release a big and balanced dataset CA8 for analogy evaluation, as well as over 100 Chinese word embeddings trained with different corpora and settings.

Based on previous works, this paper will go further into the evaluation of Chinese word embeddings, and study the correlations between intrinsic and extrinsic evaluation with representative tasks and various embeddings.

3   Intrinsic Tasks

In this paper, we propose to evaluate word embeddings with two representative intrinsic tasks: word similarity and word analogy.

3.1   Word Similarity

Word similarity is an attractive and popular task for embedding evaluation because it is computationally inexpensive and fast. In this task, the correlation coefficient between the automatic predicted results with the human labeled similarity scores is computed. This paper uses the Chinese word similarity dataset proposed by Wu and Li [26].

3.2 Word Analogy

Word analogy task, also called analogical reasoning, aims at detecting morphological and semantic relations between words. Specifically, it is to retrieve the answer of the question "a is to b as c is to ?" with vector computation. We adopt the CA8 dataset constructed by Li et al. [19], including both morphological questions and semantic questions. The questions are solved by 3COSMUL [17] objective.

4   Extrinsic Tasks

To evaluate the performance of word representations in downstream tasks, we apply them to name entity recognition and sentiment analysis. In this paper, we build a Financial NER dataset for name entity recognition, and a Book Review dataset for sentiment classification. These datasets and evaluation methods will be released at Github.

4.1   Named Entity Recognition

Named Entity Recognition (NER) is considered as a typical sequence labeling problem. In NER task, we use a hybrid BiLSTM-CRF model to detect three types of entities: Person (PER), Location (LOC) and Organization (ORG) in Chinese financial news. The texts are crawled from multiple financial news websites, including 3000 news articles (30,000 sentences in total). All the entities are manually labeled by four graduate students major in linguistics. As financial news usually involves names of companies, stocks and official agencies, we label these names as ORG in the dataset.

4.2   Sentiment Classification

Convolutional neural networks (CNNs) are effective models for sentiment and text classification. Based on Kim's [15] work, we train a simple but effective CNN model for binary sentiment classification (positive and negative). The dataset contains 40,000 reviews collected from https://book.douban.com/. Each review has a star tag rated by users from one star to five stars. It could be used to build a two-class (positive/negative)[1] classification task.

---

[1] We identify one-star and two-star reviews as negative, four-star and five-star reviews as positive. Reviews with three-star are regarded as neutral comments and thus not considered.

5 Experiments

5.1 Datasets

For intrinsic evaluation, the word similarity dataset includes 500 word pairs covering 716 unique Chinese words, which is a relatively small dataset. CA8, the word analogy dataset including 17,813 questions is a big and balanced dataset for analogical reasoning.

Table 1. Statistics of the Financial NER dataset.

	PER	LOC	ORG	Total
Training	11488	15910	29192	56590
Test	2432	3059	5874	11365
Total	13920	18969	35066	67955
Test/Total	0.1747	0.1613	0.1675	0.1672

For the NER task, We divide the dataset into training set (25000 sentences) and test set (5000 sentences). During the training of RNN, the model will be automatically validated from the training set, so there is no validation set. Table 1 shows the distribution of three types of entities in the datasets.

The data for Sentiment Classification is divided into following three sets: 10% of the 40,000 short texts are used for test, 85% for training, the remaining for validation.

5.2 Pre-trained Word Vectors

We train word embeddings with SGNS (Skip-gram with negative-Sampling) model implemented by ngram2vec toolkit[2]. Table 2 shows the hyper-parameter settings. As shown in Table 3, six large-scale corpora ranging from 1 GB to over 6 GB are used during training, including Chinese Wikipedia, Baidu-baike (an online Chinese encyclopedia), Zhihu (Chinese social QA data), People's Daily news, Sogou News and Financial News. Embedding is also trained after combining the above six corpora.

Like Li et al.'s [19] work, while training embeddings based on each corpus, we consider integrating the n-gram and characters features, which are proved

Table 2. Hyper-parameter settings for training word embeddings.

Window	Iteration	Dimension	Subsampling	Low-frequency threshold	Context distribution smoothing	Negative (SGNS)
5	5	300	1e-5	10	0.75	5

---

[2] https://github.com/zhezhaoa/ngram2vec.

effective in training word representations [29]. Specifically, we use word bigram for n-gram features, character unigram and bigram for character features. As a result, we obtain 21 embeddings for experiments.

Table 3. Seven corpora used for training word embeddings.

	Wikipedia_zh	Zhihu	Sogou News	People's Daily	Baidu-baike	Financial	Combination
Size	1.3G	2.1G	3.7G	3.9G	4.1G	6.2G	21.3G
Token	223M	384M	649M	668M	745M	1055M	4037M
Vocab size	2129K	1117K	1226K	1664K	5422K	2785K	10653K

5.3   Results and Analysis of NER

The left part of Table 5 shows the NER results of different embeddings in ascending order of corpora sizes. We will make analysis of the results from three aspects: context features, size and domain of corpus.

Context Features. As shown in Table 5, the introduction of bigram and character features has brought constant improvement of performance in most scenarios. Besides, bigram features show a more distinct advantage because after integrating it, the F1 score increases in all the cases.

Corpus Size. We can see that the embedding trained with the largest combination corpus always performs best, and best F1 scores in last four groups (from People's Daily to Combination) are increasing continuously with the growing size of corpora.

Fig. 1. Performance of different embeddings in NER task, with the best F1 score of each corpus.

Corpus Domain. If we ignore the results of the combination corpus, the performance of financial embedding achieves the best among all the groups. We speculate that the reason is not only about its size, but also its domain. As the NER dataset is constructed from financial news, the embedding trained with financial domain data should have direct and positive impacts on the recognition results. Figure 1 clearly indicates the contributions of various domains.

In order to further testify the impact of corpus size and domain, we randomly sample two smaller financial corpora from the original one, and re-evaluate their performances. One of the samples is 1.3 GB, as same as the Wikipedia corpus, because we find although Wikipedia is a much smaller corpus than financial news, but its embeddings achieve comparable results with financial data.

As shown in Table 4, financial embeddings of *word* and *word + bigram* features always outperform Wikipedia embeddings even when the size of financial corpus decreases to the same with Wikipedia (1.3 GB). The experimental results prove that both the size and domain of a corpus have important influences on embeddings.

Table 4. Comparison between wikipedia and different sizes of financial embeddings based on NER F1 scores.

	word	word+bigram	word+char
Wiki	0.7955	0.7965	0.8047
Fin._1.3G	0.7977	0.8008	0.8035
Fin._3.5G	0.7991	0.8066	0.8024
Fin._6.2G	0.8023	0.8081	0.8083

5.4   Results and Analysis of Sentiment Classification

The right part of Table 5 shows the results of sentiment classification on Book Review dataset. To keep consistent with NER task, we will also discuss the results from three aspects.

Context Features. It can be seen that character and bigram features are both advantageous to model performance. More importantly, most embeddings integrated with bigram features perform the best among three kinds of features, which is consistent with NER task. Thus these two extrinsic tasks both favors more of bigram features than character features.

Corpus Size. It is quite obvious that corpus size plays an important role in embedding performance. Firstly, Wikipedia and Baidu-baike are both online encyclopedia data, and with a bigger size, Baidu-baike gains much better performance than wikipedia, especially on *word* and *word+bigram* settings. Secondly, similar to NER task, embeddings trained with the combination corpus achieve the highest F1 scores and accuracies, indicating the size of corpus has direct and important impacts on the performance.

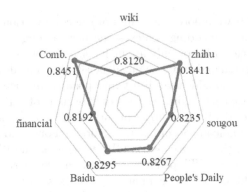

Fig. 2. Performance of different embeddings on SC task, with the best F1 score of each corpus.

Table 5. Results of Named Entity Recognition and Sentiment Classification.

		Name Entity Recognition			Sentiment Classification			
		P	R	F1	P	R	F1	Accuracy
Wiki	word	0.8194	0.7730	0.7955	0.7940	0.7883	0.7851	0.7858
	word+bigram	0.8088	0.7845	0.7965	0.7829	0.7773	0.7742	0.7749
	word+char	0.8323	0.7788	**0.8047**	0.8143	0.8133	**0.8120**	**0.8121**
Zhihu	word	0.8167	0.7722	0.7938	0.8395	0.8359	0.8337	0.8339
	word+bigram	0.8120	0.7815	0.7964	0.8416	0.8409	**0.8411**	**0.8414**
	word+char	0.8287	0.7697	**0.7981**	0.8336	0.8329	0.8317	0.8317
Sogou	word	0.8306	0.7742	0.8014	0.8178	0.8176	0.8167	0.8167
	word+bigram	0.8356	0.7726	0.8028	0.8260	0.8230	**0.8235**	**0.8242**
	word+char	0.8338	0.7750	**0.8033**	0.8216	0.8219	0.8217	0.8217
People's daily	word	0.8267	0.7700	0.7974	0.8278	0.8267	0.8254	0.8255
	word+bigram	0.8192	0.7773	**0.7977**	0.8274	0.8274	**0.8267**	**0.8267**
	word+char	0.8311	0.7612	0.7946	0.8240	0.8240	0.8233	0.8233
Baidu-baike	word	0.8335	0.7714	0.8013	0.8288	0.8279	0.8267	0.8267
	word+bigram	0.8216	0.7872	**0.8040**	0.8275	0.8274	0.8275	0.8277
	word+char	0.8273	0.7691	0.7972	0.8308	0.8305	**0.8295**	**0.8295**
Financial	word	0.8344	0.7727	0.8023	0.8152	0.8137	0.8140	0.8145
	word+bigram	0.8260	0.7910	0.8081	0.8192	0.8195	**0.8192**	**0.8192**
	word+char	0.8511	0.7697	**0.8083**	0.8152	0.8147	0.8136	0.8136
Comb.	word	0.8383	0.7795	0.8078	0.8474	0.8462	0.8448	0.8448
	word+bigram	0.8374	0.7973	**0.8169**	0.8459	0.8459	**0.8451**	**0.8451**
	word+char	0.8433	0.7851	0.8131	0.8400	0.8401	0.8400	0.8402

Corpus Domain. It can be seen that Zhihu data has a clear advantage on sentiment classification, which means the corpus domain may play more important roles on the performance of this task than that of NER. This conclusion can be reflected in at least two comparisons: (1) although Zhihu is the second smallest corpus among the corpora, the embeddings trained with zhihu data achieve significant improvements over other embeddings, e.g. nearly 7% higher than Wikipedia, and 2%-3% higher than the other news or encyclopedia corpora which are much larger than Zhihu. (2) The combination corpus is 10 times larger than Zhihu, but their results are almost the same (best F1: 0.8451 vs 0.8411, best accuracy: 0.8451 vs 0.8414).

Figure 2 clearly shows the contribution of Zhihu. One possible reason is Zhihu data is collected from a social QA website, and the Book Review data is crawled from Douban, which is also a social networking service website. Their text domains are highly similar, thus the embeddings of zhihu can make a great contribution to the classification task.

In this section, we discuss the performances of two extrinsic tasks, and find the impacts of embeddings on the tasks are not the same. More specifically, in the three kinds of context features, embeddings with *word + bigram* features perform best in both tasks. As for sizes of corpora, larger size of some corpora can improve the performance to some extent, and the largest combination corpus achieves best results in both two tasks. However, performance does not always improve with the growing size of corpora. As for domains of corpora, domain-specific corpora do have positive and significant impacts on the performance (Financial in NER and Zhihu in sentiment classification). The influence of the domain is even more important than that of size.

5.5   Results of Intrinsic Evaluation

Table 6 shows the performance of intrinsic evaluation, including analogical reasoning on morphological and semantic relations, and word similarity. It can be clearly seen that the introduction of bigram and character features brings significant and consistent improvements on all the categories of embeddings. Furthermore, character features are especially advantageous for reasoning of morphological relations. This is because word in Chinese is composed of graphical characters, known as Hanzi, which has direct influence on Chinese morphology. Thus the introduction of character features can greatly improve the performance of morphological reasoning.

Table 6. Results of intrinsic evaluation. Mor. and Sem. belong to analogical reasoning and Sim. refers to similarity.

		Wiki	Zhihu	Sogou	People's daily	Baidu-baike	Financial	Comb.
Mor.	word	0.114	0.161	0.098	0.194	0.203	0.049	0.285
	word+bigram	0.148	0.191	0.098	0.228	0.241	0.077	0.33
	word+char	**0.395**	**0.499**	**0.343**	**0.477**	**0.417**	**0.323**	**0.543**
Sem.	word	0.188	0.156	0.239	0.406	0.319	0.225	0.489
	word+bigram	0.195	0.157	0.246	**0.407**	0.325	**0.243**	**0.492**
	word+char	**0.238**	**0.173**	**0.249**	0.403	**0.412**	0.237	0.412
Sim.	word	0.388	0.476	0.472	0.461	0.462	0.354	0.503
	word+bigram	0.397	**0.489**	**0.48**	**0.477**	**0.465**	0.350	**0.519**
	word+char	**0.414**	0.438	0.468	0.469	0.407	**0.356**	0.500

As for the size of corpus, it has direct impacts on the performance, for example, Baidu-baike outperforms Wikipedia in all the evaluation measures. Corpus domain is also an important factor in intrinsic tasks. For example, vectors trained on news data (e.g. People's Daily) are beneficial to semantic reasoning, because CA8 incorporates a lot of geography questions, and the names of countries and cities have high frequencies in news data. With the largest size and varied domains, the Combination corpus performs much better than others in both analogy and similarity tasks.

5.6 Correlation Between Intrinsic Evaluation and Extrinsic Evaluation

Firstly, we can observe a lot of consistencies between intrinsic and extrinsic evaluation from above experiments.

- By introducing the character and ngram features, performances of both intrinsic and extrinsic tasks improve, but we can observe that character features are more favorable to intrinsic tasks, while ngram features prove to be more advantageous for extrinsic tasks.
- By comparing embeddings trained with corpora of different sizes and domains, we can find that larger size or similar domain can be important advantages for both intrinsic and extrinsic tasks. And the combination corpus with largest size and varied domains always performs the best.

Table 7. Correlations between evaluations. Ana. here refers to the average scores of Mor. and Sem. The strength of the correlation according to Evans (1996) [9] is : 0.00-0.19 "very weak"?, 0.20-0.39 "weak"?, 0.40-0.59 "moderate", 0.60-0.79 "strong", 0.80-1.0 "very stron".

	Inside Intrinsic Evaluation				Between Intrinsic and Extrinsic Evaluation			
	Mor. vs Sem.	Mor. vs Sim.	Sem. vs Sim.	Ana. vs Sim.	NER vs. Ana.	NER vs. Sim.	SC vs. Ana.	SC vs. Sim.
word	0.7572	0.7631	0.4931	0.6502	0.5493	0.2107	0.6464	0.7402
+bigram	0.7699	0.7012	0.4791	0.6118	0.5510	0.1534	0.4759	0.6589
+char	0.4963	0.5891	0.4317	0.5865	**-0.0658**	0.0434	0.7373	0.6456

To evaluate the correlation between these tasks objectively, we compute the correlations between above tasks by using Pearson correlation coefficient (p). To be specific, we compute not only the correlation between intrinsic and extrinsic, but also the correlation between two intrinsic tasks. We extract the F1 scores of three context features respectively in each task, and compute the coefficients between them.

From the results shown in Table 7, we can observe that in intrinsic evaluations, there is a consistent positive correlation between results of word analogy and word similarity in all the three types of context features. It is not surprising that the correlation between morphological reasoning and semantic reasoning is high, because they are both word analogy tasks. An interesting result is morphological reasoning has a higher correlation with word similarity than semantic reasoning. It is probably because both of the tasks involve word pairs that have same character morphemes.

Regarding the correlation between intrinsic and extrinsic evaluation, most coefficients show positive correlations, indicating intrinsic task can be good indicators of downstream tasks. The only exception is the *word + char* embeddings in NER task. Generally, correlations between Sentiment classification task and intrinsic tasks are stronger than those between NER and intrinsic tasks. The main reason is we use a domain-specific NER dataset for test, the performance of which is largely affected by the domain issue.

Based on above analysis, we reached a couple of interesting and useful findings for evaluation of Chinese word embeddings. Firstly, intrinsic measures are useful in predicting the performances of embeddings in downstream tasks to some extent. Secondly, each task has its favorable features. We would suggest to train word embeddings with corpus that has a similar domain with the dataset. For the same domain of corpus, the bigger, the better. Moreover, extrinsic tasks favor ngram features, while intrinsic tasks favor character features. Thus it is recommended to choose suitable embeddings for each task.

6 Conclusion

This paper conducts a comprehensive study on the correlation between intrinsic and extrinsic evaluation for word embeddings. 21 word embeddings with different corpora and context features are trained and evaluated in 4 tasks: analogy reasoning and word similarity for intrinsic evaluation, NER and Sentiment Classification for extrinsic evaluation. Experimental results prove that intrinsic and extrinsic evaluations are consistent in most cases.

Also, our study sheds some lights on how to select suitable embeddings for NLP tasks: (1) Context features can be integrated to improve the performance, and most extrinsic tasks favor ngram features, while intrinsic tasks favor character features. (2) Training Corpus is very important for the performance of word embeddings. The relevant domain is more important than size factor, especially for extrinsic tasks.

Overall, this paper presents some interesting findings for embedding evaluation, as well as several datasets which could serve as benchmarks for Chinese NLP communities. We also plan to investigate more factors that may affect the embedding performance such as different models and hyper-parameters, and to explore other downstream tasks, e.g. POS tagging and parsing.

Acknowledgements. This work is supported by the Fundamental Research Funds for the Central Universities, China Postdoctoral Science Foundation funded project (No. 2018M630095) and National Language Committee Research Program of China (No. ZDI135-42).

References

1. Bakarov, A.: A survey of word embeddings evaluation methods. arXiv preprint arXiv:1801.09536 (2018)
2. Bansal, M., Gimpel, K., Livescu, K.: Tailoring continuous word representations for dependency parsing. In: Proceedings of the 52nd Annual Meeting of the Association for Computational Linguistics (Volume 2: Short Papers), pp. 809–815 (2014)
3. Batchkarov, M., Kober, T., Reffin, J., Weeds, J., Weir, D.: A critique of word similarity as a method for evaluating distributional semantic models. In: Proceedings of the 1st Workshop on Evaluating Vector-Space Representations for NLP, pp. 7–12 (2016)
4. Camacho-Collados, J., Pilehvar, M.T., Collier, N., Navigli, R.: Semeval-2017 task 2: multilingual and cross-lingual semantic word similarity. In: Proceedings of the 11th International Workshop on Semantic Evaluation (SemEval-2017), pp. 15–26 (2017)
5. Chen, C.Y., Ma, W.Y.: Word embedding evaluation datasets and wikipedia title embedding for Chinese. In: Proceedings of the Eleventh International Conference on Language Resources and Evaluation (LREC 2018), pp. 825–831 (2018)
6. Chen, X., Xu, L., Liu, Z., Sun, M., Luan, H.B.: Joint learning of character and word embeddings. In: IJCAI, pp. 1236–1242 (2015)
7. Chiu, B., Korhonen, A., Pyysalo, S.: Intrinsic evaluation of word vectors fails to predict extrinsic performance. In: Proceedings of the 1st Workshop on Evaluating Vector-Space Representations for NLP, pp. 1–6 (2016)
8. Collobert, R., Weston, J., Bottou, L., Karlen, M., Kavukcuoglu, K., Kuksa, P.: Natural language processing (almost) from scratch. J. Mach. Learn. Res. **12**(Aug), 2493–2537 (2011)
9. Evans, J.D.: Straightforward Statistics for the Behavioral Sciences. Brooks/Cole, Boston (1996)
10. Faruqui, M., Tsvetkov, Y., Rastogi, P., Dyer, C.: Problems with evaluation of word embeddings using word similarity tasks. In: Proceedings of the 1st Workshop on Evaluating Vector-Space Representations for NLP, pp. 30–35 (2016)
11. Finkelstein, L., et al.: Placing search in context: the concept revisited. In: Proceedings of the 10th international conference on World Wide Web, pp. 406–414. ACM (2001)
12. Gao, B., Bian, J., Liu, T.Y.: Wordrep: a benchmark for research on learning word representations. arXiv preprint arXiv:1407.1640 (2014)
13. Ghannay, S., Favre, B., Esteve, Y., Camelin, N.: Word embedding evaluation and combination. In: LREC, pp. 300–305 (2016)
14. Gurnani, N.: Hypothesis testing based intrinsic evaluation of word embeddings. arXiv preprint arXiv:1709.00831 (2017)
15. Kim, Y.: Convolutional neural networks for sentence classification. arXiv preprint arXiv:1408.5882 (2014)
16. Levy, O., Goldberg, Y.: Dependency-based word embeddings. In: Proceedings of the 52nd Annual Meeting of the Association for Computational Linguistics (Vol. 2: Short Papers), vol. 2, pp. 302–308 (2014)
17. Levy, O., Goldberg, Y.: Linguistic regularities in sparse and explicit word representations. In: Proceedings of the Eighteenth Conference on Computational Natural Language Learning, pp. 171–180 (2014)
18. Levy, O., Goldberg, Y., Dagan, I.: Improving distributional similarity with lessons learned from word embeddings. Trans. Assoc. Comput. Linguist. **3**, 211–225 (2015)

19. Li, S., Zhao, Z., Hu, R., Li, W., Liu, T., Du, X.: Analogical reasoning on Chinese morphological and semantic relations. arXiv preprint arXiv:1805.06504 (2018)
20. Li, Y., Li, W., Sun, F., Li, S.: Component-enhanced Chinese character embeddings. arXiv preprint arXiv:1508.06669 (2015)
21. Mikolov, T., Chen, K., Corrado, G., Dean, J.: Efficient estimation of word representations in vector space. arXiv preprint arXiv:1301.3781 (2013)
22. Nayak, N., Angeli, G., Manning, C.D.: Evaluating word embeddings using a representative suite of practical tasks. In: Proceedings of the 1st Workshop on Evaluating Vector-Space Representations for NLP, pp. 19–23 (2016)
23. Schnabel, T., Labutov, I., Mimno, D., Joachims, T.: Evaluation methods for unsupervised word embeddings. In: Proceedings of the 2015 Conference on Empirical Methods in Natural Language Processing, pp. 298–307 (2015)
24. Tjong Kim Sang, E.F.: Introduction to the CoNLL-2002 shared task: language-independent named entity recognition. In: Proceedings of CoNLL-2002, Taipei, Taiwan, pp. 155–158 (2002)
25. Turian, J., Ratinov, L., Bengio, Y.: Word representations: a simple and general method for semi-supervised learning. In: Proceedings of the 48th Annual Meeting of the Association for Computational Linguistics, pp. 384–394. Association for Computational Linguistics (2010)
26. Wu, Y., Li, W.: Overview of the NLPCC-ICCPOL 2016 shared task: Chinese word similarity measurement. In: Lin, C.-Y., Xue, N., Zhao, D., Huang, X., Feng, Y. (eds.) ICCPOL/NLPCC -2016. LNCS (LNAI), vol. 10102, pp. 828–839. Springer, Cham (2016). https://doi.org/10.1007/978-3-319-50496-4_75
27. Xu, J., Liu, J., Zhang, L., Li, Z., Chen, H.: Improve Chinese word embeddings by exploiting internal structure. In: Proceedings of the 2016 Conference of the North American Chapter of the Association for Computational Linguistics: Human Language Technologies, pp. 1041–1050 (2016)
28. Zhai, M., Tan, J., Choi, J.D.: Intrinsic and extrinsic evaluations of word embeddings. In: AAAI, pp. 4282–4283 (2016)
29. Zhao, Z., Liu, T., Li, S., Li, B., Du, X.: Ngram2vec: Learning improved word representations from Ngram co-occurrence statistics. In: Proceedings of the 2017 Conference on Empirical Methods in Natural Language Processing, pp. 244–253 (2017)

Information Retrieval and Question Answering

Question-Answering Aspect Classification with Hierarchical Attention Network

Hanqian Wu[1,2], Mumu Liu[1,2], Jingjing Wang[3(✉)], Jue Xie[4], and Chenlin Shen[3]

[1] School of Computer Science and Engineering, Southeast University, Nanjing, China
hanqian@seu.edu.cn, liudoublemu@163.com

[2] Key Laboratory of Computer Network and Information Integration of Ministry of Education, Southeast University, Nanjing, China

[3] NLP Lab, School of Computer Science and Technology, Soochow University, Suzhou, China
djingwang@gmail.com

[4] Southeast University-Monash University Joint Graduate School, Suzhou, China

Abstract. In e-commerce websites, user-generated question-answering text pairs generally contain rich aspect information of products. In this paper, we address a new task, namely Question-answering (QA) aspect classification, which aims to automatically classify the aspect category of a given QA text pair. In particular, we build a high-quality annotated corpus with specifically designed annotation guidelines for QA aspect classification. On this basis, we propose a hierarchical attention network to address the specific challenges in this new task in three stages. Specifically, we firstly segment both question text and answer text into sentences, and then construct (sentence, sentence) units for each QA text pair. Second, we leverage a QA matching attention layer to encode these (sentence, sentence) units in order to capture the aspect matching information between the sentence inside question text and the sentence inside answer text. Finally, we leverage a self-matching attention layer to capture different importance degrees of different (sentence, sentence) units in each QA text pair. Experimental results demonstrate that our proposed hierarchical attention network outperforms some strong baselines for QA aspect classification.

Keywords: Aspect classification · Question answering
Hierarchical attention

1 Introduction

Recently, a new question-answering (QA) style reviewing form, namely customer questions and answers, has been widely adopted in many popular e-commerce platforms, such as Amazon, Taobao and JingDong. Figure 1 gives an example of QA-style reviewing form. In this new reviewing form, a customer can ask

© Springer Nature Switzerland AG 2018
M. Sun et al. (Eds.): CCL 2018/NLP-NABD 2018, LNAI 11221, pp. 225–237, 2018.
https://doi.org/10.1007/978-3-030-01716-3_19

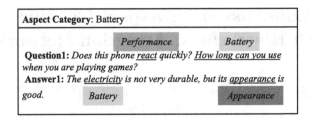

Fig. 1. A translated example of QA-style reviews from an e-commerce website.

questions about certain product which he/she wants to purchase while other customers who have bought this product can answer these questions. With the widespread of such QA-style reviewing form, the relevant research is drawing more and more attention due to its specific characteristics. On one hand, users prefer to write QA-style reviews instead of traditional reviews. On the other hand, compared with the traditional product reviews, the QA-style reviewing form is more informative and convincing. This largely avoids fake reviews and makes product reviews more reliable. Thus, aspect-based sentiment analysis for QA-style texts becomes particularly important, in which aspect classification is a critically basic subtask. However, there are no existing studies with focus on aspect classification of QA-style reviews which aims to identify the aspect category of a given QA text pair. Further, according to the corpus analysis, this new task has the following specific characteristics.

First, from Fig. 1, we could find that **Question1** involves two different aspects distributed in two different sentences, i.e., *performance* inside the first sentence and *battery* inside the second sentence, while **Answer1** involves one related aspect *battery* inside the first sentence and the other unrelated aspect *appearance* inside the second sentence. Under this circumstance, we only need to focus on identifying the matching aspect *battery* between **Question1** and **Answer1**, and ignore the unrelated and non-matching parts between **Question1** and **Answer1**. A better way to handle the above challenge is to firstly perform sentence segmentation in each QA text pair in order to segment both question and answer texts into sentences and make each sentence contain only one aspect. Then, after sentence segmentation, we construct the (sentence, sentence) units in each QA text pair to detect the in-depth aspect matching information, where each (sentence, sentence) unit is composed of two sentences from question and answer text respectively. Finally, in this study, we propose an innovative attention-based neural layer, which could match the sentences inside question and answer text, to encode the (sentence, sentence) units.

Second, for a specific aspect, the importance degree of different (sentence, sentence) units in each QA text pair may be different. For instance, the (sentence, sentence) unit, i.e., sentence *"Does this phone react quickly?"* in **Question1** and sentence *"but its appearance is good."* in **Answer1**, fails to contribute in identifying aspect *battery*, while the (sentence, sentence) unit, i.e., sentence *"How long can you use when you are playing games?"* in **Question1** and sentence *"The* *electricity is not very durable,"* in **Answer1** contributes much in identifying aspect *battery*. Therefore, a well-behaved neural network approach should be capable of capturing the importance degree of different (sentence, sentence) units in each QA text pair.

To address the above specific characteristics in QA aspect classification, in this paper, we build a high-quality annotated corpus tailored for the new QA aspect classification task and then propose a hierarchical attention network to address the new task. Specifically, we firstly perform sentence segmentation in order to segment both the question and answer texts into sentences and then construct (sentence, sentence) units in each QA text pair. Then, we leverage an innovative QA matching attention layer to encode these (sentence, sentence) units in order to match the sentences inside question and answer texts. Finally, we leverage a self-matching attention layer to capture the different importance degrees of different (sentence, sentence) units in each QA text pair. Empirical studies demonstrate that our hierarchical attention approach performs better than several baseline approaches in the task of QA aspect classification.

2   Related Work

Aspect classification, also known as aspect category classification, is regarded as a special case of text classification and often treated as a supervised classification task. Thus, approaches to text classification, such as CNN [4], LSTM [12] and so on, can be applied to aspect classification task, but very few research has been conducted on aspect classification. Toh et al. [11] train binary classifiers based on sigmoidal feedforward network for aspect category classification. Xue et al. [14] propose a multi-task learning model based on neural networks to solve the two tasks aspect category classification and aspect term extraction together.

Besides, aspect classification task is also related to the aspect extraction task. Aspect extraction aims to extract the fine-grained opinion targets from reviews, which is divided into two subtasks, i.e., extracting all aspect terms from corpus and clustering aspect terms with similar meaning into categories where each category represents an aspect. Poria et al. [7] exploit common-sense knowledge and sentence dependency trees to detect both explicit and implicit aspects from opinionated texts. Rana et al. [8] propose a two-fold rule-based method based on sequential patterns and rules mined from reviews. Supervised learning, such as Conditional Random Fields (CRF), requires more manual annotation. Shu et al. [9] propose a lifelong CRF model for aspect extraction which leverages the knowledge from many past domains to facilitate extraction for a new domain. Besides, unsupervised methods, such as Latent Dirichlet Allocation (LDA) and its variants [6], have been applied to aspect extraction. To tackle the weakness of the LDA-based methods, He et al. [3] propose a neural approach based on attention mechanism to de-emphasize irrelevant words during training and further improve the coherence of aspects.

Unlike above studies, this paper focuses on the aspect classification task for QA text pairs. To the best of our knowledge, we are the first to focus on identifying the aspect category of a QA text pair.

3   Data Collection and Annotation

We manually annotate 8,313 QA text pairs collected from *"Asking All"* in Taobao[1] which is one of the most famous e-commerce platforms in China. The QA text pairs are mainly extracted from the domain of *electronic appliances*. To ensure the high consistency of corpus annotation, we define three aspect-related annotation guidelines and annotate each QA text pair in the form of a triple of *aspect term, aspect* and *polarity* (all examples presented in this paper are translations of original Chinese texts).

Guideline1. The annotation of a QA text pair with a triple depends on whether we can extract the aspect term from the question text.

Guideline2. All aspects, i.e., aspect categories, are induced from aspect terms in our corpus. For example, aspect terms *"electricity"* and *"durability"* can be categorized into the aspect *battery*. In other words, these aspect categories are predefined based on the annotation of aspect terms. Aspects are divided into two categories, one is domain-independent aspect, i.e., the aspect does not change with the migration of domain and is applied to all domains, such as *weight, quality* and *appearance*. The other one is domain-dependent aspect, i.e., the aspect is specific to the particular domain, such as *performance, battery* and *IO*.

Guideline3. If an aspect is contained in both the question and answer texts, we need to determine the sentiment polarity of this aspect so as to annotate this QA text pair with a triple. In general, the sentiment polarity can be divided into *neutral* and *non-neutral* categories. Further, the *non-neutral* category can be subdivided into *positive, negative* and *conflict*(a mix of both positive and negative) categories. Some cases may occur in the processing of annotation, elaborated as follows:

(a) If the answer text objectively describes the aspect in the question text, the QA text pair is annotated as (*aspect term, aspect, neutral*). **E1** is an example of this category. In the question text, aspect terms *"battery"* and *"communication effect"* belong to aspects *battery* and *IO* respectively, while the valid answer is only related to the aspect *IO* and objectively describes it.

   E1: Q: How about the battery? How about the communication effect?
   A: The ring is low, but its appearance is very good.

(b) If the answer text contains emotional words that express sentiment of the aspect in the question text, such as *"bad"* and *"not good"*, the QA text pair is annotated as (*aspect term, aspect, negative*). **E2** is an example of this category in which the common aspect of question and answer texts is *battery* that is described as *"endurance"* and the answer text expresses negative sentiment to it.

---

[1] https://www.taobao.com/.

E2: Q: How do you feel about its running? What about the endurance of this phone?

A: Yes, the battery is not durable, and its memory is not as large as it is said.

(c) If there exists sentimental expressions like *"good"* and *"great"* in the answer text and they are related to specific aspect in the question text, the QA text pair is annotated as (*aspect term, aspect, positive*). **E3** is an example of this category. The aspect terms *"certified product"* and *"choppy"* with respect to aspects *certified product* and *performance* respectively are not fully referred in the answer text, which only positively responds to the question about *performance*.

E3: Q: Is it a certified product? Is your phone choppy when you rotate it?

A: My phone is not stuttering, but its screen is not very user-friendly.

(d) Given an aspect in the question text, if the answer text contains both positive and negative sentiment, the QA text pair is annotated as (*aspect term, aspect, conflict*). **E4** is an example of this category which expresses both positive and negative sentiment about the aspect term *"screen"* relating to the aspect *IO*.

E4: Q: How about this phone? How about its screen?

A: Its screen resolution is very good but its screen is a little bit small, and the battery is not durable.

We assign two annotators to label each QA text pair, and the Kappa consistency check value of the annotation is 0.81. To deal with the QA text pairs which are inconsistent annotated by two annotators, an expert is assigned to check them to ensure the quality of data annotation. After annotation of our corpus, we obtain 2,566 QA text pairs which conform to the above guidelines. In this paper, we mainly focus on identifying the aspect category of a given QA text pair. And we will release this annotated corpus if this paper is accepted.

4   Hierarchical Attention Network

In this section, we firstly describe how to segment both question and answer texts into sentences of each QA text pair. Then, we propose a hierarchical attention network to identify the aspect category of a given QA text pair. Figure 2 depicts the overall architecture of our proposed approach to solving the QA aspect classification task.

As described in Sect. 3, in a QA text pair, sentences inside question and answer text could contain different aspects. Thus, we use Stanford CoreNLP toolkit [5] to perform sentence segmentation. After sentence segmentation, we then match the sentence inside question text with the sentence inside answer text to construct (sentence, sentence) units.

The core of the proposed approach lies in the hierarchical attention network, which consists of two layers, i.e., QA matching attention layer and self-matching attention layer (see Fig. 3). We will describe the two layers respectively in the following sections.

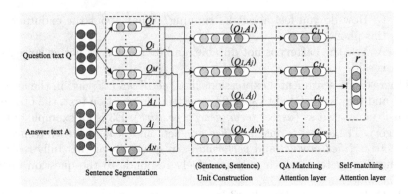

Fig. 2. The overview of our approach.

4.1   QA Matching Attention

Figure 3(a) shows the details of QA matching attention layer. In a QA text pair, assume that the question text Q has been segmented into M sentences $\{Q_1, \cdots, Q_M\}$ and each sentence contains U words. The vector representation $q_{ij} \in R^{d_w}$ denotes the j-th word of the i-th question sentence. And the answer text has been segmented into N sentences $\{A_1, \cdots, A_N\}$ and each sentence contains V words. The vector representation $a_{ij} \in R^{d_w}$ denotes the j-th word of the i-th answer sentence, where d_w represents the dimension of word embeddings in question/answer text.

Then, for (Q_i, A_j) unit, we use LSTM [12] model to encode the question sentence Q_i and the answer sentence A_j, where $i \in [1, M]$, and $j \in [1, N]$, and obtain the hidden state matrix $H_{Q_i} = [h_{i1}, \cdots, h_{iU}]$ of Q_i and the hidden state matrix $H_{A_j} = [h_{j1}, \cdots, h_{jV}]$ of A_j, i.e.,

$$H_{Q_i} = \text{LSTM}(Q_i) \tag{1}$$

$$H_{A_j} = \text{LSTM}(A_j) \tag{2}$$

where $H_{Q_i} \in R^{N_w \times d_h}$, $H_{A_j} \in R^{N_w \times d_h}$, N_w is the number of words in a sentence inside question/answer text and d_h is the size of hidden layers.

Further, the following formulas are applied to compute the attention weight vector α_{ij} between H_{Q_i} and H_{A_j} to concentrate on such words in question sentence Q_i and answer sentence A_j with respect to the annotated aspect and capture the matching information between them.

$$M_{i,j} = \tanh(W_{ij} \cdot (H_{Q_i}^T \cdot H_{A_j}) + b_{ij}) \tag{3}$$

$$\alpha_{i,j} = \text{softmax}(W_e^T \cdot M_{i,j}) \tag{4}$$

where $M_{i,j} \in R^{N_w \times N_w}$, $\alpha_{ij} \in R^{N_w}$. W_{ij} and W_e are the weight matrices, b_{ij} is the bias and $\cdot$ denotes the dot product between matrices.

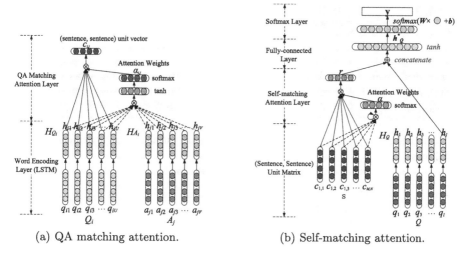

(a) QA matching attention.   (b) Self-matching attention.

Fig. 3. The overall architecture of hierarchical attention model.

From **Guideline1** and **Guideline2** in Sect. 3, the question text is important for QA aspect classification. Thus, we obtain the (sentence, sentence) unit vector $c_{i,j} \in R^{d_h}$ of (Q_i, A_j) unit with the hidden state matrix H_{Q_i} of the question sentence Q_i based on the weights, i.e.,

$$c_{i,j} = H_{Q_i} \cdot \alpha_{i,j}^T \tag{5}$$

Finally, we obtain (sentence, sentence) unit vector set $C = \{c_{1,1}, \cdots, c_{i,j}, \cdots, c_{M,N}\}$, where $|C|$ is $M * N$.

4.2 Self-matching Attention

In this layer, we introduce another attention mechanism to automatically determine the importance of each (Q_i, A_j) unit, where $i \in [1, M]$, and $j \in [1, N]$, as shown in Fig. 3(b).

First, we concatenate multiple (sentence, sentence) unit vectors $\{c_{1,1}, \cdots, c_{i,j}, \cdots, c_{M,N}\}$ together into a new (sentence, sentence) unit matrix $s \in R^{N_c \times d_h}$, where N_c is $M * N$, i.e.,

$$s = [c_{1,1}, \cdots, c_{M,N}] \tag{6}$$

Further, we perform self-matching to obtain the importance degree vector $\alpha \in R^{N_c}$ of different (sentence,sentence) units by the following formulas,

$$M = \tanh(W_s \cdot (s^T \cdot s) + b_s) \tag{7}$$

$$\alpha = \text{softmax}(W_f^T \cdot M) \tag{8}$$

where $M \in R^{N_c \times N_c}$, W_s and W_f are the weight matrices, b_s is the bias and $\cdot$ denotes the dot product between matrices.

After obtaining self-matching attention weights, we can get the attention representation $r \in R^{d_h}$ of (sentence, sentence) units as follows,

$$r = s \cdot \alpha^T \tag{9}$$

Besides, from **Guideline1** and **Guideline2** in Sect. 3, the aspect mainly exists in the question text. Therefore, we use LSTM model to encode the original question text Q that is not split and get its hidden state $H_Q = [h_1, \cdots, h_l]$. And we then concatenate the attention representation r of (sentence, sentence) units with the last hidden vector h_l of Q to generate the final representation h_Q^* of Q as follows,

$$h_Q^* = \tanh(W_p r + W_x h_l) \tag{10}$$

where $H_Q \in R^{N_q \times d_h}$, and N_q represents the number of words in the question text. $h_Q^* \in R^{d_h}$, W_p and W_x are the weight matrices.

Finally, we put the final representation h_Q^* into a *softmax* layer to obtain the conditional probability distribution:

$$y = \text{softmax}(W h_Q^* + b) \tag{11}$$

where W and b are parameters for the *softmax* layer. Based on that, the label with the highest probability stands for the predicted aspect of a QA text pair.

4.3   Model Training

Cross-entropy loss function is used to train our model end-to-end for classification. Given a set of training data S_{Q_t}, S_{A_t} and y_t, where S_{Q_t} is the t-th question text, S_{A_t} is the corresponding answer text, and y_t is the ground-truth aspect for a QA text pair (S_{Q_t}, S_{A_t}), if we represent this model as a black-box function $\phi(S_Q, S_A)$ whose output is a vector representing the probability of aspects, then the goal of training is to minimize the loss function:

$$J(\theta) = -\sum_{t=1}^{N_s} \sum_{k=1}^{K} y_t^k \cdot \log \phi(S_{Q_t}, S_{A_t})) + \frac{l}{2} ||\theta||_2^2 \tag{12}$$

where N_s is the number of training samples, K is the number of aspects for classification and l is a L_2 regularization to bias parameters.

In the equation above, parameters in our model are optimized by *Adagrad* optimizer [1]. All the matrix and vector parameters are initialized with uniform distribution $[-\sqrt{6/(r + c')}, \sqrt{6/(r + c')}]$, where r and c' are rows and cols of the matrix respectively [2]. Besides, the dropout strategy is used in LSTM layer to avoid over-fitting.

5   Experimentation

5.1   Experimental Settings

- **Data Settings:** Considering the imbalanced distribution of data, we omit the aspect categories which contain less than 50 QA text pairs. In our experiments, there are 7 aspects and 2,427 QA text pairs in total, as shown in

Table 1. The distribution of experimental data.

Aspect category	Number of QA text pairs
Performance	548
Battery	230
IO	908
Function	111
Quality	165
Certified product	370
Computation	95

Table 1. We also set aside 10% from the training data as the development data to tune learning algorithm parameters.

- **Word Representation:** *Wording Embedding* is used to initialize the words of our dataset and we use skip-gram model to pre-train the word embeddings using 320 thousand QA text pairs crawled from *"Asking All"* in Taobao.

- **Evaluation Metrics:** The performance is evaluated by using *Accuracy* and *Macro-F1* (F) which is calculated by the formula $F = \frac{2PR}{P+R}$, where the overall precision P and recall R are averaged on the precision/recall scores of all categories. Furthermore, t-test is used to evaluate the significance of the performance difference between two approaches [15].

- **Hyper-parameters:** In our experiment, all out-of-vocabulary words are initialized by sampling from the uniform distribution $U(-0.01, 0.01)$. The dimensions of word embedding and LSTM hidden states are set to be 100. The other hyper-parameters are tuned according to the development data. Specifically, the learning rate is 0.02. The dropout rate is set to 0.25.

5.2 Experimental Results

For a thorough comparison, we implement the following baseline approaches to aspect classification to evaluate the performance of our proposed approach. Note that all these approaches employ the same representations.

- **CNN(A):** This approach is a basic baseline approach proposed by Kim et al. [4] which takes answer texts as input of CNN.
- **CNN(Q):** This approach is a basic baseline approach which takes question texts as input of CNN.
- **CNN(Q+A):** This approach is a basic baseline approach which takes the concatenation of question and answer texts as the input of CNN.
- **LSTM(A):** This baseline approach puts answer texts into the input layer of LSTM proposed by Tang et al. [10].
- **LSTM(Q):** This baseline approach puts question texts into the input layer of LSTM.

- **LSTM(Q+A):** This baseline approach puts the concatenation of question and answer texts into the input layer of LSTM.
- **Hierarchical LSTM:** This is a question classification approach proposed by Xia et al. [13], which uses a hierarchical LSTM model to encode the question texts.
- **LSTM-Attention:** This baseline approach introduces attention mechanism to capture the aspect matching information between question texts and answer texts proposed by Wang et al. [12].
- **QA Matching Attention:** This is our proposed approach which only employs a QA matching attention layer without a self-matching attention layer.
- **Hierarchical Attention:** This is our proposed approach which employs both QA matching attention layer and self-matching attention layer.

Table 2 gives the performance comparison of different approaches. From Table 2, we can find that:

First, according to the first six approaches, all approaches training using question texts perform better than those training using answer texts, which indicates that question texts contain richer aspect information than answer texts and this conforms to annotation guidelines **Guideline1** and **Guideline2** in Sect. 3.

Second, when using the concatenation of question and answer texts as input, both **CNN** and **LSTM** methods could obtain better performance, which indicates that answer texts can assist question texts to further improve the performance of QA aspect classification.

Third, we can find that approaches based on **LSTM** perform better than those based on **CNN**. Therefore, LSTM is more suitable for QA aspect classification task than CNN.

Therefore, in the last four approaches, LSTM model is used to encode in regardless of question or answer texts. Moreover, in our proposed approaches **QA**

Table 2. *Accuracy* and *Macro-F1* on QA aspect classification.

Approaches	Accuracy	Macro-F1
CNN(A) (Kim et al. [4])	0.575	0.294
CNN(Q)	0.744	0.585
CNN(Q+A)	0.771	0.595
LSTM(A) (Tang et al. [10])	0.663	0.466
LSTM(Q)	0.779	0.668
LSTM(Q+A)	0.800	0.678
Hierarchical LSTM (Xia et al. [13])	0.825	0.733
LSTM-Attention (Wang et al. [12])	0.827	0.729
QA Matching Attention (ours)	0.842	0.751
Hierarchical Attention (ours)	**0.873**	**0.804**

Matching Attention and **Hierarchical Attention**, we take both question and answer texts as input. We could find that the performance of **Hierarchical LSTM** method is better than **LSTM** but worse than **LSTM-Attention** method, and the accuracy of **LSTM-Attention** method is 4.8% higher than that of **LSTM** method, which indicate that attention mechanism is a good choice to capture the information related to the specific aspect.

Besides, our proposed **QA Matching Attention** approach achieves the improvement of 1.5% (Accuracy) and 2.2% (*Macro-F1*) compared to **LSTM-Attention** method, which highlights the importance of employing the aspect matching information between the sentence inside question text and the sentence inside answer text.

Noted that, our proposed **Hierarchical Attention** approach using both QA Matching Attention layer and self-matching attention layer achieves the best performance than all other approaches. The accuracy and *Macro-F1* of **Hierarchical Attention** approach are 3.1% and 5.3% higher than **QA Matching Attention** only using QA Matching Attention layer respectively, which indicates that using self-matching attention layer can effectively capture the different importance degrees of different (sentence, sentence) units in a QA text pair and thus could further boost the performance of QA aspect classification. Significance test shows that this improvement is significant ($p - value < 0.05$).

5.3   Error Analysis

Based on the analysis of misclassified QA text pairs, we can find some main reasons for misclassification as follows:

(1) The distribution of experimental data is imbalanced (see Table 1), so the most of misclassified QA text pairs tend to be predicted as the aspect categories accounting for a large rate in experimental data. According to statistics, in our proposed approach, the predicted aspect of 22.95% of misclassified QA text pairs is *IO* and 16.39% is *performance*.
(2) Some manually annotated aspect terms are ambiguous. We can factitiously determine their different meanings and categorize them into correct categories, while it is difficult for the well-trained machine. For instance, the aspect term *sound quality* is related to the aspect *IO*, but the machine may regard it relating to the aspect category *quality*.
(3) The word segmentation toolkit cannot accurately segment out-of-vocabulary words, such as *mhl* and *wifi*, which may also affect the performance of QA aspect classification.

6   Conclusion

In this paper, we propose a hierarchical attention network for aspect classification of QA text pairs. The main idea of the proposed approach is to segment question and answer texts into sentences and then extend feature representations of question texts with the two layers of QA matching attention and self-matching attention. Experimental results demonstrate that our proposed approach outperforms some strong baseline methods using neural networks.

In our future work, we would like to perform joint learning on both aspect classification and aspect term extraction tasks to further improve the performance of QA aspect classification.

Acknowledgements. This work is supported in part by Industrial Prospective Project of Jiangsu Technology Department under Grant No. BE2017081 and the National Natural Science Foundation of China under Grant No. 61572129.

References

1. Duchi, J., Hazan, E., Singer, Y.: Adaptive subgradient methods for online learning and stochastic optimization. J. Mach. Learn. Res. **12**(Jul), 2121–2159 (2011)
2. Glorot, X., Bengio, Y.: Understanding the difficulty of training deep feedforward neural networks. In: Proceedings of the 13th International Conference On Artificial Intelligence and Statistics, pp. 249–256 (2010)
3. He, R., Lee, W.S., Ng, H.T., Dahlmeier, D.: An unsupervised neural attention model for aspect extraction. In: Proceedings of the 55th Annual Meeting of the Association for Computational Linguistics, vol. 1, pp. 388–397 (2017)
4. Kim, Y.: Convolutional neural networks for sentence classification. In: Proceedings of the 2014 Conference on EMNLP, pp. 1746–1751 (2014)
5. Manning, C., Surdeanu, M., Bauer, J., Finkel, J., Bethard, S., McClosky, D.: The stanford corenlp natural language processing toolkit. In: Proceedings of 52nd Annual Meeting of ACL: System Demonstrations, pp. 55–60. ACL (2014)
6. Mukherjee, A., Liu, B.: Aspect extraction through semi-supervised modeling. In: Proceedings of the 50th Annual Meeting of ACL, pp. 339–348. ACL (2012)
7. Poria, S., Cambria, E., Ku, L., Gui, C., Gelbukh, A.: A rule-based approach to aspect extraction from product reviews. In: Proceedings of the Second Workshop on Natural Language Processing for Social Media, pp. 28–37 (2014)
8. Rana, T.A., Cheah, Y.: A two-fold rule-based model for aspect extraction. Expert Syst. Appl. **89**, 273–285 (2017)
9. Shu, L., Xu, H., Liu, B.: Lifelong learning CRF for supervised aspect extraction. In: Proceedings of the 55th Annual Meeting of the Association for Computational Linguistics, vol. 2, pp. 148–154 (2017)
10. Tang, D., Qin, B., Feng, X., Liu, T.: Effective LSTMs for target-dependent sentiment classification. In: Proceedings of COLING 2016, the 26th International Conference on Computational Linguistics: Technical Papers, pp. 3298–3307 (2016)
11. Toh, Z., Su, J.: NLANGP: supervised machine learning system for aspect category classification and opinion target extraction. In: International Workshop on Semantic Evaluation, pp. 496–501 (2015)
12. Wang, Y., Huang, M., Zhao, L., et al.: Attention-based LSTM for aspect-level sentiment classification. In: Proceedings of the 2016 Conference on Empirical Methods in Natural Language Processing, pp. 606–615 (2016)
13. Xia, W., Zhu, W., Liao, B., Chen, M., Cai, L., Huang, L.: Novel architecture for long short-term memory used in question classification. Neurocomputing **299**, 20–31 (2018)

14. Xue, W., Zhou, W., Li, T., Wang, Q.: MTNA: a neural multi-task model for aspect category classification and aspect term extraction on restaurant reviews. In: Proceedings of the 8th International Joint Conference on Natural Language Processing, pp. 151–156. Asian Federation of Natural Language Processing (2017)
15. Yang, Y., Liu, X.: A re-examination of text categorization methods. In: International ACM SIGIR Conference on Research and Development in Information Retrieval, pp. 42–49 (1999)

End-to-End Task-Oriented Dialogue System with Distantly Supervised Knowledge Base Retriever

Libo Qin, Yijia Liu, Wanxiang Che$^{(\boxtimes)}$, Haoyang Wen, and Ting Liu

Research Center for Social Computing and Information Retrieval,
Harbin Institute of Technology, Harbin, China
{lbqin,yjliu,car,hywen,tliu}@ir.hit.edu.cn

Abstract. Task-oriented dialog systems usually face the challenge of querying knowledge base. However, it usually cannot be explicitly modeled due to the lack of annotation. In this paper, we introduce an explicit KB retrieval component (*KB retriever*) into the seq2seq dialogue system. We first use the *KB retriever* to get the most relevant entry according to the dialogue history and KB, and then apply the copying mechanism to retrieve entities from the *retrieved KB* in decoding time. Moreover, the *KB retriever* is trained with distant supervision, which does not need any annotation efforts. Experiments on Stanford Multi-turn Task-oriented Dialogue Dataset shows that our framework significantly outperforms other sequence-to-sequence based baseline models on both automatic and human evaluation.

Keywords: Task-oriented dialog systems · Sequence-to-sequence Knowledge base

1 Introduction

Task-oriented dialogue system, which helps users to achieve specific goals with natural language, attracts more and more research attention. With the sequence-to-sequence (seq2seq) approaches being successfully applied in text generation [1,10,14,15,21,21], several works tried to model the task-oriented dialogue as the seq2seq generation of response from the dialogue history [4,5,22]. This kind of modeling scheme frees the task-oriented dialogue system from the manually designed pipeline modules and heavy annotation labor for these modules.

Different from typical text generation, the success of serving users in the task-oriented dialogue system largely relies on the success of querying knowledge base (KB). Taking the dialogue in Fig. 1 for example, to answer the driver's query on the gas station, the dialogue system is required to pick out the entry that has the "gas station" (the fourth row in this table). To tackle the KB query challenges, Eric and Manning [4] use an additional copy mechanism to retrieve entities both in KB and dialogue history. Eric et al. [5] further proposed key-value retrieval

© Springer Nature Switzerland AG 2018
M. Sun et al. (Eds.): CCL 2018/NLP-NABD 2018, LNAI 11221, pp. 238–249, 2018.
https://doi.org/10.1007/978-3-030-01716-3_20

Address	Distance	POI type	POI	Traffic info
638 Amherst St	3 miles	grocery store	Sigona Farmers Market	car collision nearby
269 Alger Dr	1 miles	coffee or tea place	Cafe Venetia	car collision nearby
5672 barringer street	5 miles	certain address	5672 barringer street	no traffic
200 Alester Ave	2 miles	gas station	Valero	road block nearby
899 Ames Ct	5 miles	hospital	Stanford Childrens Health	moderate traffic
481 Amaranta Ave	1 miles	parking garage	Palo Alto Garage R	moderate traffic
145 Amherst St	1 miles	coffee or tea place	Teavana	road block nearby
409 Bollard St	5 miles	grocery store	Willows Market	no traffic

Driver: Address to the gas station.
Car: Valero is located at 200 Alester Ave.
Driver: OK , please give me directions via a route that avoids all heavy traffic.
Car: Since there is a road block nearby, I found another route for you and I sent it on your screen.
Driver: Awesome thank you.

Fig. 1. An example of a task-oriented dialogue that incorporates a knowledge base.

network which incorporates world knowledge into its dialogue utterances via attention over the key-value entries of the underlying knowledge base.

Besides using soft attention to model the interaction between dialogue history and KB entries, a component that directly retrieves the KB was used in dialogue pipeline [9, 23]. However, such component is generally considered intractable for the seq2seq dialogue system because probabilistically modeling calls for annotated data which are absent in the seq2seq settings. Past decades witness the success of the distant supervision in information extraction [11,12,25,26], which induces the training signal from a set of heuristic on the existing KB. Inspired by this line of research, we explore the possibility of introducing an explicit KB retrieval component into the seq2seq dialogue system and train this component with distant supervision.

In this paper, we propose a novel seq2seq dialogue system that explicitly queries KB and uses the queried result to generate the response. A KB retrieving component (retriever) is proposed to model the interaction between dialogue history and the KB and its trained with a novel distant supervision algorithm. In practice, KB retriever first gets the most relevant entry given dialogue history and KB, and then perform column attention to get retrieved KB cell based on the selected entry while decoding time. Finally, the retrieved KB cell is then fed into a copy network to generate the final response. Our method represents a shift in perspective compared to existing work, we not only follow the basic method of task-oriented dialog based on seq2seq model, but also explicitly model a KB retrieving component into the basic seq2seq framework. Moreover, the KB retrieving component is trained with a novel distant supervision which doesn't need heavy annotation. Experiments on Stanford Multi-turn, Dialogue Dataset [5] verify the effectiveness of our method by significantly outperforming the baseline in both the automatic and human evaluation.

Our contributions can be summarized as follows:

- We propose a *KB retriever* based seq2seq model in task-oriented dialogue systems, which can greatly improve the ability of entire system interacting with and querying the knowledge base.

- In our framework, the *KB retriever* is trained with a novel distant supervision which does not need heavy human annotation.
- Experiments on a publicly available dataset show that our approach significantly and consistently outperforms all baselines.

2   Related Work

Historically, task-oriented dialog systems have been built as pipelines of separately trained modules. A typical pipeline design contains four components: (1) a user intent classifier, (2) a belief tracker, (3) a dialogue policy maker and a (4) response generator. Recently, the powerful distributed representation ability of neural networks makes task-oriented dialogue system end-to-end possible. Wen et al. [22] built a system that connects classic pipeline modules by a policy network. It queries KB by a database operator which is consistent with the most likely belief state. However, their modules like belief tracker still needs to be trained separately before end-to-end training. Unlike their work, our framework use an explicit KB retriever to extract useful information from a knowledge base, without the need for explicit training of belief or intent trackers. Other dialogue agents can also interface with the database by augmenting their output action space with predefined API calls [2,9,13,27]. While Dhingra et al. [3] applied a soft-KB lookup on an entity-centric knowledge base to compute the probability of that the user knows the values of slots, and has tried to model the posterior distributions over all slots. However, our framework does not require any slots information. Eric and manning [4] use an additional copy mechanism to retrieve entities both in KB and dialogue history. Eric et al. [5] further introduced retrieval from key-value KB based seq2seq model. The key difference between our work and their work is that they query the KB only by attention-based method while our model proposes an explicit *KB retriever* component to query KB into a seq2seq framework. Inspired to those works of the distant supervision in information extraction [11,12,25,26]. we train our *KB retriever* component with distant supervision and collect the training data only by history dialogue and the existing KB, which doesn't need heavy human annotation.

3   Method

In this section, we describe our framework for task-oriented dialogue system. Our framework consists of a *KB retriever* that takes the encoded dialogue history along with the representation of all KB entries as input and returns the most possible KB entry (*retrieved KB*) (Sect. 3.3), and an *encoder-decoder* framework that takes the *retrieved KB* and an attentively represented dialogue history and use a copy network [6] to determine the next generated token.

3.1   Problem Definition and Notation

Dialogue History. Given a dialogue between a user (u) and a system (s), we follow Eric and Manning [4,5] and represent the k-turned *dialogue utterances*

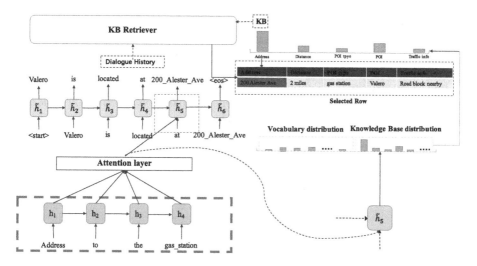

Fig. 2. Given with dialogue history and KB, the *KB Retriever* return the *Retrieved KB Row*. For each time-step of decoding, the cell state is used to compute an attention over the encoder states and a separate column attention over the column of *Retrieved KB Row*. The attention over the encoder is used to generate a context vector which is combined with the cell state to get a distribution over the normal vocabulary. The hierarchical attention over the column of the KB become the logits for their associated entity in a now augmented vocabulary that we argmax over.

as $\{(u_1, s_1), (u_2, s_2), ..., (u_k, s_k)\}$. At the ith turn of the dialogue, we aggregate dialogue context which consists of the tokens of $(u_1, s_1, ..., s_{i-1}, u_i)$ and use $\mathbf{x} = (x_1, x_2, ..., x_m)$ to denote the whole *dialogue history* word by word, where m is the number of tokens in the dialogue.

Knowledge Base. In this paper, we assume to have the assessment of a relational database-like KB T, which consists of several rows and five columns. Each column is associated with a attribute name f.

Sequence-to-Sequence Task-Oriented Dialogue. We define the seq2seq task-oriented dialogue as finding the most likely response sequence according to the input dialogue history and KB. Formally, it is defined as

$$p(\mathbf{y} \mid \mathbf{x}, T) = \prod_{t=1}^{n} p(y_t \mid y_1, ..., y_{t-1}, \mathbf{x}, T)$$

where y represent an output token.

3.2 Vanilla Sequence-to-Sequence Task-Oriented Dialogue System

Eric and Manning [5] proposed the vanilla seq2seq task-oriented dialogue system. In their model, a long short term memory (LSTM, [7]) is used to encode the dialogue history **x**. More specifically, the tokens in **x** are mapped to vectors with embedding function ϕ^{emb}. The vectors are then fed into LSTM to produce context-sensitive hidden representations $(h_1, h_2, ..., h_m)$, by repeatedly applying the recurrence $h_i = \text{LSTM}\left(\phi^{emb}\left(x_i\right), h_{i-1}\right)$.

LSTM is also used to represent the partially generated output sequence $(y_1, y_2, ..., y_{t-1})$ as $(\tilde{h}_1, \tilde{h}_2, ..., \tilde{h}_t)$. For the generation of next token y_t, their model first calculates an attentive representation $\tilde{h}'_t$ of the dialogue history as

$$u_i^t = w^T \tanh(W_2 \cdot \tanh(W_1 \cdot [h_i, \tilde{h}_t]))$$
$$a_i^t = \text{softmax}(u_i^t)$$
$$\tilde{h}'_t = \sum_{i=1}^{m} a_i^t \cdot h_i$$

Finally, a concatenation of the hidden representation of outputted sequence $\tilde{h}_t$ and the attentive dialogue history representation $\tilde{h}'_t$ are projected to the vocabulary space by U as

$$o_t = U \cdot [\tilde{h}_t, \tilde{h}'_t]$$
$$p(y_t \mid y_1, ..., y_{t-1}, \mathbf{x}, T) = \text{softmax}(o_t)$$

where $\mathcal{V}$ is the vocabulary and $y \in \mathcal{V}$.

Seq2Seq Task-Oriented Dialogue with Copy Net. To enable the network to generate the entry in KB, Eric and Manning [5] also proposed an augmented decoder that decodes over the combination of vocabulary and candidate entries in KB. In [5], the logit o_t is expanded with a KB-attention score v^t as

$$o_t = U \cdot [\tilde{h}_t, \tilde{h}'_t] + v^t$$

where o_t's dimensionality is $|\mathcal{V}| + |\mathcal{E}|$. In v^t, lower $|\mathcal{V}|$ is zero and the rest is $|\mathcal{E}|$ attention scores. Our major difference with Eric and Manning [5] is that we don't use attention scores of the whole KB to augment o_t but the scores of one concrete row (*retrieved KB*) of the relational KB.

3.3 KB Retriever

As described in Sect. 1, our goal is to examine the possibility for directly retrieves the KB. To accomplish this goal, we propose a *KB retriever*.

Dialogue History Representation. We encode the dialogue history by adopting the neural bag-of-words (BoW). Each token in the dialogue history is mapped into a vector by another embedding function $\phi^{emb'}(x)$ and the dialogue history representation E_d is computed as summing these vectors: $E_d = \sum_{i=1}^{m} \phi^{emb'}(x_i)$

KB Encoder. In this section, we describe how we encode the KB table. Each KB cell is represented as the cell value v embedding $\phi^{\text{value}}(v)$. The representation can be formalized as

$$c = \phi^{\text{value}}(v)$$

The representation of a row of KB C_k is denoted as $C_k = [c_{k,1}, ..., c_{k,m}]$, where m represents number of column attributes.

KB Retriever. Past decades witness the success of the memory network [20] in some reasoning tasks [17,24]. Inspired by those works, we follow the structure of memory network to explore the deep correlation of the dialogue history and the every KB row, hoping to help us reasoning and find the KB row most relevant to the Dialogue history. In practice, we consider the dialogue history as the query which mentioned in [20] and regard KB as the information should be stored in memory. We model the retrieval process as a hierarchical classification over KB, which first select the row, then select the column.

For the Row Selection. We take the encoding of dialogue history E_d and the table encoding as input, which are fed into multi-hop memory network to get the relevance score of every row in the KB. Finally, we select the row that corresponds to the maximum score. Below we describe how to get the probability distribution of each row through the memory network given the dialog history and table encode. In our model, we give a row entity of KB set $C_1, ..., C_i$ to be stored in memory. The entire set of $[c_{k,1}, ..., c_{k,m}]$ are converted into memory vectors m_k of dimension d computed by embedding each KB cell $c_{k,i}$ and sums the resulting vectors: $m_k = \Sigma_j c_{k,j}$. Then, we compute the match between E_d and each memory m_k by taking the inner product followed by a softmax:

$$p_k = \text{softmax}\left(E_d^T m_k\right)$$

where $\text{softmax}(z_i) = e^{z_i} / \sum_j e^{z_j}$. Defined in this way p is a probability vector over the row entities of KB set. Each C_i has a corresponding output vector z_i given in the simplest case by another embedding matrix M. The response vector from the memory o is then a sum over the transformed inputs z_i, weighted by the probability vector from the input:

$$o = \sum_i p_i z_i$$

In the single layer case, the sum of the output vector o and the dialogue history representation E_d is then passed through a final weight matrix W and a softmax to produce the predicted logits:

$$\tilde{a} = \text{softmax}\left(W\left(o + E_d\right)\right)$$

In our framework, we also explore the multi-hop memory network. The memory layers are stacked in the following way:

- The input to layers above the first is the sum of the output o^k and the input E_d^k from layer k.

– At the top of the network, the input to W also combines the input and the output of the top memory layer: $\tilde{a} = \text{softmax}\left(W\left(o^k + E_d^k\right)\right)$.

where $\tilde{a}$ represents the predicted row logits of KB which is used to query KB. Its dimension size is the number of KB's row rather than the size of word vocabulary which is the difference between our model and [20]. Based on $\tilde{a}$, we select the row with the largest probability value as the *retrieved KB*. Moreover, we use the adjacent type of weight typing to reduce the number of parameters and use Temporal Encoding to improve the performance of the *KB retriever*.

For the Column Selection. After getting the *retrieved KB*, we perform column attention in decoding time to select column of KB. We use the decoder hidden state $(\tilde{h}_1, \tilde{h}_2, ..., \tilde{h}_t)$ to compute an attention score with the embedding of column attribute name. The attention logits then become the logits of the column be selected based *retrieved KB*. Finally, we use a copy network to determine the next generated token. Similar to [5], the final logit o_t is expanded with a KB-attention score v^t as

$$o_t = U \cdot [\tilde{h}_t, \tilde{h'}_t] + v^t$$

where o_t's dimensionality is $|\mathcal{V}| + |\mathcal{E}|$. In v^t, lower $|\mathcal{V}|$ is zero and the rest is $|\mathcal{E}|$ attention scores. We just use attention scores of one concrete row of the *retrieved KB*, not use the whole KB to augment o_k, which is the key difference between our work and Eric and Manning [5]. This description seeks to capture the intuition that when in response to the query *Address to the gas station* in Fig. 1, our *KB retriever* have selected the fourth KB row which includes the correct response entries *Valero* and *200 Alester Ave*. Therefore, our model only put an attention weight on the *retrieved KB* rather than the whole KB, which can improve the performance of response results. We provide a visualization of the whole framework in Fig. 2.

4   Data Collection for Training the Retriever with Distant Supervision

In this section, we talk about how we collect the training data for the *KB retriever*. Different from other works that need heavy human annotation, we only use the dialogue history and the existing KB to collect our training data.

Given with dialogue history $(x_1, x_2, ..., x_m)$ and the KB C_i (i represents the row index of KB), we can use a simple match algorithm to collect training data. For every row of KB, take the kth row $([c_{k,1}, ..., c_{k,m}])$ for example, we judge whether the cells $(c_{k,1}, ..., c_{k,m})$ of each row of KB have appeared in the dialogue history $(x_1, x_2, ..., x_m)$. If they match, the counter is incremented by one, and then we get a match score for each row. Finally, we select the row corresponding to the largest match score as our selected row.

Take the dialogue and KB in Fig. 1 for example, we show how we get the match score of the fourth row of KB. First, we initialize every row's match counter to zero. Then, for every cell in the fourth row(200 Alester Ave, 2 miles, gas station, Valero, road block nearby), we find those cells (200 Alester Ave, gas station, Valero) can be matched in the dialogue history. So we change the value of counter to three. After getting each row's match score, we select the row corresponding to the largest match score as our selected row. The intuition is that we believe the knowledge base with the largest number of matched entities in the dialogue history is the supported KB row in most of the time. Through the above steps, we get the training data for training the retriever.

5   Experiments

In this section, we first introduce the details of the experiments and then present results from both automatic and human evaluation. Then we provide results and analyses of automatic evaluation and human evaluation. Besides, we present ablation test to evaluate and analyze the function of different components in our framework.

5.1   Experiment Setting

We choose a KB-rich domain from Stanford Multi-turn Multi-domain Task-oriented Dialogue Dataset [5], which is point-of-interest navigation.

Our framework is trained separately in these two stages, using the same train/validation/test split sets as [5]. We do not map the entities in dialogue into its canonical form as what [5] have done, since our framework extract entities directly from KB. And we evaluate our framework on exact entities as well. In the first stage, we applied three hops and weight typing to train memory network for positioning row of KB. In the second stage, we trained our main framework by an end-to-end approach. Our framework is trained using the Adam optimizer [8]. The learning rate is 10^{-3}. We applied dropout [19] to the input and the output of LSTM, with a dropout rate at 0.75. We add the weight decay on the model. The coefficient of weight decay is $5 * 10^{-6}$. The embedding size and all hidden size are 200. The number of epochs for pretraining memory network is 100 for and the number of hops is 3.

5.2   Baseline Models

We provide several baseline models for comparing the performance of our whole framework:

- **Copy-Augmented Sequence-to-Sequence Network.** This model is adapted from [4]. It augments a sequence-to-sequence architecture with encoder attention, with an additional attention-based hard-copy mechanism over the KB entities mentioned in the encoder context.

- **Key-Value Retrieval Network.** This model is adapted from [5]. It utilizes key-value forms to represent KBs. Key representations are used for an attention-based value retrieval. Note that in the original paper, they simplified the task by mapping the expression of entities to a canonical form using named entity recognition (NER) and linking.

5.3   Automatic Evaluation

In this section, we provide two different automatic evaluations to compare with other baseline models. The results and analyses are provided in the following sections.

Evaluation Metrics:

- **BLEU.** We use the BLEU metric, commonly employed in evaluating machine translation systems [16], which has also been used in past literature for evaluating dialogue systems both of the chatbot and task-oriented variety [6,18,22]. Hence, we include BLEU score in our evaluation (i.e. using Moses multi-bleu.perl script).

- **Entity F1.** We micro-average and macro-average the entire set of system responses and compare the entities in plain text. The entities in each gold system response are selected by a predefined entity list. This metric evaluates the ability to generate relevant entities from the provided KBs and to capture the semantics of the dialogue flow [4,5].

Results and Analyses. Experiment results are illustrated in Table 1. The results show that our model outperforms other models in all automatic evaluation metrics. Compared to KV Net, we achieve 2.85 improvements on BLEU score and 20.5 improvements on Micro F1. And compared to Copy Net, we achieve 2.88 improvements on BLEU score and 26.3 improvements Macro F1. The results in navigation show our model's capability to generate more natural and meaningful response than the Seq2Seq baseline models.

We also find that the KV Net's results are lower than that reported by [5]. We address this to the differences in the preprocessing, model training and evaluation metrics. In spite of the difference of evaluation metrics that we evaluate on exact entities rather than their canonical forms, the Micro F1 score of our model still outperforms what [5] reported, which is 41.3 in navigation domain and which is evaluated on canonical forms.

Ablation. In this section, we perform several ablation experiments to evaluate different components in our framework on the navigation domain. The results are shown in Table 2. The results demonstrate the strong impact that components of our model to the final performance.

Copying mechanism enables our framework to retrieve entities directly from KBs. Without copying mechanism, such retrieval is infeasible and our framework cannot produce values in KBs. The results show that it introduces more variability to the generation process if we do not use copying mechanism.

KB retriever first retrieve the KB row most relevant to Dialogue history is the key difference between our model with other baselines. It can effectively reduce the scale of KB while in decoding time, which can improve the performance of the generation.

Table 1. Automatic evaluation on test data. Best results are shown in bold. Generally, our framework significantly outperforms other models in all automatic evaluation metrics.

Model	BLEU	Micro F1	Macro F1
Seq2Seq with attention	8.32	17.5	15.6
Copy Net	8.67	23.7	20.8
KV Net	8.70	29.5	24.9
Our model	**11.55**	**50.0**	**42.8**

Table 2. Ablation experiment on navigation domain. -copy refers to a framework without copying. -*KB retriever* refers to a framework without *KB retriever*.

Model	BLEU	Micro F1	Macro F1
Our model	**11.55**	**50.0**	**42.8**
-copying	8.9	26.0	22.7
-*KB retriever*	10.42	33.2	30.1

Table 3. Human evaluation of responses based on random selected previous dialogue history in test dataset.

Model	Correct	Fluent	Humanlike
Copy Net	4.01	4.58	4.51
KV Net	4.23	4.68	4.56
Our model	**4.66**	**4.81**	**4.78**

5.4 Human Evaluation

In this section, we provide human evaluation on our framework and other baseline models. We generated all responses in test dataset. These responses are based on distinct dialogue history. We hire many human experts, and they were asked to judge the quality of their responses according to correctness, cooperativeness, and humanlikeness on a scale from 3 to 5. And each judgment indicates a relative score compared to the standard response from test data. The results are illustrated in Table 3. The results show that our framework outperforms other baseline models on all metrics. The most significant improvement is from correctness, indicating that our model generates more accurate information that the users want to know.

6 Conclusion

In this work, we explore the possibility of introducing an explicit KB retrieval component (*KB retriever*) into the seq2seq dialogue system. Our framework performed an explicit *KB retriever* to lookup over the knowledge base, and applied the copying mechanism to retrieve entities from the *retrieved KB* while decoding. Besides, the *KB retriver* component is trained with distant supervision, which does not need heavy human annotation. Experiments showed that our model outperforms other competitive Seq2Seq models on both automatic and human evaluation metrics. In the future, we would like to jointly model the *KB retriever* and seq2seq framework in an end-to-end training method.

Acknowledgements. We are grateful for helpful comments and suggestions from the anonymous reviewers. This work was supported by the National Key Basic Research Program of China via grant 2014CB340503 and the National Natural Science Foundation of China (NSFC) via grant 61632011 and 61772153.

References

1. Bahdanau, D., Cho, K., Bengio, Y.: Neural machine translation by jointly learning to align and translate. arXiv preprint arXiv:1409.0473 (2014)
2. Bordes, A., Boureau, Y.L., Weston, J.: Learning end-to-end goal-oriented dialog. arXiv preprint arXiv:1605.07683 (2016)
3. Dhingra, B., et al.: End-to-end reinforcement learning of dialogue agents for information access. arXiv preprint arXiv:1609.00777 (2016)
4. Eric, M., Manning, C.D.: A copy-augmented sequence-to-sequence architecture gives good performance on task-oriented dialogue. arXiv preprint arXiv:1701.04024 (2017)
5. Eric, M., Manning, C.D.: Key-value retrieval networks for task-oriented dialogue. arXiv preprint arXiv:1705.05414 (2017)
6. Gu, J., Lu, Z., Li, H., Li, V.O.: Incorporating copying mechanism in sequence-to-sequence learning. arXiv preprint arXiv:1603.06393 (2016)
7. Hochreiter, S., Schmidhuber, J.: Long short-term memory. Neural Comput. **9**(8), 1735–1780 (1997)
8. Kingma, D.P., Ba, J.: Adam: a method for stochastic optimization. arXiv preprint arXiv:1412.6980 (2014)
9. Li, X., Chen, Y.N., Li, L., Gao, J.: End-to-end task-completion neural dialogue systems. arXiv preprint arXiv:1703.01008 (2017)
10. Luong, M.T., Pham, H., Manning, C.D.: Effective approaches to attention-based neural machine translation. arXiv preprint arXiv:1508.04025 (2015)
11. Min, B., Grishman, R., Wan, L., Wang, C., Gondek, D.: Distant supervision for relation extraction with an incomplete knowledge base. In: Proceedings of the 2013 Conference of the North American Chapter of the Association for Computational Linguistics: Human Language Technologies, pp. 777–782 (2013)
12. Mintz, M., Bills, S., Snow, R., Jurafsky, D.: Distant supervision for relation extraction without labeled data. In: Proceedings of the Joint Conference of the 47th Annual Meeting of the ACL and the 4th International Joint Conference on Natural Language Processing of the AFNLP: Volume 2-Volume 2, pp. 1003–1011. Association for Computational Linguistics (2009)
13. Möller, S., et al.: Memo: towards automatic usability evaluation of spoken dialogue services by user error simulations. In: Ninth International Conference on Spoken Language Processing (2006)
14. Nallapati, R., Xiang, B., Zhou, B.: Sequence-to-sequence RNNs for text summarization (2016)

15. Nallapati, R., Zhou, B., Gulcehre, C., Xiang, B., et al.: Abstractive text summarization using sequence-to-sequence RNNs and beyond. arXiv preprint arXiv:1602.06023 (2016)
16. Papineni, K., Roukos, S., Ward, T., Zhu, W.J.: BLEU: a method for automatic evaluation of machine translation. In: Proceedings of the 40th Annual Meeting on Association for Computational Linguistics, pp. 311–318. Association for Computational Linguistics (2002)
17. Pavez, J., Allende, H., Allende-Cid, H.: Working memory networks: augmenting memory networks with a relational reasoning module. arXiv preprint arXiv:1805.09354 (2018)
18. Ritter, A., Cherry, C., Dolan, W.B.: Data-driven response generation in social media. In: Proceedings of the Conference on Empirical Methods in Natural Language Processing, pp. 583–593. Association for Computational Linguistics (2011)
19. Srivastava, N., Hinton, G., Krizhevsky, A., Sutskever, I., Salakhutdinov, R.: Dropout: a simple way to prevent neural networks from overfitting. J. Mach. Learn. Res. **15**(1), 1929–1958 (2014)
20. Sukhbaatar, S., Weston, J., Fergus, R., et al.: End-to-end memory networks. In: Advances in Neural Information Processing Systems, pp. 2440–2448 (2015)
21. Sutskever, I., Vinyals, O., Le, Q.V.: Sequence to sequence learning with neural networks. In: Advances in Neural Information Processing Systems, pp. 3104–3112 (2014)
22. Wen, T.H., et al.: A network-based end-to-end trainable task-oriented dialogue system. arXiv preprint arXiv:1604.04562 (2016)
23. Williams, J.D., Asadi, K., Zweig, G.: Hybrid code networks: practical and efficient end-to-end dialog control with supervised and reinforcement learning. arXiv preprint arXiv:1702.03274 (2017)
24. Xiong, C., Merity, S., Socher, R.: Dynamic memory networks for visual and textual question answering. In: International Conference on Machine Learning, pp. 2397–2406 (2016)
25. Xu, W., Hoffmann, R., Zhao, L., Grishman, R.: Filling knowledge base gaps for distant supervision of relation extraction. In: Proceedings of the 51st Annual Meeting of the Association for Computational Linguistics (Volume 2: Short Papers), vol. 2, pp. 665–670 (2013)
26. Zeng, D., Liu, K., Chen, Y., Zhao, J.: Distant supervision for relation extraction via piecewise convolutional neural networks. In: Proceedings of the 2015 Conference on Empirical Methods in Natural Language Processing, pp. 1753–1762 (2015)
27. Zhao, T., Eskenazi, M.: Towards end-to-end learning for dialog state tracking and management using deep reinforcement learning. arXiv preprint arXiv:1606.02560 (2016)

Attention-Based CNN-BLSTM Networks for Joint Intent Detection and Slot Filling

Yufan Wang, Li Tang, and Tingting He[(⊠)]

School of Computer, Central China Normal University, Wuhan 430079, China
{yufan_wang, lt}@mails.ccnu.edu.cn,
tthe@mail.ccnu.edu.cn

Abstract. Dialogue intent detection and semantic slot filling are two critical tasks in nature language understanding (NLU) for task-oriented dialog systems. In this paper, we present an attention-based encoder-decoder neural network model for joint intent detection and slot filling, which encodes sentence representation with a hybrid Convolutional Neural Networks and Bidirectional Long Short-Term Memory Networks (CNN-BLSTM), and decodes it with an attention-based recurrent neural network with aligned inputs. In the encoding process, our model firstly extracts higher-level phrase representations and local features from each utterance using convolutional neural network, and then propagates historical contextual semantic information with a bidirectional long short-term memory network layer architecture. Accordingly, we could obtain sentence representation by merging the two architectures mentioned above. In the decoding process, we introduce attention mechanism in long short-term memory networks that can provide additional sematic information. We conduct experiment on dialogue intent detection and slot filling tasks with standard data set Airline Travel Information System (ATIS). Experimental results manifest that our proposed model can achieve better overall performance.

Keywords: Nature language understanding · Slot filling · Intent detection Attention model

1 Introduction

Intelligent dialogue system is a core technology in artificial intelligence and will become a friendly human-computer interaction approach. Building dialogue systems usually requires great effort, because the training of NLU and speech generation modules are essential to conduct conversations [1]. NLU, as a critical component of the dialogue system, is a major problem in current intelligent voice interaction and human-machine dialogue. The purpose of NLU is to convert user dialogue text into a form that computer can understand. Typically, NLU system involves the identification of user intent from natural language queries and the extraction of semantic constituents. Generally, the two tasks are referred as intent detection and slot filling [2].

Usually, intent detection and slot filling are performed separately. Intent detection can be abstracted as a classification problem. Slot filling can be abstracted as a sequence labeling problem. There are some traditional methods based on statistics used

M. Sun et al. (Eds.): CCL 2018/NLP-NABD 2018, LNAI 11221, pp. 250–261, 2018.
https://doi.org/10.1007/978-3-030-01716-3_21

for both tasks. However, the traditional methods are not capable of learning the deep semantic information. Deep neural networks have brought new inspiration to various natural language processing tasks, including intent detection and slot filling. In intent detection task, convolutional neural networks (CNNs) and recurrent neural networks (RNNs) are two popular models. While RNN and encoder-decoder have achieved good results in slot filling [2, 3]. Nowadays, there are many joint models used to solve intent detect and slot filling tasks in the meantime. And the joint models achieved good performance in two tasks.

Encoder-decoder neural network is used to solve the slot filling and performs very well in [2]. Generally, sentence representation obtained from encoder based RNN only considers the chronological features of sentences but ignores the importance of local semantic information of sentences. In task of slot filling, however, local semantic information has important influence on identifying the slot correctly. Fortunately, convolution layer is specialized in capturing local semantic information [4], and it can get higher-level representation of sentences. Therefore, we explore CNN results input into the RNN for encoding and get the final sentence representation. In order to keep the original chronological order in the RNN input, we rearrange the output of convolution layer in the relative order of the original sentence. Intuitively, each word between the input and output in the slot filling task is a one-to-one correspondence. Attention mechanism and alignment could obtain corresponding relationship. The sentence representation obtained by encoder can also be used for intention detection. The model joint the two tasks can make them influence and promote each other [5]. In this paper, a new joint model is presented to solve intent detection and slot filling tasks. In the new joint model, a combination of CNN and bidirectional long short-term memory networks (BLSTM) encodes a dialogue sentence into a dense vector that imply features of the sentence. Then, this vector is used to initialize the decoder state. The decoder uses long short-term memory networks (LSTM) cell as basic RNN. User intent and slot labels can be generated by decoding. In addition, an attention-based RNN with aligned inputs is used to decode. Experimental results manifest that our proposed joint model can achieve better overall performance.

The remainder of the paper is organized as follows. Section 2 discusses related work. The details of our model and the proposed methods are described in Sect. 3. Section 4 describes experimental setup. In Sect. 5, the experiment results and analysis are reported. Finally, Sect. 6 draws conclusions and discusses future work.

2   Related Work

In natural language processing, NLU has been a hot research field. NLU typically involves two major tasks—intent detection and slot filling. Intent detection can be abstracted as a classification problem. There are some popular classifiers based on supervised machine-learning methods like Support Vector Machine (SVM) [6], Nave Bayes [7] and so on. Deep learning methods recently have been shown to give excellent performance in natural language processing field. Grabes et al. [8] compared feedforward, recurrent, and gated—such as LSTM and GRU (Gated Recurrent Unit)— networks for intent classification task. Xiao et al. [9] proposed a neural network architecture that connect both the convolution and recurrent layers for intent detection. Slot filling can be abstracted as a sequence annotation problem. The popular traditional approaches include Maximum Entropy Markov Model (MEMM) [10], Conditional Random Field (CRF) [11]. Recently, Aliannejadi et al. [12] proposed graph-based semi-supervised conditional random fields for spoken language understanding using unaligned data. Although traditional machine learning methods have achieved good results, these methods require good feature engineering and additional semantic resources. In recent years, RNN, CNN, as well as their variations or combinations are used to solve slot filling problems. Xu et al. [13] proposed using CNN based triangular CRF for joint intent detection and slot filling. Yao et al. [14] proposed a regression model on top of the LSTM unnormalized scores for slot filling. Vu et al. [15] proposed bi-directional recurrent neural network with ranking loss for slot filling. Kurata et al. [16] proposed leveraging sentence-level information with encoder LSTM for slot filling. Zhu et al. [17] proposed encoder-decoder with focus-mechanism for slot filling. And the combination of LSTM and CRF achieved good results in slot filling.

Joint model for intent detection and slot filling has also been proposed last several years. Guo et al. [18] proposed using a recursive neural network that learns hierarchical representations of the dialogue for the joint task. Specially, an encoding-decoding joint model is used to solve intent detection and slot filling [19]. Liu [2] proposed an attention-based encoder-decoder neural network models for intent detection and slot filling. Zhang et al. [5] proposed using Bidirectional Gated Recurrent Unit (BGRU) for the joint task. Weigelt et al. [20] proposed jointly modeling intent identification and slot filling with contextual and hierarchical information. Such joint models simplify intent detection and slot filling systems, as only one model needs to be trained and deployed. In our work, we have made some changes on the encoder-decoder architecture and integrated CNN with BLSTM as the encoder. Combining the strengths of CNN and LSTM in encoder, we propose a joint model for intent detection and slot filling.

3 Methodology

Our model is inspired by the works of Liu et al. [2], Zhou et al. [21], and Yao et al. [22]. The architecture of the Encode-decoder model is shown in Fig. 1. The main structure includes: encoder (input layer, convolutional layer, window feature sequence and a bidirectional recurrent structure of LSTM) and decoder (LSTM network with attention mechanism and aligned inputs). In terms of encoder, the structure which combines CNN and BLSTM was proved to be more powerful than single RNN in our experiment. In particular, we choose the BLSTM network. Compared to vanilla RNN, BLSTM has the ability to better model long-term dependencies. As for decoder, on one channel, the result of intent detection can be obtained through a softmax function; on the other channel, a LSTM network with attention mechanism and aligned inputs can be used for slot filling.

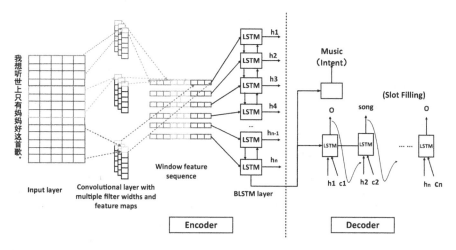

Fig. 1. The architecture of the encoder-decoder model is shown above. The word embedding is used as input to the model. Rearrange the feature maps obtained from convolution layer then feed them to BLSTM. The last state of the encoder is used to initialize the decoder state. An attention-based RNN with aligned inputs is used to decode.

3.1 Encoder

The normal encoder uses LSTM as the encoding network. The improved encoder structure is shown in Fig. 1. In this section, we will introduce the functions and calculation methods for each layer of encoder.

Embedding Layer. In the first step, the dialogue text needs to be converted into a feature vector matrix as the input layer of the model. This model uses embedding technology to convert each word feature into a word embedding vector. We can get the word embedding vectors which we need by using word embedding tools (e.g. Word2vec [23] by google). So we get dialogue text vectors of input $X = [x_1, x_2, x_3, \ldots, x_n]$.

Convolutional Layer. The CNN model has been proved to perform well on some tasks in the NLP field. Convolutional layer of CNN can encode significant information contained in input data with significantly fewer parameters [24]. Convolutional layer is equivalent to a sliding window, which can get the contextual characteristics within the local window of the current word. The parameters of each filter are shared across all the windows. In intent detection and slot filling tasks, we find that a good or comparable performance can be achieved with only one convolutional layer. Multiple filters with differently initialized weights are used to improve the model's learning capability. Generally, multiple filter sizes perform better than a single size. The number of filters k is determined to use experience value. The convolution operation is as follows:

$$s_i = h\left(W_S^T \cdot x_{i:i+m-1} + b_s\right) \tag{1}$$

where $W_S \in \mathbb{R}^{m \times d}$, d is the word embeddings' dimension, and m is the size of convolution window. The vector $x_{i:i+m-1}(x_i \in X = [x_1, x_2, x_3 \ldots x_n])$ represents a window of m words starting from the i-th word to $(i+m-1)$-th word, and the $x_i \in \mathbb{R}^d$, the term b_s is a bias vector. $h(\cdot)$ denotes a nonlinear activation function. The outputs of the feature maps is: $S = [s_1, s_2, s_3 \ldots s_n]$, where n is the number of convolution windows, s_i is the result of every convolution.

Window Feature Sequence. To learn sequential correlations from higher-level sequence representations, we do not feed the output of the convolutional layer directly into the BLSTM. Instead, we add a structure, namely the window feature sequence layer, before feeding the output into the BLSTM. We get the outputs $(S = [s_1, s_2, s_3 \ldots s_n])$ of the convolution layer and rearrange it in the relative order of the original sentence to get the vector $V = [v_1, v_2, v_3 \ldots v_l]$. The window feature sequence is connected by the elements corresponding to the i-th dimension of each feature map after convolution.

$$v_i = f_1^i \oplus f_2^i \oplus f_3^i \oplus \cdots f_{n-1}^i \oplus f_n^i \tag{2}$$

Here, f_n^i denotes the elements corresponding to the i-th dimension of the s_n. The length of feature map and the input length of convolution layer are equal to l. So a total of l combinatorial vectors can be obtained. Each sentence representative is transformed into successive window features. Typically, the pooling operation would be performed after the convolutional layer, but the discontinuous feature selection sampling will destroy the sequential information of the sentence. While, the window feature sequence layer can effectively maintain the sequential information, which contributes to the effective extraction of various features of a given sentence.

Sentence Representation with BLSTM. RNN is able to propagate historical information via a chain-like neural network architecture. Therefore, we choose to use RNN network after the convolution layer instead of max-pooling. In RNN, gradient vanishing and gradient exploding are two problems which are difficult to solve when propagating backwards in the time series [25]. Graves A et al. [26] designed LSTM in order to overcome the problem. An adaptive gating mechanism is introduced to help LSTM unit to keep the previous state and remember the extracted features of the current input data.

A basic LSTM neural network consists of four modules: input gate i_t, forget gate f_t, output gate o_t and memory cell c_t. The following formulas show how to calculate each gate and memory cell unit:

$$i_t = \sigma(W_i \cdot v_t + U_i \cdot h_{t-1} + b_i) \tag{3}$$

$$f_t = \sigma\left(W_f \cdot v_t + U_f \cdot h_{t-1} + b_f\right) \tag{4}$$

$$o_t = \sigma(W_o \cdot v_t + U_o \cdot h_{t-1} + b_o) \tag{5}$$

$$g_t = tanh\left(W_g \cdot v_t + U_g \cdot h_{t-1} + b_g\right) \tag{6}$$

Here $v_t(v_t \in V = [v_1, v_2, v_3 \ldots v_l])$ denotes the current input, and h_{t-1} denotes the previous hidden state. W, U are weight matrices for input and output, respectively, and b is the corresponding bias.

$$c_t = i_t \otimes g_t + f_t \otimes c_{t-1} \tag{7}$$

$$h_t = o_t \otimes tanhc_t \tag{8}$$

Where g_t is the new memory content, c_t is updated by partially forgetting the existing memory and adding a new memory content. Here we use tanh as our non-linear activation function.

According to the LSTM correlation formula, we can deduce the formula of the forward and the backward LSTM. For each timestep t in the input sequence, we will get two kinds of sentence representations $\overleftarrow{h_t}, \overrightarrow{h_t}$. We can obtain the outputs of each timestep in the BLSTM layer:

$$h_t = \left[\overleftarrow{h_t}, \overrightarrow{h_t}\right] \tag{8}$$

The encoder output of sequence: $H = [h_1, h_2, h_3 \ldots h_n]$

3.2   Decoder

The decoder for intent detection and the decoder for slot filling share the same encoder. During model training, we add the two loss functions together as the final cost function. Costs from both decoders are back-propagated to update the model parameters.

Decoder for Intent Detection. We add an additional decoder for intent detection task that shares the same encoder with slot filling decoder. For intent detection, we can get the last output o_0 of encoder, which encodes information of the entire source sequence.

$$y_i = W \cdot o_0 + b \tag{9}$$

where o_0 is also initial decoder state, we use the softmax function to convert the output data into probabilities p_i as follow:

$$p_i = \frac{\exp(y_i)}{\sum_{k=1}^{n} \exp(y_k)} \tag{10}$$

We get the category of max probabilities as the intention of dialogue text

Attention-Based Decoder for Slot Filling. In our model, attention mechanism is introduced to improve the model effect. The attention mechanism can automatically select some important information of word sequence. And attention mechanism let the task-processing system focus more on finding useful information related to the current output in the input data. The ultimate goal of the attention model is to help a framework, such as encoder-decoder, to better learn the interrelationships between multiple content modes and to better represent this information. o_i is the decoder state of the slot label predicted in time step i. The formula is the calculation process of decoder state o_i:

$$o_i = f(o_{i-1}, y_{i-1}, h_i, c_i) \tag{11}$$

where o_{i-1} is the previous decoder state, and y_{i-1} is the previous emitted label. h_i is the encoder hidden state to introduce aligned information. And the context vector c_i is computed as a weighted sum of the encoder states $H = [h_1, h_2, h_3 \ldots h_n]$ as follow:

$$\alpha_i = \frac{\exp(\eta(h_i, o_{i-1}))}{\sum_{i=1}^{n} \exp(\eta(h_i, o_{i-1}))} \tag{12}$$

and

$$\eta(h_i, q) = W h_i + U o_{i-1} \tag{13}$$

$$c_i = \Sigma_{i=1}^{n} \alpha_i h_i \tag{14}$$

where W and U. are parameters. Compared to the decoder without using attention mechanism, the context vector c_i. provides more information for decoder.

4 Experiments

4.1 Data Set

In order to verify the validity of our model, we conduct dialogue intent detection and slot filling on Airline Travel Information Systems (ATIS) dataset [27], which is widely used in NLU research. ATIS is split into 4978 utterances for training and 893 utterances for test. There are 127 slot labels and 22 intent types in the data set. And the ATIS data set has an average of 15 words per sentence. F1-score is used as evaluation metric for slot filling, and accuracy is used as evaluation metric for the task of intent detection.

4.2 Model Training

We explore the effects of different hyper parameters in the proposed method: word embedding size, learning rate, number of filters, batch size, window size for convolution, the hidden layer size and so on.

The hyper-parameter settings of the neural networks may have a great impact on the final results of the experiment, especially the filter sizes in CNN. We discuss the four different sizes of 2, 3, [2, 3, 5] and [2, 4, 5] about convolution kernels, and the sizes of [2, 3, 5] performs the best. The default forget gate bias is set to 1 in LSTM network [28]. Adam optimization algorithm is adopted to calculate the model. In this work, other hyper-parameters are shown in Table 1.

Table 1. Hyper-parameters of our model

Embedding size	64
Hidden units	128
Learning rate	0.001
Loss regularization	0.001
Window size	[2, 3, 5]
Filter number	128
Dropout	0.5

4.3 Compared Methods

Standard neural network model LSTM and the CRF model are chosen for the comparative experiment. The specific method for each model is designed as follows

- CRF: We choose the standard CRF as baseline. The input is the n-grams in a context window. The baseline was demonstrated in the ATIS dataset.
- CNN-CRF: The architecture combination of CNN and CRF, which exploits the dependency between intents and slots. The two tasks to share the same features extracted through CNN layers.
- Deep LSTM: We select a deep LSTM network which consists of multiple layers of LSTMs for comparison.
- RecNN: The model adapted recursive neural networks for joint training. To improve the result on slot filling, the Viterbi algorithm was introduced.
- Bi-RNN with attention: We select bidirectional recurrent neural network with attention mechanism for comparison.
- Bi-RNN with Ranking Loss: We select bidirectional recurrent neural network for comparison. Furthermore, the model use ranking loss function to train the model.
- Encoder-laber Deep LSTM: encodes the whole input sequence into a fixed length vector and then uses this encoded vector as the initial state of another LSTM for sequence labeling.
- Encoder + Decoder: We choose BLSTM as encoder and a unidirectional LSTM as decoder. Attention mechanism and aligned inputs are introduced in decoder.
- Encoder + Decoder (Focus): We choose a standard bidirectional LSTM as encoder and a unidirectional LSTM as decoder. Focus mechanism which is for better alignment is introduced in decoder.

5 Experimental Results and Analysis

Table 2 shows the results of our joint model and previous reported models. Compared with other standard neural networks on the same dataset, our model outperforms both on precision of intent detection and F1-score of slot filling. We discuss the advantage of using BLSTM-CNN as an encoder to encode the input sequence for intent detection and slot filling in task-oriented dialog systems in the next experimental analysis.

Table 2. Comparison with published results on ATIS.

Model	ID precision	SF F1
CRF	–	92.94%
Simple RNN	–	94.11%
CNN-CRF [13]	–	94.35%
LSTM	–	94.85%
Deep LSTM [14]	–	95.08%
Bi-RNN with ranking loss [15]	–	95.56%
Encoder-labeler Deep LSTM [16]	–	95.66%
BLSTM-LSTM (focus) [17]	–	95.79%
RecNN [18]	95.40%	93.22%
Bi-GRU & max-pooling [5]	96.98%	95.61%
BRNN (Attention) [2]	96.65%	95.78%
Encoder-decoder [2]	96.57%	95.87%
Our model	**97.17%**	**97.76%**

5.1 Our Model vs. Other Joint Model

In Table 3, compared with the previous joint model, the F1-score of slot filling and precision of intent detection are both higher in our model

Table 3. Joint model result on AITS

Model	ID precision	SF F1
Encoder-decoder	96.57%	95.87%
Our model (only attention)	96.87%	95.77%
Our model (only aligned)	96.99%	96.89%
Our model (attention and aligned)	**97.17%**	**97.76%**

In the slot filling task, the results of the various models we discussed show that: First, slot filling models with no alignment information do not perform well. The reason is that the alignment between the words and labels is explicit. But the model may not be able to learn the alignment from the training data. Second, we can see that the F1 score and accuracy of the attention-based models are higher than models with no attention mechanism. By investigating the attentional mechanism learned by the model, we found that there are differences in the distribution of attentional weights among words in the source sequence. The reason is that attention mechanism can better consider the importance of contextual information. Third, the attention-based encoder-decoder model with aligned inputs has a lower F1 score than the attention-based encoder-decoder model based on CNN-BLSTM. Because combining the bidirectional LSTM with the convolutional structure enables the model to extract comprehensive information, namely historical, future and local context of any position in a sequence. In the intent detection task, the accuracy of our model also exceeds the standard Bidirectional Recurrent neural networks. The attention-based encoder-decoder model performs better in that the model extracts the deep local semantic information of the dialogue text and the salient features of implicit category information through convolution layer. We infer that our model could find the balance between two loss functions of two subtasks. This is an important reason for our model to achieve better overall performance.

5.2 Separate Model vs. Joint Model

Table 4 shows the experiment results of joint model and separate model. First, we give the definition of joint model and separate model. The joint model we propose can simultaneously get the both results of intent detection and slot filling. The separate model is only for one task. For intent detection, the separate model is the same as the joint model in encoder, but only intent detection task is concerned in decoder. Slot filling is also the same. The joint model outperforms the single model on both tasks, indicating that joint training is effective. We add the loss functions of the two tasks as the loss function of the joint model. Through a unified loss function and a common representation, the model learns the correlation between the two tasks to promote each other. Because of the information sharing of two channels, our joint model outperforms the pipeline method with single channel. The joint model could make a balance between the two loss functions of two subtasks. In addition, the model shares the model parameters to speed up the training procedure.

Table 4. The result of model Joint and separate on AITS

Model	ID precision	SF precision	SF F1
Intent detection only	96.51%	–	–
Slot filling only	–	97.36%	97.03%
Join model	**97.17%**	**97.94%**	**97.76%**

5.3 Filter Sizes Setting

In the convolutional layer, the local feature of the dialogue text is extracted by the filter, which is similar to the N-gram feature. Different sizes of the filter bring about different local feature maps. So, the different size of filters in the convolutional layer has an important effect on convolutional results. Generally, multiple filter sizes perform better than a single size.

In Table 5, we found that single convolutional layer with filter size [2, 3, 5] outperforms the other cases in our experiments. Comparing 3, [2, 3, 5] with 2, [2, 4, 5], we can see that 3, [2, 3, 5] performs better in slot filling and intent detection, which proves that tri-gram features play an important role in capturing local features in our task. We speculate that BLSTM could learn better semantic sentence representation from the sequences of tri-gram features.

Table 5. The result of different size of filters in the convolution layer

Filter size	ID precision	SF precision	SF F1
2	96.37%	97.20%	96.69%
3	96.48%	97.52%	96.91%
[2, 4, 5]	96.57%	97.60%	97.13%
[2, 3, 5]	**97.17%**	**97.94%**	**97.76%**

6 Conclusions and Future Work

In this paper, we discuss the feasibility of joint model which uses a combination of CNN and BLSTM as an encoder and uses an attention-based LSTM with aligned inputs as decoder for intent detection and slot filling. In our model, the semantics of the context can be captured by bidirectional LSTM neural network. And the local important latent semantic factors can be selected by convolution layer. Experiments show that our model which combines the advantages of BLSTM and CNN can achieve better performance than baseline models.

Due to the particularity of the dialogue text, there are still many problems in the task, such as new word feature misrecognition and other issues. Consequently, how to construct effective word embeddings and solve new words problem in dialogue utterance will be the future work of our study.

Acknowledgments. This research is supported by the Fundamental Research Funds for the Central Universities (CCNU18JCK05), the National Natural Science Foundation of China (61532008), the National Science Foundation of China (61572223), the National Key Research and Development Program of China (2017YFC0909502).

References

1. Shen, B., Inkpen, D.: Speech intent recognition for robots (2017)
2. Liu, B., Lane, I.: Attention-based recurrent neural network models for joint intent detection and slot filling (2016)
3. Liu, B., Line, I.: Recurrent neural network structured output prediction for spoken language understanding (2015)
4. Kim, Y.: Convolutional neural networks for sentence classification. Eprint Arxiv (2014)
5. Zhang, X., Wang, H.: A joint model of intent determination and slot filling for spoken language understanding, pp. 5690–5694 (2016)

6. Haffner, P., Tur, G., Wright, J.H.: Optimizing SVMs for complex call classification (2003)
7. Chen, J., Huang, H., Tian, S., et al.: Feature selection for text classification with Naive Bayes. Expert Syst. Appl. Int. J. **36**(3), 5432–5435 (2009)
8. Graves, A., Jaitly, N., Mohamed, A.R.: Hybrid speech recognition with deep bidirectional LSTM (2014)
9. Xiao, Y., Cho, K.: Efficient character-level document classification by combining convolution and recurrent layers (2016)
10. Xiao, J., Wang, X., Liu, B.: The study of a nonstationary maximum entropy Markov model and its application on the pos-tagging task. ACM Trans. Asian Lang. Inf. Process. **6**(2), 7 (2007)
11. Raymond, C., Riccardi, G.: Generative and discriminative algorithms for spoken language understanding (2007)
12. Aliannejadi, M., Kiaeeha, M., Khadivi, S., et al.: Graph-based semi-supervised conditional random fields for spoken language understanding using unaligned data (2017)
13. Xu, P., Sarikaya, R.: Convolutional neural network based triangular CRF for joint intent detection and slot filling (2014)
14. Yao, K., Peng, B., Zhang, Y., et al.: Spoken language understanding using long short-term memory neural networks (2015)
15. Vu, N.T., Gupta, P., Adel, H., et al.: Bi-directional recurrent neural network with ranking loss for spoken language understanding (2016)
16. Kurata, G., Xiang, B., Zhou, B., et al.: Leveraging sentence-level information with encoder LSTM for natural language understanding (2016)
17. Zhu, S., Yu, K.: Encoder-decoder with focus-mechanism for sequence labelling based spoken language understanding (2017)
18. Guo, D., Tur, G., Yih, W.T., et al.: Joint semantic utterance classification and slot filling with recursive neural networks (2015)
19. Liu, B., Lane, I.: Joint online spoken language understanding and language modeling with recurrent neural networks (2016)
20. Weigelt, S., Hey, T., Landhäußer, M.: Integrating a dialog component into a framework for spoken language understanding (2018)
21. Zhou, C., Sun, C., Liu, Z., et al.: A C-LSTM neural network for text classification. Comput. Sci. **1**(4), 39–44 (2015)
22. Yao, K., Peng, B., Zhang, Y., et al.: Spoken language understanding using long short-term memory neural networks. In: IEEE – Institute of Electrical & Electronics Engineers, pp. 189–194 (2014)
23. Word2vec Homepage. http://code.google.com/archive/p/word2vec/
24. Yin, W., Schütze, H., Xiang, B., et al.: ABCNN: attention-based convolutional neural network for modeling sentence pairs (2015)
25. Morin, F., Bengio, Y.: Hierarchical probabilistic neural network language model. Aistats (2005)
26. Graves, A., Mohamed, A.R., Hinton, G.: Speech recognition with deep recurrent neural networks (2013)
27. Hemphill, C.T., Godfrey, J.J., Doddington, G.R.: The ATIS spoken language systems pilot corpus. In: Proceedings of the Darpa Speech & Natural Language Workshop, pp. 96–101 (1990)
28. Jozefowicz, R., Zaremba, W., Sutskever, I.: An empirical exploration of recurrent network architectures. In: International Conference on Machine Learning, pp. 2342–2350. JMLR.org (2015)

Multi-Perspective Fusion Network for Commonsense Reading Comprehension

Chunhua Liu[1], Yan Zhao[1], Qingyi Si[1], Haiou Zhang[1], Bohan Li[1], and Dong Yu[1,2(✉)]

[1] Beijing Language and Culture University, Beijing, China
chunhualiu596@gmail.com, zhaoyan.nlp@gmail.com,
xk17sqy@126.com, hozhangel@126.com, yudong_blcu@126.com,
bohanli.lavida@gmail.com
[2] Beijing Advanced Innovation for Language Resources of BLCU, Beijing, China

Abstract. Commonsense Reading Comprehension (CRC) is a significantly challenging task, aiming at choosing the right answer for the question referring to a narrative passage, which may require commonsense knowledge inference. Most of the existing approaches only fuse the interaction information of choice, passage, and question in a simple combination manner from a *union* perspective, which lacks the comparison information on a deeper level. Instead, we propose a Multi-Perspective Fusion Network (MPFN), extending the single fusion method with multiple perspectives by introducing the *difference* and *similarity* fusion. More comprehensive and accurate information can be captured through the three types of fusion. We design several groups of experiments on MCScript dataset [11] to evaluate the effectiveness of the three types of fusion respectively. From the experimental results, we can conclude that the difference fusion is comparable with union fusion, and the similarity fusion needs to be activated by the union fusion. The experimental result also shows that our MPFN model achieves the state-of-the-art with an accuracy of 83.52% on the official test set.

Keywords: Commonsense Reading Comprehension
Fusion Network · Multi-perspective

1 Introduction

Machine Reading Comprehension (MRC) is an extremely challenging topic in natural language processing field. It requires a system to answer the question referring to a given passage.In real reading comprehension, the human reader can fully understand the passage with the prior knowledge to answer the question. To directly relate commonsense knowledge to reading comprehension, SemEval2018 Task 11 defines a new sub-task called Commonsense Reading Comprehension, aiming at answering the questions that requires both commonsense knowledge

© Springer Nature Switzerland AG 2018
M. Sun et al. (Eds.): CCL 2018/NLP-NABD 2018, LNAI 11221, pp. 262–274, 2018.
https://doi.org/10.1007/978-3-030-01716-3_22

and the understanding of the passage. The challenge of this task is how to answer questions with the commonsense knowledge that does not appear in the passage explicitly. Table 1 shows an example of CRC.

Table 1. An example of CRC.

Passage: It was night time and it was time to go to bed. The boy wanted to keep playing. I told him that after he got ready for bed I would read a story to him. He dawdled a bit but finally started getting ready for bed. First of all he had to take a bath. He splashed in the tub and split water all over the floor. Next he dried off in a big, fluffy blue towel. Then he brushed his teeth with his special Star Wars toothbrush. Next he dressed in his Star Wars underwear and then put on his Star Wars pajamas. His dad and I tucked him into his bed that was made with Star Wars sheets. He said his prayers. Next was story time. I pulled out his favorite book about (you guessed it) Star Wars. He gradually dozed off dreaming about Anakin Skywalker and a galaxy far, far away
Q1: Did they sleep in the same room as their parents?
A. Yes, they all slept in one big loft   B. No they have their own room
Q2: Why didn't the child go to bed by themselves?
A. The child wanted to watch a Star Wars movie.   B. The child wanted to continue playing

Most studies on CRC task are neural network based (NN-based) models, which typically have the following characteristics. Firstly, word representations are augmented by additional lexical information. Secondly, the interaction process is usually implemented by the attention mechanism, which can provide the interaction representations like choice-aware passage, choice-aware question, and question-aware passage. Thirdly, the original representations and interaction representations are fused together and then aggregated by a Bidirectional Long Short-Term Memory Network (BiLSTM) [4] to get high-order semantic information. Fourthly, the final output based on their bilinear interactions.

The NN-based models have shown powerfulness on this task. However, there are still some limitations. Firstly, the two fusion processes of passage and question to choice are implemented separately, until producing the final output. Secondly, the existing fusion method used in reading comprehension task is usually implemented by concatenation [2,24], which is monotonous and cannot capture the partial comparison information between two parts. Studies on Natural Language Inference (NLI) have explored more functions [1,10], such as element-wise subtraction and element-wise multiplication, to capture more comparison information, which have been proved to be effective.

In this paper, we introduce a Muti-Perspective Fusion Network (MPFN) to tackle these limitations. The model can fuse the choice with passage and question simultaneously to get a multi-perspective fusion representation. Furthermore, inspired by the element-wise subtraction and element-wise multiplication function used in [1], we define three kinds of fusion functions from multiple perspectives to fuse choice, choice-aware passage, and choice-aware question. The three fusions are union fusion, difference fusion, and similarity fusion. Note that, we name the concatenation fusion method as union fusion in this paper, which collects the global information. The difference fusion and the similarity fusion can discover the different parts and similar parts among choice, choice-aware passage, and choice-aware question respectively.

MPFN comprises an encoding layer, a context fusion layer, and an output layer. In the encoding layer, we employ a BiLSTM as the encoder to obtain context representations. To acquire better semantic representations, we apply union fusion in the word level. In the context fusion layer, we apply union fusion, difference fusion, and similarity fusion to obtain a multi-perspective fusion representation. In the output layer, a self-attention and a feed-forward neural network are used to make the final prediction.

We conduct experiments on MRScript dataset released by [11]. Our single and ensemble model achieve the accuracy of 83.52% and 84.84% on the official test set respectively. Our main contributions are as follows:

- We propose a general fusion framework with two-layer fusion, which can fuse the passage, question, and choice simultaneously.
- To collect multi-perspective fusion representations, we define three types of fusions, consisting of union fusion, difference fusion, and similarity fusion.
- We design several groups of experiments to evaluate the effectiveness of the three types of fusion and prove that our MPFN model outperforms all the other models.

2   Related Work

MRC has gained significant popularity over the past few years. Several datasets have been constructed for testing the comprehension ability of a system, such as *MCTest* [15], *SQuAD* [14], *BAbI* [22], *TriviaQA* [6], RACE [8], and NewsQA [17]. Each dataset focuses on one specific aspect of reading comprehension. Particularly, the MCScript [11] dataset concerns answering the question which requires using commonsense knowledge.

Many architectures on MRC follow the process of representation, attention, fusion, and aggregation [5, 16, 20, 24, 25, 27]. BiDAF [16] fuses the passage-aware question, the question-aware passage, and the original passage in context layer by concatenation, and then uses a BiLSTM for aggregation. The fusion levels in current advanced models are categorized into three types by [5] , including word-level fusion, high-level fusion, and self-boosted fusion. They further propose a FusionNet to fuse the attention information from bottom to top to obtain a fully-aware representation for answer span prediction.

On SemEval2018 Task 11, most of the models use the attention mechanism to build interactions among the passage, the question, and the choice [2,3,18,23]. The most competitive models are [2,18], and both of them employ concatenation fusion to integrate the information. [18] utilizes choice-aware passage and choice-aware question to fuse the choice in word level. In addition, they apply the question-aware passage to fuse the passage in context level. Different from [18], both the choice-aware passage and choice-aware question are fused into choice in the context level in [2] , which is the current state-of-the-art result on the MCSript dataset.

On NLI task, fusing the premise-aware hypothesis into the hypothesis is an effective and commonly-used method. [12,19] leverage the concatenation of the hypothesis and the hypothesis-aware premise to help improve the performance of their model. The element-wise subtraction and element-wise multiplication between the hypothesis and the hypothesis-aware premise are employed in [1] to enhance the concatenation.

Almost all the models on CRC only use the union fusion. In our MPFN model, we design another two fusion methods to extend the perspective of fusion. We evaluate the MPFN model on MRC task and achieve the state-of-the-art result.

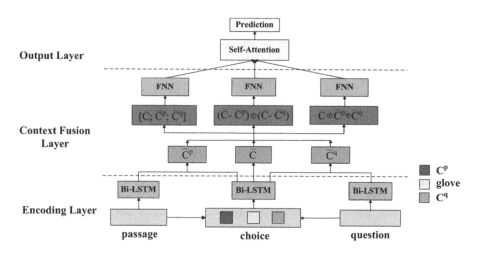

Fig. 1. Architecture of our MPFN model.

3   Model

The overview of our Multi-Perspective Fusion Network (MPFN) is shown in Fig. 1. Given a narrative passage about a series of daily activities and several corresponding questions, a system requires to select a correct choice from two options for each question. In this paper, we denote $\mathbf{p} = \{\mathbf{p}_1, \mathbf{p}_2, ..., \mathbf{p}_{|\mathbf{p}|}\}$ as the passage, $\mathbf{q} = \{\mathbf{q}_1, \mathbf{q}_2, ..., \mathbf{q}_{|\mathbf{q}|}\}$ as a question, $\mathbf{c} = \{\mathbf{c}_1, \mathbf{c}_2, ..., \mathbf{c}_{|\mathbf{c}|}\}$ as one of the candidate choice, and a true label $y^* \in \{0, 1\}$. Our model aims to compute a probability for each choice and take the one with higher probability as the prediction label. Our model consists of three layers: an encoding layer, a context fusion layer, and an output layer. The details of each layer are described in the following subsections.

3.1   Encoding Layer

This layer aims to encode the passage embedding p, the question embedding q, and the choice embedding c into context embeddings. Specially, we use a one-layer BiLSTM as the context encoder.

$$\bar{c}_i = \text{BiLSTM}(c, i), \qquad\qquad i \in [1, 2, \cdots, |c|] \qquad (1)$$

$$\bar{p}_j = \text{BiLSTM}(p, j), \qquad\qquad j \in [1, 2, \cdots, |p|] \qquad (2)$$

$$\bar{q}_k = \text{BiLSTM}(q, k), \qquad\qquad k \in [1, 2, \cdots, |q|] \qquad (3)$$

The embeddings of p, q and c are semantically rich word representations consisting of several kinds of embeddings. Specifically, the embeddings of passage and question are the concatenation of the Golve word embedding, POS embedding, NER embedding, Relation embedding and Term Frequency feature. And the embeddings of choice comprise the Golve word embedding, the choice-aware passage embedding, and choice-aware question embedding . The details about each embedding are follows:

Glove Word Embedding. We use the 300-dimensional Glove word embeddings trained from 840B Web crawl data [13]. The out-of-vocabulary words are initialized randomly. The embedding matrix are fixed during training.

POS&NER Embedding. We leverage the Part-of-Speech (POS) embeddings and Named-Entity Recognition (NER) embeddings. The two embeddings are randomly initialized and updated during training.

Relation Embedding. Relations are extracted form ConceptNet. For each word in the choice, if it satisfies any relation with another word in the passage or the question, the corresponding relation will be taken out. If the relations between two words are multiple, we just randomly choose one. The relation embeddings are generated in the similar way of POS embeddings.

Term Frequency. Following [18], we introduce the term frequency feature to enrich the embedding of each word. The calculation is based on English Wikipedia.

Choice-Aware Passage Embedding. The information in the passage that is relevant to the choice can help encode the choice [21]. To acquire the choice-aware passage embedding c_i^p, we utilize dot product between non-linear mappings of word embeddings to compute the attention scores for the passage [9].

$$c_i^p = Attn(c_i, \{p_j\}_1^{|p|}) = \sum_{j=1}^{|p|} \alpha_{ij} p_j \qquad (4)$$

$$\alpha_{ij} \propto exp(S(c_i, p_j)), \quad S(c_i, p_j) = ReLU(Wc_i)^T ReLU(Wp_j) \qquad (5)$$

Choice-Aware Question Embedding. The choice relevant question information is also important for the choice. Therefore, we adopt the similar attention way as above to get the choice-aware question embedding $c_i^q = Attn(c_i, \{q_k\}_1^{|q|})$.

The embeddings delivered to the BiLSTM are the concatenation the above components, where $p_j = [p_j^{glove}, p_j^{pos}, p_j^{ner}, p_j^{rel}, p_j^{tf}]$, $c_i = [c_i^{glove}, c_i^p, c_i^q]$, and $q_k = [q_k^{glove}, q_k^{pos}, q_k^{ner}, q_k^{rel}, q_k^{tf}]$.

3.2   Context Fusion Layer

This is the core layer of our MPFN model. In this layer, we define three fusion functions, which consider the union information, the different information, and the similar information of the choice, passage, and question.

Since we have obtained the choice context $\bar{c}_i$, the passage context $\bar{p}_j$, and the question context $\bar{q}_k$ in the encoding layer, we can calculate the choice-aware passage contexts $\tilde{c}_i^p$ and choice-aware question contexts $\tilde{c}_i^q$. Then we deliver them together with the choice contexts $\bar{c}_i$ to the three fusion functions.

Choice-Aware Passage Context. In this part, we calculate the choice-aware passage representations $\tilde{c}_i^p = \sum_j \beta_{ij} \bar{p}_j$. For model simplification, here we use dot product between choice contexts and passage contexts to compute the attention scores β_{ij}:

$$\beta_{ij} = \frac{exp(\bar{c}_i^T \bar{p}_j)}{\sum_{j'=1}^{|p|} exp(\bar{c}_i^T \bar{p}_{j'})} \tag{6}$$

Choice-Aware Question Context. In a similar way as above, we get the choice-aware question context $\tilde{c}_i^q = \sum_j \beta_{ik} \bar{q}_k$. The β_{ik} is the dot product of the choice context $\bar{c}_i$ and question context $\bar{q}_k$.

Multi-Perspective Fusion. This is the key module in our MPFN model. The goal of this part is to produce multi-perspective fusion representation for the choice $\bar{c}_i$, the choice-aware passage $\tilde{c}_i^p$, and the choice-aware question $\tilde{c}_i^q$. In this paper, we define fusion in three perspectives: *union*, *difference*, and *similarity*. Accordingly, we define three fusion functions to describe the three perspectives. The outputs and calculation of the three functions are as follows:

$$u_i = [\bar{c}_i; \tilde{c}_i^p; \tilde{c}_i^q], \tag{7}$$
$$d_i = (\bar{c}_i - \tilde{c}_i^p) \odot (\bar{c}_i - \tilde{c}_i^q), \tag{8}$$
$$s_i = \bar{c}_i \odot \tilde{c}_i^p \odot \tilde{c}_i^q, \tag{9}$$

where ; , $-$, and $\odot$ represent concatenation, element-wise subtraction, and element-wise multiplication respectively. And u_i, d_i, and s_i are the representations from the union, difference and similarity perspective respectively.

The union perspective is commonly used in a large bulk of tasks [5,12,24]. It can see the whole picture of the passage, the question, and the choice by concatenating the $\tilde{c}_i^p$ and $\tilde{c}_i^q$ together with c_i . While the difference perspective captures the different parts between choice and passage, and the difference parts between choice and question by $\bar{c}_i - \tilde{c}_i^p$ and $\bar{c}_i - \tilde{c}_i^q$ respectively. The $\odot$ in difference perspective can detect the two different parts at the same time and emphasize them. In addition, the similarity perspective is capable of discovering the similar parts among the passage, the question, and the choice.

To map the three fusion representations to lower and same dimension, we apply three different FNNs with the ReLU activation to u_i, d_i, and s_i. The final output g_i is the concatenation of the results of the three FNNs, which represents a global perspective representation.

$$g_i = [f^u(u_i), f^d(d_i), f^s(s_i)] \tag{10}$$

3.3 Output Layer

The output layer includes a self-attention layer and a prediction layer. Following [26], we summarize the global perspective representation $\{g_i\}_1^{|c|}$ to a fixed length vector r. We compute the $r = \sum_{i=1}^{|c|} b_i g_i$, where b_j is the self-weighted attention score:

$$b_i = \frac{exp(Wg_i)}{\sum_{i'=1}^{|c|} exp(Wg_{i'})} \tag{11}$$

In the prediction layer, we utilize the output of self-attention r to make the final prediction.

4 Experiments

4.1 Experimental Settings

Data. We conduct experiments on the MCScript [11], which is used as the official dataset of SemEval2018 Task11. This dataset constructs a collection of text passages about daily life activities and a series of questions referring to each passage, and each question is equipped with two answer choices. The MCScript comprises 9731, 1411, and 2797 questions in training, development, and test set respectively. For data preprocessing, we use spaCy[1] for sentence tokenization, Part-of-Speech tagging, and Name Entity Recognition. The relations between two words are generated by ConceptNet.

Parameters. We use the standard cross-entropy function as the loss function. We choose Adam [7] with initial momentums for parameter optimization. As for hyper-parameters, we set the batch size as 32, the learning rate as 0.001, the dimension of BiLSTM and the hidden layer of FNN as 123. The embedding size of Glove, NER, POS, Relation are 300, 8, 12, 10 respectively. The dropout rate of the word embedding and BiLSTM output are 0.386 and 0.40 respectively.

---

[1] https://github.com/explosion/spaCy.

Table 2. Experimental results of models

Model	Test (%acc)
SLQA	79.94
Rusalka	80.48
HMA Model (single) [2]	80.94
TriAN (single) [18]	81.94
MPFN (single)	**83.52**
(jiangnan) (ensemble) [23]	80.91
MITRE (ensemble) [3]	82.27
TriAN (ensemble) [18]	83.95
HMA Model (ensemble) [2]	84.13
MPFN (ensemble)	**84.84**

4.2   Experimental Results

Table 2 shows the results of our MPFN model along with the competitive models on the MCScript dataset. The TriAN achieves 81.94% in terms of test accuracy, which is the best result of the single model. The best performing ensemble result is 84.13%, provided by HMA, which is the voting results of 7 single systems.

Our single MPFN model achieves 83.52% in terms of accuracy, outperforming all the previous models. The model exceeds the HMA and TriAN by approximately 2.58% and 1.58% absolute respectively. Our ensemble model surpasses the current state-of-the-art model with an accuracy of 84.84%. We got the final ensemble result by voting on 4 single models. Every single model uses the same architecture but different parameters.

4.3   Discussion of Multi-Perspective

To study the effectiveness of each perspective, we conduct several experiments on the three single perspectives and their combination perspective. Table 3 presents their comparison results. The first group of models are based on the three single perspectives, and we can observe that the union perspective performs best compared with the difference and similarity perspective. Moreover, the union perspective achieves 82.73% in accuracy, exceeding the TriAN by 0.79% absolute. We can also see that the similarity perspective is inferior to the other two perspectives.

The second group of models are formed from two perspectives. Compared with the single union perspective, combining the difference perspective with the union perspective can improve 0.11%. Composing union and similarity fusion together doesn't help the training. To our surprise, the combination of similarity perspective and difference perspective obtains 83.09% accuracy score.

Table 3. Test accuracy of multi-perspective

Perspective	MPFN	MPFN+BiLSTM
U	82.73	82.73
D	82.27	81.77
S	81.55	80.59
DU	82.84	82.16
SU	82.48	82.87
SD	83.12	83.09
SDU	**83.52**	82.70

Table 4. Encoding inputs ablation study.

Model	Test (%acc)
MPFN	**83.52**
w/o POS	82.70
w/o NER	82.62
w/o Rel	81.98
w/o TF	81.91
w/o C^p	81.62
w/o C^q	82.16
w/o C^p & C^q	81.66

The last model is our MPFN model, which performing best. The final result indicates that composing the union perspective, difference perspective, and similarity perspective together to train is helpful.

Many advanced models employ a BiLSTM to further aggregate the fusion results. To investigate whether a BiLSTM can assist the model, we apply another BiLSTM to the three fusion representations in Formula 10 respectively and then put them together. The results are shown in the second column in Table 3, which indicate that the BiLSTM does not help improve the performance of the models.

4.4 Encoding Inputs Ablation

In the section, we conduct ablation study on the encoding inputs to examine the effectiveness each component. The experiment results are listed in Table 4.

From the best model, if we remove the POS embedding and NER embedding, the accuracy drops by 0.82% and 0.9%. Without Relation embedding,

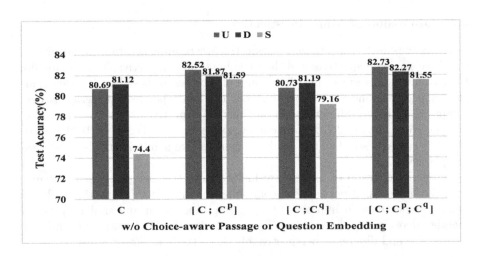

Fig. 2. Influence of word-level interaction.

the accuracy drops to 81.98%, revealing that the external relations are helpful to the context fusions. Without Term Frequency, the accuracy drops by approximately 1.61%. This behavior suggests that the Term Frequency feature has a powerful capability to guide the model.

After removing the C^p, we find the performance degrades to 81.62%. This demonstrates that information in the passage is significantly important to final performance. If we remove C^q from the MPFN, the accuracy drops to 82.16%. If we remove the word level fusion completely, we will obtain an 81.66% accuracy score. These results demonstrate that each component is indispensable and the bottom embeddings are the basic foundations of the top layer fusions.

4.5   Influence of Word-Level Interaction

In this section, we explore the influence of word-level interaction to each perspective. Figure 2 reports the overall results of how each perspective can be affected by the lower level interaction. The C^p and the C^q represent the choice-aware passage embedding and the choice-aware question embedding respectively. We can observe that the results of $[C; C^p]$, $[C; C^q]$, and $[C; C^p; C^q]$ are all higher than the result of C alone, indicating the effectiveness of word embedding interaction.

Both the union fusion and difference fusion can achieve more than 80% accuracy, while the similarity fusion is very unstable. We also observe that the difference fusion is comparable with the union fusion, which even works better than the union fusion when the information of C^p is not introduced into the input of encoding. The similarity fusion performs poorly in C and $[C; C^q]$, while yielding a huge increase in the remaining two groups of experiments, which is an interesting phenomenon. We infer that the similarity fusion needs to be activated by the union fusion.

In summary, we can conclude that integrate the information of C^p into C can greatly improve the performance of the model. Combining C^q together with C^p can further increase the accuracy.

4.6   Visualization

In this section, we visualize the union and difference fusion representations and show them in Fig. 3. And, we try to analyze their characteristics and compare them to discover some connections. The values of similarity fusion are too small to observe useful information intuitively, so we do not show it here. We use the example presented in Table 1 for visualization, where the question is *Why didn't the child go to bed by themselves?* and the corresponding True choice is *The child wanted to continue playing*.

The left region in Fig. 3 is the union fusion. The most intuitive observation is that it captures comprehensive information. The values of *child, wanted, playing* are obvious higher than other words. This is consistent with our prior cognition, because the concatenation operation adopted in union fusion does not lose any content. While the difference union shows in the right region in Fig. 3 focuses on some specific words. By further comparison, we find that the difference fusion

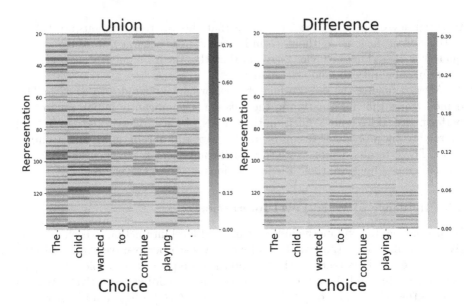

Fig. 3. Visualization of fusions

can pay attention to the content ignored by the union fusion. What's more, the content acquired by the union would not be focused by the difference again. In other words, the union fusion and difference fusion indeed can emphasize information from the different perspective.

5   Conclusion

In this paper, we propose the Multi-Perspective Fusion Network (MPFN) for the Commonsense Reading Comprehension (CMC) task. We propose a more general framework for CRC by designing the difference and similarity fusion to assist the union fusion. Our MPFN model achieves an accuracy of 83.52% on MCScript, outperforming the previous models. The experimental results show that union fusion based on the choice-aware passage, the choice-aware question, and the choice can surpass the TriAN and HMA model. The difference fusion performs stably, which is comparable with the union fusion. We find that the word-level union fusion can significantly influence the context-level fusion. The choice-aware passage word embedding can activate the similarity fusion. We find that combining the similar parts and the difference parts together can obtain the best performance among the two-perspective models. By taking the three types of fusion methods into consideration, our MPFN model achieves a state-of-the-art result.

Acknowledgements. This work is funded by Beijing Advanced Innovation for Language Resources of BLCU, the Fundamental Research Funds for the Central Universities in BLCU (17PT05), the Natural Science Foundation of China (61300081), and the Graduate Innovation Fund of BLCU (No. 18YCX010).

References

1. Chen, Q., Zhu, X., Ling, Z.H., Wei, S., Jiang, H., Inkpen, D.: Enhanced LSTM for natural language inference. In: Proceedings of the 55th Annual Meeting of the Association for Computational Linguistics (Volume 1: Long Papers), pp. 1657–1668. Association for Computational Linguistics, Vancouver, July 2017. http://aclweb.org/anthology/P17-1152
2. Chen, Z., Cui, Y., Ma, W., Wang, S., Liu, T., Hu, G.: HFL-RC system at SemEval-2018 task 11: hybrid multi-aspects model for commonsense reading comprehension. arXiv preprint arXiv:1803.05655 (2018)
3. Merkhofer, E., Henderson, J., Bloom, D., Strickhart, L., Zarrella, G.: MITRE at SemEval-2018 task 11: commonsense reasoning without commonsense knowledge (2018)
4. Hochreiter, S., Schmidhuber, J.: Long short-term memory. Neural Comput. **9**(8), 1735–1780 (1997). https://doi.org/10.1162/neco.1997.9.8.1735
5. Huang, H., Zhu, C., Shen, Y., Chen, W.: FusionNet: fusing via fully-aware attention with application to machine comprehension. CoRR abs/1711.07341 (2017)
6. Joshi, M., Choi, E., Weld, D.S., Zettlemoyer, L.: TriviaQA: a large scale distantly supervised challenge dataset for reading comprehension. In: Proceedings of the 55th Annual Meeting of the Association for Computational Linguistics. Association for Computational Linguistics, Vancouver, July 2017
7. Kingma, D.P., Ba, J.: Adam: a method for stochastic optimization. CoRR abs/1412.6980 (2014)
8. Lai, G., Xie, Q., Liu, H., Yang, Y., Hovy, E.H.: RACE: large-scale reading comprehension dataset from examinations. In: EMNLP (2017)
9. Lee, K., Kwiatkowski, T., Parikh, A.P., Das, D.: Learning recurrent span representations for extractive question answering. CoRR abs/1611.01436 (2016). http://arxiv.org/abs/1611.01436
10. Mou, L., et al.: Natural language inference by tree-based convolution and heuristic matching. In: Proceedings of the 54th Annual Meeting of the Association for Computational Linguistics (Volume 2: Short Papers), pp. 130–136. Association for Computational Linguistics, Berlin, August 2016. http://anthology.aclweb.org/P16-2022
11. Ostermann, S., Modi, A., Roth, M., Thater, S., Pinkal, M.: MCScript: a novel dataset for assessing machine comprehension using script knowledge. In: Proceedings of the Eleventh International Conference on Language Resources and Evaluation (LREC 2018), 7–12 May 2018, Miyazaki, Japan. European Language Resources Association (ELRA) (2018)
12. Parikh, A., Täckström, O., Das, D., Uszkoreit, J.: A decomposable attention model for natural language inference. In: Proceedings of the 2016 Conference on Empirical Methods in Natural Language Processing, pp. 2249–2255. Association for Computational Linguistics (2016). https://doi.org/10.18653/v1/D16-1244, http://www.aclweb.org/anthology/D16-1244
13. Pennington, J., Socher, R., Manning, C.D.: Glove: global vectors for word representation. In: EMNLP (2014)
14. Rajpurkar, P., Zhang, J., Lopyrev, K., Liang, P.: Squad: 100, 000+ questions for machine comprehension of text. CoRR abs/1606.05250 (2016)
15. Richardson, M., Burges, C.J.C., Renshaw, E.: MCTest: a challenge dataset for the open-domain machine comprehension of text. In: EMNLP (2013)

16. Seo, M.J., Kembhavi, A., Farhadi, A., Hajishirzi, H.: Bidirectional attention flow for machine comprehension. CoRR abs/1611.01603 (2016)
17. Trischler, A., et al.: NewsQA: a machine comprehension dataset. In: Rep4NLP@ACL (2017)
18. Wang, L., Sun, M., Zhao, W., Shen, K., Liu, J.: Yuanfudao at SemEval-2018 task 11: three-way attention and relational knowledge for commonsense machine comprehension. In: SemEval@NAACL-HLT, pp. 758–762. Association for Computational Linguistics (2018)
19. Wang, S., Jiang, J.: Learning natural language inference with LSTM. In: Proceedings of the 2016 Conference of the North American Chapter of the Association for Computational Linguistics: Human Language Technologies, pp. 1442–1451. Association for Computational Linguistics, San Diego, June 2016. http://www.aclweb.org/anthology/N16-1170
20. Wang, W., Yang, N., Wei, F., Chang, B., Zhou, M.: Gated self-matching networks for reading comprehension and question answering. In: Proceedings of the 55th Annual Meeting of the Association for Computational Linguistics (Volume 1: Long Papers), pp. 189–198. Association for Computational Linguistics (2017). https://doi.org/10.18653/v1/P17-1018, http://www.aclweb.org/anthology/P17-1018
21. Weissenborn, D., Wiese, G., Seiffe, L.: FastQA: a simple and efficient neural architecture for question answering. CoRR abs/1703.04816 (2017). http://arxiv.org/abs/1703.04816
22. Weston, J., Bordes, A., Chopra, S., Mikolov, T.: Towards AI-complete question answering: a set of prerequisite toy tasks. CoRR abs/1502.05698 (2015)
23. Xia, J.: Jiangnan at SemEval-2018 task 11: deep neural network with attention method for machine comprehension task (2018)
24. Xiong, C., Zhong, V., Socher, R.: DCN+: mixed objective and deep residual coattention for question answering. In: International Conference on Learning Representations (2018). https://openreview.net/forum?id=H1meywxRW
25. Xu, Y., Liu, J., Gao, J., Shen, Y., Liu, X.: Towards human-level machine reading comprehension: reasoning and inference with multiple strategies. CoRR abs/1711.04964 (2017)
26. Yang, Z., Yang, D., Dyer, C., He, X., Smola, A., Hovy, E.: Hierarchical attention networks for document classification. In: NAACL, pp. 1480–1489. Association for Computational Linguistics, San Diego, June 2016. http://www.aclweb.org/anthology/N16-1174
27. Zhu, H., Wei, F., Qin, B., Liu, T.: Hierarchical attention flow for multiple-choice reading comprehension. In: AAAI (2018)

Text Classification and Summarization

A Hierarchical Hybrid Neural Network Architecture for Chinese Text Summarization

Yunheng Zhang[1], Leihan Zhang[1(✉)], Ke Xu[1], and Le Zhang[2]

[1] State Key Laboratory of Software Development Environment, Beihang University, Beijing, China
zhangleihan@gmail.com
[2] School of Economics and Management, Beijing University of Posts and Telecommunications, Beijing, China

Abstract. Using sequence-to-sequence models for abstractive text summarization is generally plagued by three problems: inability to deal with out-of-vocabulary words, repetition in summaries and time-consuming in training. The paper proposes a hierarchical hybrid neural network architecture for Chinese text summarization. Three mechanisms, hierarchical attention mechanism, pointer mechanism and coverage mechanism, are integrated into the architecture to improve the performance of summarization. The proposed model is applied to Chinese news headline generation. The experimental results suggest that the model outperforms the baseline in ROUGE scores and the three mechanisms can improve the quality of summaries.

Keywords: Abstractive text summarization
Hierarchical attention mechanism · Pointer mechanism
Coverage mechanism

1 Introduction

Text summarization is to generate a brief and coherent summary to represent the key ideas of the text. The methods of text summarization can be broadly classified into two categories: extractive summarization and abstractive summarization. The models of extractive summarization extract the segments from the original to compose the summary. In contrast, the abstractive models generate a compressed paraphrase of the main ideas of texts, potentially using words which don't exist in the source text. The extractive models can guarantee the grammaticality of summarization, while the abstractive models have more sophisticated abilities such as paraphrasing and generalization [11].

Many researches have been concentrating on abstractive summarization and the sequence-to-sequence models have been successfully introduced into the abstractive summarization [9]. Based on the framework of sequence-to-sequence

M. Sun et al. (Eds.): CCL 2018/NLP-NABD 2018, LNAI 11221, pp. 277–288, 2018.
https://doi.org/10.1007/978-3-030-01716-3_23

models, some mechanisms, such as attention mechanism [3,10], pointer mechanism [11] and coverage mechanism [11], have been proposed to improve the quality of summarization. Attention mechanism is usually used to solve the sequence-to-sequence tasks [3,10]. The weaknesses of these models are that they can't deal with out-of-vocabulary (OOV) words and usually repeat words in summaries. Pointer mechanism was proposed to copy words from the original [4,11]. Inspired by the coverage model of Tu et al. [14], See et al. [11] proposed coverage mechanism to reduce the repetition in the output.

The above models and mechanisms can improve the quality of abstractive summarization, but usually require much time for training. In order to improve the efficiency of summarization, a hierarchical hybrid neural network architecture is proposed. The hierarchical structure can shorten the input for the encoder and reduce the need for time. Attention mechanism, pointer mechanism and coverage mechanism are modified to integrate into the hierarchical structure. Then we apply the proposed model to generate headlines for the Chinese news from Sina society news. Experimental results show that the proposed hierarchical hybrid neural network is remarkably effective in Chinese text summarization.

2   Related Work

Abstractive text summarization is a challenging problem and a few distinguished works have been achieved. Sequence-to-sequence model have been successful in many problems like machine translation [1] and abstractive text summarization [10]. Sutskever et al. [12] used encoder-decoder model to solve the sequence-to-sequence tasks. Bahdanau et al. [1] proposed the attention mechanism for the machine translation. Then, the attention-based model was introduced into the sentence summarization [10] and Chopra et al. [3] extended the model with LSTM. Guo et al. [5] used the encoder-decoder model for headline generation. Ma et al. [7] proposed a model for Chinese social media text summarization and the length of these texts is usually less than 140 characters.

Based on the encoder-decoder model, Vinyals et al. [15] proposed the pointer networks. The decoder picks the elements from the input sequence and copies them to form an output sequence. The traditional encoder-decoder model can't generate the elements which don't appear in the training data, but the pointer networks can handle the input sequence with OOV words. The advantage of the pointer networks is very helpful for abstractive text summarization. A few researchers [4,9,11] applied the pointer mechanism from the pointer networks to abstractive text summarization and proved that the models are able to copy the rare or unseen key words from the original instead of generating the imprecise words. Coverage is usually applied to dampen repeated attention. See et al. [11] proposed coverage mechanism to alleviate the repetition in summaries.

In order to handle the long documents better, the hierarchical document structure is used to the document representation. The hierarchical model can integrate the information on the word level and the sentence level. Tang et al. [13] fed the word representations through a CNN or LSTM to get the sentence representation, and then fed the sentence representations through a gated RNN to get the document representation. Yang et al. [16] applied the two-level attention mechanisms to the construction of the sentence representation and the document representation. Nallapati et al. [9] used the sentence-level attention to re-scale the corresponding word-level attention and further redefined the importance of the words according to the importance of the sentence.

3   Proposed Architecture

3.1   Vanilla Sequence-to-Sequence Model with Vanilla Attention Mechanism

The vanilla sequence-to-sequence model [15] contains an encoder and a decoder. Generally, the encoder can understand the input sequence and represent the sequence with a vector. The decoder can produce the output sequence according to the input representation.

The recurrent neural network (RNN) is used to process the input sequence or generate the output sequence as an encoder or a decoder. The gated recurrent unit (GRU) is a gated RNN variant [2] which outperforms the traditional RNN. A GRU contains a reset gate and an update gate. At the encoding time t, the reset gate r_t and the update gate z_t are computed as follows:

$$r_t = \sigma(W_r x_t + U_r h_{t-1} + b_r) \tag{1}$$

$$z_t = \sigma(W_z x_t + U_z h_{t-1} + b_z) \tag{2}$$

where x_t is the input, h_{t-1} is the previous hidden state, σ is the sigmoid function and W_r, U_r, b_r, W_z, U_z and b_z are learnable parameters. The new hidden state h_t is computed as follows:

$$\widetilde{h}_t = \tanh(W_h x_t + r_t \odot (U_h h_{t-1}) + b_h) \tag{3}$$

$$h_t = z_t h_{t-1} + (1 - z_t)\widetilde{h}_t \tag{4}$$

where W_h, U_h and b_h are learnable parameters. x_t and h_{t-1} are deemed as the inputs for each iteration of GRU. And let $h_t = f(x_t, h_{t-1})$ denote the whole process of Eqs. (1), (2), (3) and (4) to simplify the below description.

We use the GRU as the encoder in the sequence-to-sequence model. $d = \{w_1, w_2, \ldots, w_n\}$ represents the input sequence of the tokens in the whole document, and h_n represents the last hidden state of the encoder GRU. In addition, we use the '[UNK]' token to represent any OOV word. The document representation d_e is computed as shown below:

$$d_e = W_e h_n + b_e \tag{5}$$

where W_e and b_e are learnable parameters.

In the decoder, another GRU generates the tokens one by one to form the output sequence, namely the summary. At the decoding time t, the hidden state h'_t is computed by the previous hidden state h'_{t-1} of the GRU, the last output

y_{t-1} in the generation step and the context vector c_t, and h'_t is fed through a linear layer to produce the vocabulary distribution P_v:

$$e_{ti} = g(h'_t, h_i) \tag{6}$$

$$\alpha_t = \text{softmax}(e_t) \tag{7}$$

$$c_t = \sum_i \alpha_{ti} h_i \tag{8}$$

$$h'_t = f([y_{t-1}, c_t], h'_{t-1}) \tag{9}$$

$$P_v = \text{softmax}(W_v[h'_t, c_t] + b_v) \tag{10}$$

where W_v and b_v are learnable parameters. The function g is the combine of the linear function and the tanh function. At the first decoding time step, h'_0 is the final output of the encoder, namely the document representation d_e.

Furtherly, the final probability of the word y in the fixed vocabulary is the corresponding probability in the vocabulary distribution:

$$P(y) = P_v(y) \tag{11}$$

The architecture of vanilla sequence-to-sequence model with vanilla attention mechanism is displayed in Fig. 1.

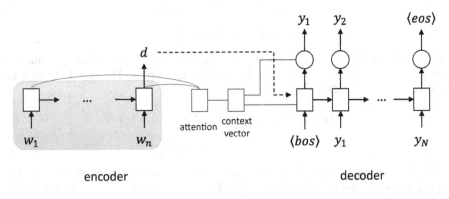

Fig. 1. The architecture of vanilla sequence-to-sequence model with vanilla attention mechanism.

For training, the training objective is the minimization of the total loss L as shown below:

$$L = -\frac{1}{N} \sum_{t=1}^{N+1} \log P(y_t^* | y_1^*, \ldots, y_{t-1}^*, x; \theta) \tag{12}$$

where $y_N^* = \{y_1^*, y_2^*, \ldots, y_t^*\}$ is the ground truth sequence, x is the corresponding input sequence and θ represents the learnable parameters in the whole model. For testing, we pick the token with maximum probability to serve as the optimal result at each decoding time step.

3.2 Hierarchical Sequence-to-Sequence Model

Hierarchical sequence-to-sequence model is a variant of the vanilla sequence-to-sequence model. The module of decoder is unchanged and the hierarchical encoder is introduced to process the Chinese documents. The advantage of this structure is that it can shorten the input length for each GRU. In addition, the structure can also help the model understand the text on the word level and the sentence level.

A Chinese document usually consists of some sentences, and each sentence consists of some words. Therefore, the hierarchical document structure mainly includes two parts. For the first part, the sequence $s_i = \{w_{i1}, w_{i2}, \ldots, w_{in}\}$ is inputted to a GRU named word encoder, where w_{ij} represents the jth word in the corresponding i^{th} sentence. The sequence $h_s = \{h_{s1}, h_{s2}, \ldots, h_{sm}\}$ is fed through a linear layer to get the output s_i as the sentence representation, where h_{si} represents the last hidden state of word encoder of the ith sentence:

$$s_i = W_{se}h_{si} + b_{se} \tag{13}$$

where W_{se} and b_{se} are learnable parameters. For the other part, the sequence $d = \{s_1, s_2, \ldots, s_n\}$ is inputted to another GRU named sentence encoder and the last hidden state h_{dn} is fed through a linear layer to get the document representation d_e:

$$d_e = W_{de}h_{dn} + b_{de} \tag{14}$$

where W_{de} and b_{de} are learnable parameters.

3.3 Hierarchical Attention Mechanism

Attention mechanism enables neural network models to focus on the important information of the original at each decoding time step. However, because the hierarchical encoder has two encoding parts and the encoding of the sentences in the document is separate, the traditional attention mechanism is not suitable for the hierarchical document structure. In order to utilize the two encoding processes and make the hierarchical model focus on the different word-level and sentence-level information during the decoding, we apply the hierarchical attention mechanism. The sentence-level attention is based on the hidden state of the sentence encoder. It makes the decoder have the ability to identify the key sentences and produce the relevant tokens. The sentence-level context vector c_{st} is computed as follows:

$$e_{sti} = g_s(h'_t, h_{si}) \tag{15}$$

$$\alpha_{st} = \text{softmax}(e_{st}) \tag{16}$$

$$c_{st} = \sum_i \alpha_{sti}h_{si} \tag{17}$$

where i is the ID of the sentence.

Then, the word-level attention is applied to help the decoder pay attention to the key words in the document. Because it is independent to compute the word-level context vectors for the different sentences, so the word-level attention is cooperated with the sentence-level attention. For each sentence in the document, the word-level context vector c_{wt} is computed as follows:

$$e_{wtij} = g_w(h'_t, h_{wij}) \tag{18}$$

$$\alpha_{wti} = \text{softmax}(e_{wti}) \tag{19}$$

$$\alpha_{wtj} = \alpha_{sti}\alpha_{wtij} \tag{20}$$

$$c_{wt} = \sum_{i,j} \alpha_{wtj}h_{wij} \tag{21}$$

where j is the ID of the word in the corresponding sentence. The biggest difference from the sentence-level attention is that the weight of the word-level attention is also influenced by the sentence-level attention of the corresponding sentence, which means that the importance of the word is related to the word-level attention and the sentence-level attention.

To realize the hierarchical attention mechanism, we put c_{st} and c_{wt} into Eqs. (9) and (10) and respectively change them to new equations as shown below:

$$h'_t = f([y_{t-1}, c_{st}, c_{wt}], h'_{t-1}) \tag{22}$$

$$P_v = \text{softmax}(W_v[h'_t, c_{st}, c_{wt}] + b_v) \tag{23}$$

In this way, the decoder can focus on the important words and sentences when generating the tokens to form the summary of the document.

3.4   Pointer Mechanism

The traditional sequence-to-sequence model can only generate the summaries with the fixed generation vocabulary and it can't produce OOV words. Therefore, pointer mechanism is introduced to the decoding step to enable the decoder to copy the key words from the original as part of the summary. Inspired by the idea of pointer-generator network [11], we use the generation probability p to control the decoder to generate a word or directly copy an existing word from the source. The generation probability p seems like a switch and can be calculated by:

$$p = \sigma(w_p^T[y_{t-1}, c_{st}, c_{wt}, h'_{t-1}] + b_p) \tag{24}$$

where w_p and b_p are learnable parameters.

According to Eqs. (22) and (23), we can get the vocabulary distribution P_v. It represents the generative probabilities of the words in the fixed vocabulary. Due to the introduction of pointer mechanism, we should consider the copying probability for the words existing in the fixed vocabulary and coming from the original. The word-level attention distribution α_{wt} means the importance of the words in the original and can also represent the copying probability in some sense.

Therefore, the copying distribution P_c is computed by the word-level attention distribution α_{wt}:

$$P_c(y) = \sum_{j:w_j=y} \alpha_{wtj} \tag{25}$$

That is to say, for the word y in the document, the corresponding copying probability is the sum of all the word-level attention weights of the words which are the same as the word y. And the copying probability will be zero if the word y doesn't exist in the document.

Then, the vocabulary distribution P_v and the copying distribution P_c are combined to get the final probability of the word y as follows:

$$P(y) = pP_v(y) + (1 - p)P_c(y) \tag{26}$$

In fact, the final probability is a linear interpolation between the vocabulary distribution and the copying distribution. In this way, the model can generate a word from the fixed vocabulary or copy a word from the original at each decoding time step. Pointer mechanism can help the decoder produce the words which are not in the vocabulary and make the summaries containing the key information from the particular documents.

3.5 Coverage Mechanism

According to See et al. [11], the coverage mechanism is helpful for the model to alleviate the repetition in the output. The main idea of the coverage mechanism is to prevent the model from focusing on the same word repeatedly. Then, the coverage mechanism is introduced into the hierarchical model. The architecture of the model is shown in Fig. 2. Coverage vector c'_t reflects the all previous total word-level attentions of the model and it can affect the next word-level attention distribution:

$$c'_t = \sum_{t'=0}^{t-1} \alpha_{wt'} \tag{27}$$

$$e_{wtij} = g_w(h'_t, h_{wij}, c'_{tj}) \tag{28}$$

Generally, paying attention to the same words or phrases repeatedly can result in the repetition of words in the summaries. Therefore, it is reasonable to alleviate the repetition by restricting the repeated attention on the word-level content. And because the sentences usually consist of some words or phrases, it is not necessary to restrict the attention on the sentence-level content.

In addition, the loss L in this model should be added the coverage loss and be changed as shown below:

$$L = -\frac{1}{N} \sum_{t=1}^{N+1} \log P(y_t^*|y_1^*, \ldots, y_{t-1}^*, x; \theta) + \lambda \sum_i \min(\alpha_{wt}, c'_t) \tag{29}$$

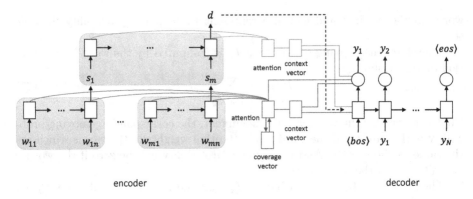

Fig. 2. The hierarchical hybrid architecture with hierarchical attention mechanism, pointer mechanism and coverage mechanism.

4   Experiments

4.1   Dataset

The proposed hierarchical hybrid architecture is designed for relative long Chinese document summarization, so a proper dataset is necessary for evaluation. However, to the best of our knowledge, there is no standard dataset for long Chinese documents. Hence, we construct a dataset by collecting the Sina society news[1]. The body texts and the corresponding headlines of Sina society from June 11, 2012 to August 31, 2017 were collected. We kept the news whose body texts are not less than 60 characters and finally got 41,792 valid pairs for our experiments. We randomly selected 33,434 pairs as the training set and remained 8,358 pairs as the test set.

4.2   Implementation

Firstly, the input documents were split into sentences according to punctuation. And the sentences and headlines were tokenized by jieba[2]. Statistically, the average length of the input is about 649 tokens and the average length of the output is about 12 tokens. Then, the word representations were derived by using word2vec model [8] and the embedding dimension was set to 128. There are 359,043 different tokens in the corpus, and the tokens with frequency less than 20 were filtered. At last, 52,614 tokens were remained to build the fixed vocabulary.

We used PyTorch[3] to implement the deep learning model. The hidden state dimension of the encoder and the decoder was set to 512. For training, we used Adam with an initial learning rate of 0.001 and a batch size of 8. We applied

---

[1] http://news.sina.com.cn/society/.

[2] https://pypi.org/project/jieba/.

[3] https://pytorch.org/.

gradient clipping with a maximum gradient norm of 2. And the training samples were shuffled in each epoch. The output length was limited to 30 tokens. We set $\lambda = 1$ for Eq. (29). Then the model was evaluated with the F1-score of ROUGE-1, ROUGE-2 and ROUGE-L metrics [6]. In addition, the quantity of OOV words was also analyzed. At last, the running time for the proposed model was compared with the baseline.

4.3   Results

As shown in Table 1, the performance of our model was compared with the baseline. In detail, the vanilla sequence-to-sequence model with vanilla attention mechanism (*seq2seq + attn*) is selected as baseline. The hierarchical sequence-to-sequence model with hierarchical attention mechanism is denoted *h-h-attn*. The hierarchical sequence-to-sequence model with hierarchical attention mechanism and pointer mechanism is denoted *h-h-attn + pointer*. And the hierarchical sequence-to-sequence model with hierarchical attention mechanism, pointer mechanism and coverage mechanism is denoted as *h-h-attn + pointer + coverage*. Experimental results in Table 1 suggest the outstanding performance of all the three mechanisms in the model. Hierarchical sequence-to-sequence model with hierarchical attention mechanism outperforms the baseline by focusing on the information on the word level and the sentence level during the decoding stage. And the pointer mechanism enables the model to copy the rare or unseen words as part of the summaries. Hierarchical sequence-to-sequence model with hierarchical attention mechanism, pointer mechanism and coverage mechanism outperforms all the other models in the three metrics of ROUGE-1, ROUGE-2 and ROUGE-L, which proves the effectiveness of our model.

Table 1. The ROUGE scores on the test set.

Model	ROUGE-1	ROUGE-2	ROUGE-L
seq2seq + attn	14.88	1.75	13.40
h-h-attn	15.89	2.05	14.34
h-h-attn + pointer	20.87	2.65	18.17
h-h-attn + pointer + coverage	**21.97**	**3.59**	**19.27**

Table 2. The percentage of OOV words in generated summaries.

Model	OOV percentage
seq2seq + attn	13.3%
h-h-attn	13.2%
h-h-attn + pointer	8.8%
h-h-attn + pointer + coverage	8.7%

Pointer mechanism can help the model copy the rare or unseen words from the source texts and reduce the appearance of OOV words. The percentage of OOV words was computed in the output sequences and the results are shown in Table 2. As can be seen, the models with pointer mechanism can produce summaries with less OOV words, which can further improve the quality of summarization.

In order to evaluate the efficiency of the hierarchical mechanism, we also compared the running time for the above models. As shown in Table 3, the time represents the sum of the preprocessing time, the training time and the test time. In general, the hierarchical model with hierarchical attention mechanism is a little faster than the vanilla sequence-to-sequence model with vanilla attention mechanism.

Table 3. The running time for various models in our experiments.

Model	Time spent
seq2seq + attn	37.2 h
h-h-attn	35.3 h
h-h-attn + pointer	82.0 h
h-h-attn + pointer + coverage	95.2 h

Table 4. The summaries generated by the proposed models for one document in the test dataset.

source text: ...刘师傅说，...，他接到了"巡视组"的电话，...，并说要查账核对...刘师傅的儿子说，父亲两次被骗转账1.79万余元... ...Mr. Liu said, ..., he picked the phone from "inspection group", ..., and they claimed to audit his account ...Mr. Liu's son said, his father was cheated to transfer more than 17.9 thousand yuan ...
ground truth: 男子轻信"巡视组查账" 手机操作被骗1.79万元 A man believed in that "inspection group audited his account", and he was cheated out of 17.9 thousand yuan by operating the phone
***seq2seq + attn*:** 男子被冒名贷款后转走走走走走走走走走走走走 An imposter of a man got a loan and the money was transferred transferred transferred
***h-h-attn*:** 男子网购[UNK]万元现金被骗 骗子称[UNK][UNK] A man was cheated out of [UNK] yuan in cash when he shopped online, and cheater claimed that [UNK] [UNK]
***h-h-attn + pointer*:** 男子手机被骗 男子骗损失损失 A man was cheated on phone, a man cheated loss loss
***h-h-attn + pointer + coverage*:** 男子轻信"巡视组" 被骗近万 被骗近万 A man believed in "inspection group", and was cheated out of about ten thousand yuan, and was cheated out of about ten thousand yuan

At last, we also made case analysis to investigate the contributions of each mechanism. As shown in Table 4, the results show that the hierarchical sequence-to-sequence model and three mechanisms can obviously improve the quality of the summaries. Hierarchical attention mechanism makes the model focus on the source text and generate more related content. Pointer mechanism reduce the appearance of the '[UNK]' token. And coverage mechanism alleviates the repetition in the output.

5   Conclusion and Future Work

In this work, we proposed a hierarchical hybrid neural network architecture for Chinese text summarization. The hierarchical document structure was utilized to shorten the length of the input sequences for the encoder, and the hierarchical attention mechanism, pointer mechanism and coverage mechanism were incorporated to generate the rational and informative summaries. And experiments on Chinese news with long texts proved the efficiency of the proposed model. In the future, we will try to incorporate the extractive and abstractive methods to deal with the summarization for long texts.

References

1. Bahdanau, D., Cho, K., Bengio, Y.: Neural machine translation by jointly learning to align and translate. arXiv e-prints abs/1409.0473, September 2014
2. Cho, K., et al.: Learning phrase representations using RNN encoder-decoder for statistical machine translation. In: Proceedings of the 2014 Conference on Empirical Methods in Natural Language Processing (EMNLP), pp. 1724–1734. Association for Computational Linguistics (2014)
3. Chopra, S., Auli, M., Rush, A.M.: Abstractive sentence summarization with attentive recurrent neural networks. In: Proceedings of the 2016 Conference of the NAACL: Human Language Technologies, pp. 93–98. Association for Computational Linguistics (2016)
4. Gu, J., Lu, Z., Li, H., Li, V.O.: Incorporating copying mechanism in sequence-to-sequence learning. In: Proceedings of the 54th Annual Meeting of the Association for Computational Linguistics (Volume 1: Long Papers), pp. 1631–1640. Association for Computational Linguistics (2016)
5. Guo, Y., Huang, H., Gao, Y., Lu, C.: Conceptual multi-layer neural network model for headline generation. In: Sun, M., Wang, X., Chang, B., Xiong, D. (eds.) CCL/NLP-NABD -2017. LNCS (LNAI), vol. 10565, pp. 355–367. Springer, Cham (2017). https://doi.org/10.1007/978-3-319-69005-6_30
6. Lin, C.Y.: Rouge: a package for automatic evaluation of summaries. In: Marie-Francine Moens, S.S. (ed.) Text Summarization Branches Out: Proceedings of the ACL-04 Workshop, pp. 74–81. Association for Computational Linguistics, Barcelona, July 2004
7. Ma, S., Sun, X., Xu, J., Wang, H., Li, W., Su, Q.: Improving semantic relevance for sequence-to-sequence learning of Chinese social media text summarization. In: Proceedings of the 55th Annual Meeting of the Association for Computational Linguistics (Volume 2: Short Papers), pp. 635–640. Association for Computational Linguistics (2017)

8. Mikolov, T., Sutskever, I., Chen, K., Corrado, G., Dean, J.: Distributed representations of words and phrases and their compositionality. In: Proceedings of the 26th International Conference on Neural Information Processing Systems, NIPS 2013, vol. 2, pp. 3111–3119. Curran Associates, Inc., Red Hook (2013)

9. Nallapati, R., Zhou, B., dos Santos, C., Gulcehre, C., Xiang, B.: Abstractive text summarization using sequence-to-sequence RNNs and beyond. In: Proceedings of the 20th SIGNLL Conference on Computational Natural Language Learning, pp. 280–290. Association for Computational Linguistics (2016)

10. Rush, A.M., Chopra, S., Weston, J.: A neural attention model for abstractive sentence summarization. In: Proceedings of the 2015 Conference on Empirical Methods in Natural Language Processing, pp. 379–389. Association for Computational Linguistics (2015)

11. See, A., Liu, P.J., Manning, C.D.: Get to the point: summarization with pointer-generator networks. In: Proceedings of the 55th Annual Meeting of the Association for Computational Linguistics (Volume 1: Long Papers), pp. 1073–1083. Association for Computational Linguistics (2017)

12. Sutskever, I., Vinyals, O., Le, Q.V.: Sequence to sequence learning with neural networks. In: Proceedings of the 27th International Conference on Neural Information Processing Systems, NIPS 2014, vol. 2, pp. 3104–3112. MIT Press, Cambridge (2014)

13. Tang, D., Qin, B., Liu, T.: Document modeling with gated recurrent neural network for sentiment classification. In: Proceedings of the 2015 Conference on Empirical Methods in Natural Language Processing, pp. 1422–1432. Association for Computational Linguistics (2015)

14. Tu, Z., Lu, Z., Liu, Y., Liu, X., Li, H.: Modeling coverage for neural machine translation. In: Proceedings of the 54th Annual Meeting of the Association for Computational Linguistics (Volume 1: Long Papers), pp. 76–85. Association for Computational Linguistics (2016)

15. Vinyals, O., Fortunato, M., Jaitly, N.: Pointer networks. In: Advances in Neural Information Processing Systems 28, pp. 2692–2700. Curran Associates, Inc., Red Hook (2015)

16. Yang, Z., Yang, D., Dyer, C., He, X., Smola, A., Hovy, E.: Hierarchical attention networks for document classification. In: Proceedings of the 2016 Conference of the NAACL: Human Language Technologies, pp. 1480–1489. Association for Computational Linguistics (2016)

TSABCNN: Two-Stage Attention-Based Convolutional Neural Network for Frame Identification

Hongyan Zhao[1,2(✉)], Ru Li[1,3(✉)], Fei Duan[1(✉)], Zepeng Wu[1(✉)], and Shaoru Guo[1(✉)]

[1] School of Computer and Information Technology, Shanxi University,
Taiyuan 030006, China
{liru, fduan}@sxu.edu.cn, 289977640@qq.com,
guoshaoru0928@163.com
[2] School of Computer Science and Technology,
Taiyuan University of Science and Technology, Taiyuan 030024, China
hongyanzhao@tyust.edu.cn
[3] Key Laboratory of Ministry of Education for Computation Intelligence
and Chinese Information Processing, Shanxi University, Taiyuan 030006, China

Abstract. As an essential sub-task of frame-semantic parsing, Frame Identification (FI) is a fundamentally important research topic in shallow semantic parsing. However, most existing work is based on sophisticated, hand-crafted features which might not be compatible with FI procedure. Besides that, they usually heavily rely on available natural language processing (NLP) toolkits and various lexical resources. Thus existing methods with hand-crafted features may not achieve satisfactory performance. In this paper, we propose a two-stage attention-based convolutional neural network (TSABCNN) to alleviate this problem and capture the most important context features for FI task. In order to dynamically adjust the weight of each feature, we build two levels of attention over instances at input layer and pooling layer respectively. Furthermore, the proposed model is an end-to-end learning framework which does not need any complicated NLP toolkits and feature engineering, and can be applied to any language. Experiments results on FrameNet and Chinese FrameNet (CFN) show the effectiveness of the proposed approach for the FI task.

Keywords: Frame identification · FrameNet · Convolutional neural network

1 Introduction

As the core task of Natural Language Processing (NLP), shallow semantic parsing abandons the complexity of deep components and relationships and has attracted great attention. In recent years, more semantic knowledge bases such as WordNet, PropBank, and HowNet have been built and widely used in the shallow semantic parsing task. Among these semantic knowledge engineering projects, FrameNet (Baker et al. 1998) is a rich linguistic resource containing considerable expert knowledge about lexical and predicate-argument semantics, and frame-semantic parsing has been proven to be an effective way that extracts a shallow semantic structure from text.

© Springer Nature Switzerland AG 2018
M. Sun et al. (Eds.): CCL 2018/NLP-NABD 2018, LNAI 11221, pp. 289–301, 2018.
https://doi.org/10.1007/978-3-030-01716-3_24

According to the theory of frame semantics (Fillmore,1982), one semantic frame represents an event or scenario, and possesses a set of targets (namely lexical units or predicate) that can evoke the semantic scenario and some frame elements (or semantic roles) that participate in the event (Hermann et al. 2015). Most work on frame-semantic parsing (Das et al. 2010; Das et al. 2014) has divided the task into two subtasks: (1) the first one is frame identification, which identifies the most suitable semantic frame for a given target in a sentence; (2) the second one is argument identification (or semantic role labeling), which performs semantic role labeling for the identified frame. However, current researches on frame-semantic parsing mostly focus on argument identification for given target and its frame (Carreas et al. 2008), skipping the frame identification step, which leads to the failure to automatically implement frame-semantic analysis task. This is also the main reason why frame-semantic parsing can't be widely used in many NLP tasks. We argue that the first subtask is an essential step in the frame-semantic parsing task. In this paper, we focus on the FI for given targets.

At present, FI task is treated as a multi classification task, and virtually all of the state-of-the-art approaches for this task are based on sophisticated, hand-crafted features, such as conditional random fields (CRF), support vector machine (SVM), maximum entropy (ME). In addition, extracting these features usually heavily rely on available NLP toolkits and various lexical resources, which might lead to the error propagation. Thus pre-existing methods with hand-crafted features may not achieve satisfactory performance.

In order to reduce the manual labor in feature extraction, recently, deep learning is used to learn features in many NLP tasks, and convolutional neural networks (CNN) have shown to be efficient to capture syntactic and semantic context features between words within a sentence for NLP tasks such as sentence modeling (Kalchbrenner et al. 2014), sentence classification (Kim 2014) and relation classification (Zeng et al. 2014), event extraction(Chen et al. 2015), question answering (Dong et al. 2015), short text ranking (Severyn et al. 2015), text chunks matching (Yin and Schütze, 2015).

FI can also be considered as a sentence-classification task for a marked target. In this paper, we propose a two-stage attention-based convolutional neural network (TSABCNN) to alleviate this problem and capture the most important context features for FI task. In order to dynamically adjust the weight of each feature, we build two levels of attention mechanism over instances at input layer and pooling layer respectively. Furthermore, the proposed model is an end-to-end learning framework which does not need any complicated NLP toolkits and feature engineering, and can be applied to any language. Experiments results on FrameNet and Chinese FrameNet (CFN) show the effectiveness of the proposed approach for the FI task.

Our main contributions are: (1) we analyze the problem of exiting models on the task of FI, and propose an end-to-end FI method based on CNN, which does not need any complicated NLP toolkits and feature engineering; (2) In order to make the weight of important features bigger, we introduce a supervised attention based FI model on input layer and pooling layer respectively. (3) we improve the performance of FI and achieve better performance than the baselines.

2 Related Work

Since Gildea and Jurafsky (2002) pioneered semantic role labeling (SRL), particularly followed by the CONLL2004 (Carreras, 2004) and CONLL2005 (Carreras and Màrquez, 2005) treat SRL as a shared evaluation task, frame-semantic analysis has received a boost in attention. The FrameNet lexicon contains abundant linguistic information about lexical items and predicate-argument structures. In a frame-analyzed sentence, predicates evoking frames are known as targets, and a word or phrase filling a role is known as an argument. Figure 1 shows frame-semantic annotations for a sentence. In this figure, the target buy.V evokes the Commerce_buy frame. Buyer and Goods are some arguments (semantic roles) for this frame.

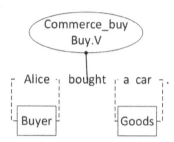

Fig. 1. Frame-semantic annotations for a sentence

Most early work on frame-semantic parsing used a supervised machine learning method. Fleischman, Kwon and Hovy (2003) used maximum entropy models to identify arguments and their roles for a given frame. Erk et al. (2005) used the traditional word disambiguation method to conduct experiment on German FrameNet frame disambiguation. The LTH system presented by Johansson and Nugues (2007) achieved the best performance in the SemEval 2007 task of identifying frame. They adopted a series of SVMs to classify the frame for a given target, associating unseen lexical items to frames and identifying and classifying a word or phrase as various semantic roles. Adrian Bejan and Hathaway (2007) selected 556 ambiguous target words which can evoke two or more semantic frames and have more than five annotated sentences for each frame. They trained a multi-classifier for ambiguous targets from the FrameNet lexicon.

Recently, a tool called SEMAFOR was presented (Das et al. 2010), with a probabilistic models for FI that used a latent-variable log-linear model to capture frames for unseen targets. The feature set of this model leads to better performance on the SemEval 2007 data. The FrameNet project released a new version of annotating data in 2010. Das et al. using a two-stage statistical model (Das et al. 2014) on this dataset, improved their prior work and set the new state of the art. A few salient aspects of this updated version of SEMAFOR involved handling unseen targets by using a graph-based semi-supervised learning approach and a dual decomposition algorithm. Subsequently, Hermann et al. (2014) presented a novel model using distributed representations of the word context and dependency path for better FI, outperforming the aforementioned SEMAFOR.

Unfortunately, above work is mostly based on sophisticated, hand-crafted features. Besides that, they usually heavily rely on available natural language processing (NLP) toolkits and various lexical resources.

3 TSABCNN Model

The aim of FI is to choose a correct frame for the given target in a sentence. Because targets are usually ambiguous, a target may arouse multiple frames, which poses challenges for FI tasks. Traditional statistical machine learning method is based on sophisticated, hand-crafted features. In recent years, CNN has been proved to be able to learn more advanced contextual features, and has been applied to many NLP tasks (Chen et al. 2015; Li et al. 2015; Severyn et al. 2015; Yin and Schütze, 2015; Wenpeng Yin et al. 2016). Inspired by Wenpeng Yin, we here propose a novel general two-stage attention-based convolution neural network model for FI. Our network architecture is shown as Fig. 2. The input sentence is transformed into a vector by looking up word embedding. To acquire the word order, we here combine word embedding with relative position vector between the word and target. In input layer, an attention mechanism is used to acquire relevance of words and the target. In order to obtain context information such as n-gram, we use multiple kinds of filters with different region size

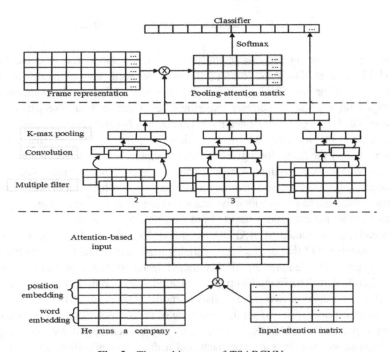

Fig. 2. The architecture of TSABCNN.

(Zhang and Wallace 2017), and next is a maximum pool layer. Another attention is used to determine more important n-gram features for FI. Finally, a *softmax* classifier is used, and the highest scoring frame is regarded as the best frame. Our model will be further described in the remainder of this section.

3.1 Word Representation

Before entering the network, each word w_i is mapped into a real vector $v_i \epsilon R^d$ (d denote the word embedding dimension) by looking up the word embedding table V, which can be trained by word2vec (Mikolov et al. 2013) model. In addition, in order to embed the position information of a word in a sentence for FI, we introduce the relative distance between $i - th$ word and the target marked by a word position embedding P (Collobert et al. 2011). For example, given sentence in Fig. 2, the relative distances of "He" and "company" to "runs" are -1 and 2 respectively. Every relative distance is randomly initialized a position vector p_i, and the dimensionality of the word position vector is q. Finally, combining the word embeddings v_i and the relative position embedding p_i, the word feature (WF) is represented as w_i^f. Thus, a given sentence or the sequence of n words can be encoded into a matrix as follows:

$$S = \left[w_1^f, w_2^f, \ldots, w_n^f \right] w_i^f \in R^{(d+q)} \tag{1}$$

3.2 Input-Attention Mechanism

Attentive neural networks have been successfully applied to natural language process (NLP) tasks such as machine translations, question answering, relation extraction, and sentence pairs modeling (Bahdanau et al. 2014; Hermann et al. 2015; Zhou et al. 2016; Yin et al. 2016). In this subsection, we propose an input-attention mechanism for FI. Previous work has focused on the alignment of the input and output sequences, e.g. the alignment of input language and target language in machine translation. To make our model automatically learn the more useful features for the FI task, we propose a novel idea of applying the two-stage attention mechanism to heterogeneous objects, respectively input-attention and pooling-attention. The attention matrix on the input-layer intends to give higher weights to those words related to the target, and guide the convolution layer to learn more useful and higher level features. As is shown in Fig. 2, the word "company" is very significant to FI, then the higher weight is allocated to it. Here, we define the input-layer Attention as a diagonal matrix A. It denote that the relevance of word w_i and target t in the given sentence. Formally, we define the diagonal attention matrix A as:

$$A_{ii} = f(w_i, t) \tag{2}$$

The function f can be computed by different ways. We here exploit $\frac{1}{0.0001 + |w_i - t|}$ to initialize the matrix, where $|w_i - t|$ is Euclidean Distance of between $i - th$ word in a sentence and target word t, and A_{ii} is updated during the network training process. The relevance degree of $i - th$ word in the sentence and its target word is defined as:

$$a_i = \frac{exp(A_{ii})}{\sum_{k=1}^{n} exp(A_{kk})} \tag{3}$$

with n being the length of the sentence. The attention-based input is represented as follows:

$$X = [a_1 w_1, a_2 w_2, \ldots, a_n w_n] \tag{4}$$

subsequently, X is fed into convolution layer of the model.

3.3   CNN Architecture with Multiple-Size Filters

FI is a complicated task. In prior work, a variety of context features (e.g. n-gram features) were extracted to solve this problem. Apparently, it is difficult to obtain these information only by word features. In order to enable the model to learn advanced features, following Collobert and Weston (2008), we regard the sentence matrix as a special matrix, and perform a convolution operation on it by multi-filters of different sizes. As columns represent discrete units (namely word or phrase) and a sentence has its inherent sequential structure, it is meaningful to use filters with the same height as the WF dimension. Therefore, we only choose the width of filters, namely adjusting the number of words jointed. In the subsequent section of the paper, the region size of the filter refers only to its width.

Given a filer with a region size m, it is a weight matrix W_f including $m \cdot k$ (k is the dimension of a word feature) parameters to be trained. The attention-based input matrix of the sentence X is fed into convolution layer. We use a sub-matrix $X[:, i, :, j]$ to represent the mapping from sentence i-th column to sentence j-th column. Input matrix X is folded by the above mentioned filters, and at the same time wide convolution method is used, and phrase-level features are generated. More formally, a feature map h_i is generated from the window of words $X_{[:,i:i+m-1]}$ as follows:

$$h_i = \sigma\left(W_f^T X_{[:,i:i+m-1]} + b_f\right) \tag{5}$$

where b_f is the bias of the convolution layer.

Through convolution layer, the output length is $n_{out} = n_{in} + m - 1$ (n_{in} is the input length of the convolution).

3.4 K-Max-Pooling and Pooling-Attention

The higher-level phrase features are generated by the filter windows on convolution layer. Some features are very important features for the target task, but others are not relevant to the task. To assign greater weight to those important features, we hereby propose a novel attention-based pooling method to extract some crucial features for the FI task. We firstly use K-max pooling on every feature map, which helps to obtain a fixed size matrix, reduces dimensionality, and keeps important features and global information about position. Combined with all K-max features, the context of the target t is represented as:

$$S_p = \left[c_1^t, c_2^t, c_3^t, \ldots, c_l^t\right] \tag{6}$$

with c_i^t being the i- **th** K-max feature vector, and l being the number of filters. Secondly, we use an pooling-attention strategy to determine the importance of the context features of target that encoded by convolutional kernel and $K - \max$ pooling. We create a correlation matrix C that obtains relative connections between the context feature of target and frames embedding W_F, $C = S_p U W_F$, where U is a weight matrix to be learned by the network, and the frame embedding is represented by the mean of the word embedding of the frame name. Then we normalize the matrix C, and obtain pooling-attention matrix A^p as:

$$A_{i,j}^p = \frac{C_{i,j} - minvalue}{maxvalue - minvalue} \tag{7}$$

Finally, to highlight important context features, we multiply this pooling-attention matrix with S_p to get S_o. It is denoted as follows:

$$S_o = A^p \cdot S_p \tag{8}$$

3.5   Regularization and Classification

In order to overcome over fitting, following the works of Hinton et al. (2012) and Kim (2014), we execute a dropout regularization for S_o and produce the dropout vector S_d. What should be noted is that dropout is only performed during the training phase. The dropout vector S_d is input into a standard neural network by fully connected method, which use a weight matrix W_c as model parameters. Finally, a softmax layer is used to implement the classification and output the vector o. The vector o of i-th dimension represents the probability of the i-th frame classified, computed as follows:

$$p(f_i|x, \theta) = \frac{e^{o^i}}{\sum_{j=1}^{F} e^{o^j}} \tag{9}$$

Where θ is a set of all network parameter to be learned, and F is the frame number.

3.6   The Network Training

For all training examples $\left(x^{(i)}, y^{(i)}\right)$, we create the log-likelihood about the parameter θ as the objective function. It is denoted as follows:

$$L(\theta) = \sum_{i=1}^{N} logp\left(y^{(i)}|x^{(i)}, \theta\right) \tag{10}$$

With N is the number of all training examples, and we train θ by maximizing $L(\theta)$.

4   Experiments

In this section, we firstly introduce the datasets and evaluation metrics, experimental setting, and then we present our experiments and results obtained in FI.

4.1   Datasets and Evaluation Metrics

Datasets. We evaluate our model on English FrameNet (FN) and Chinese FrameNet (CFN) respectively. For English FrameNet, we use the FrameNet 1.5 release which is the full-text annotations and was used by Das et al. (2014). We use the same test data as Das et al. containing 4,458 targets. There are 19,582 targets in training data.

In experiment with Chinese FrameNet dataset. we select 25,000 annotated sentences in CFN exemplar sentences database as the training data set, which contained 1,567 targets and 180 frames. In order to compare with previous work in Chinese frame identification (CFI), we use two test sets. The first one, named tc1, consisted of 5000 sentences with marked targets that have not appeared in the training set. The second one, named tc2, used the same data set as Li et al. (2010) at the Coling 2010 Conference. This data set contains 7 different Chinese ambiguous targets. For each target, sentences were collected from Sogo Corpus and Contemporary Chinese Corpus of Beijing University and 940 sentences are selected for training data and 128 for test data.

Evaluation Metrics. We use the accuracy $= \frac{b}{r}$ to evaluate our model, where b is the number of correct frames identified, and r is the total number of frames identified.

4.2   Experimental Setting

In our experiment, word embedding is 100 dimensions, trained on Wikipedia by the skip-gram model (Mikolov et al. 2013). Position embedding is 50 dimension and initialized randomly. The filter matrixes on Convolution layer and the other weight matrixes are initialized randomly, following a Gaussian distribution. All biases are initialized to 0. Hyper-parameters are tuned on the development dataset. The results of the final values of hyper-parameter are shown in Table 1.

Table 1. Hyper-parameter setting.

Description	Value
Filter window size	2, 3, 4
Filters number of each size	100
Filter height	150
Dropout rate	0.5
Batch size	64
K-max	3
Initial learning rate	0.01

Table 2. The result of English FrameNet

Model	SEMAFOR Lexicon			Full Lexicon		
	All	Ambiguous	Unseen	All	Ambiguous	Unseen
Das et al.supervised	82.97	69.27	23.08	–	–	–
Das et al.best	83.60	69.19	42.67	–	–	–
LOG-LINEAR WORDS	84.53	70.55	27.27	87.33	70.55	–
LOG-LINEAR EMBEDDING	83.94	70.26	27.97	86.74	70.26	–
WSABIE EMBEDDING	86.49	73.39	46.15	88.41	73.10	–
CNN	86.13	80.35	70.46	86.54	80.43	70.66
CNN + multifilter	86.56	81.12	72.31	87.23	81.43	72.32
CNN + Input-att	86.45	81.45	73.67	86.55	81.53	73.85
CNN + Input-att + multifilter	87.25	82.03	72.64	87.75	82.47	72.13
CNN + Pooling-att	87.13	81.97	72.48	87.43	82.39	72.65
CNN + Pooling-att + multifilter	87.55	82.77	74.65	88.13	83.29	75.77
TSABCNN	**89.72**	**83.07**	**75.12**	**91.4**	**83.78**	**76.34**

4.3   Experimental Results and Analysis

To show the effectiveness of our proposed method, several state-of-the-art methods are selected as baseline for comparison on English FrameNet and CFN.

English FrameNet baseline are shown as follows:

Das et al.: A semi-supervised learning method was used to improve upon a supervised latent-variable log-linear model (Das et al. 2014).

Hermann et al.: Distributed representations of predicates and their syntactic context were used (Hermann et al. 2014).

Chinese FrameNet baseline are shown as follows:

Li et al.: A tree-structed conditional random field model was used to solve Chinese FI based on Dependency Parsing (Li et al. 2010).

Zhao et al.: BP neural network was used to learn the context features representation of a given target, and the selection of a frame for a given target (Zhao et al. 2016).

In addition, in order to analyze the effectiveness of each component of our neural network architecture, we use the ablation experiment on English FrameNet data set and CFN data set respectively, and compare with the above methods. The results are shown in Tables 2, 3 and 4.

Table 3. The comparision result with Zhao et al. of CFN

Model	All	Ambiguous	Unseen
Zhao et al.	79.64	74.37	67.21
CNN	82.56	78.34	71.87
CNN + multifilter	83.34	78.95	72.25
CNN + Input-att	83.36	78.82	72.30
CNN + Input-att + multifilter	84.43	79.54	73.15
CNN + Pooling-att	84.13	79.34	72.10
CNN + Pooling-att + multifilter	85.3	80.34	73.41
TSABCNN	**86.8**	**81.76**	**73.97**

Table 4. The comparision result with Li et al..

Model	Accuracy
Li et al.	81.46
CNN	83.57
CNN + multifilter	84.56
CNN + Input-att	84.58
CNN + Input-att + multifilter	85.65
CNN + Pooling-att	85.27
CNN + Pooling-att + multifilter	87.18
TSABCNN	**88.87**

Table 2 presents the accuracy for the state-of-the-art models on English FrameNet, and the comparision results with our proposed methods. Test data set is the SEMAFOR Lexicon and Full Lexicon. We present the results on all targets, ambiguous targets that evoke Multiple frames and unseen targets in the FrameNet lexicon or training data. As shown in Table 2, a high performance achieved on all targets and ambiguous word by traditional feature-based methods (Li et al. 2010), but a very low accuracy is obtained on unseen targets. Hermann et al. (2014) used distributed representation of targets and their syntactic context performs better results on three test datasets than that of Das. For unseen targets, although the accuracy rate increased, it was far from being used. Simultaneously, we can see the performance is greatly improved on these datasets by a convolutional neural networks, especially on ambiguous targets and unseen targets, which show that deep learning can automatically produce some useful features. By ablation studies, multi filters and attention mechanism can also improve the accuracy of FI, which shows multi filters can learn diversity features and attention mechanism can make the important features get more weight. Consider all components, TSABCNN model proposed in this paper achieves the best performance, and strongly outperform other methods not only on all targets, but also ambiguous targets and unseen targets. On all target on FrameNet 1.5 release, our approach achieves 91.4 accuracy, and beyond the best baseline nearly 4 point.

Table 3 shows the results on Chinese tc1 dataset. The results are consistent with the results in Table 2. From Table 3, we can see accuracy on the CFN is slightly lower than that of FrameNet. It may be caused by the following factors: (1) Language differences. Chinese is more flexible and more ambiguous than English. (2) The scale of CFN corpus is much smaller than that of FN, and the neural network learning needs a large corpus support. Table 4 shows the results on Chinese tc2. On this dataset, CNN modelexceed that of Li et al. by 2.11 accuracy. It proves CNN can learn more useful context features for FI. While multi-filter, input-att, pooling-att are successively added to the model, Performance has a different degree of improvement. Finally, our proposed model, TSABCNN achieves 88.87 accuracy, and is higher 7.41 than the baseline.

5   Conclusion

In this paper, we propose a novel two-stage attention-based CNN model for FI. Our model utilizes CNN to automatically learn more important features for FI. We base attention mechanism to give higher weight to more important features. The experimental result shows that our model achieves better performance on three FI tasks than the baselines. Our model is not only suitable for the frame recognition of the all targets but also has a good effect on that of the unseen targets and the ambiguous targets. Furthermore, it is not limited to a language, so it is a general FI model. In the future work, we will focus on the research into joint identification of frames and arguments, and realize automatic labeling of semantic roles on FrameNet.

Acknowledgements. This work is supported by National Natural Science Foundation of China (No. 61772324, No. 61673248), and Shanxi Province Postgraduate Joint Training Base Talent Training Project (No. 2018JD01, No. 2018JD02).

References

Baker, C.F., Fillmore, C.J., Lowe, J.B.: The Berkeley FrameNet project. In: Meeting of the Association for Computational Linguistics and, International Conference on Computational Linguistics, vol. 47, pp. 86–90 (1998)

Fillmore, C.J.: Frame semantics. In: Linguistics in the Morning Calm, pp. 111–138 (1982)

Hermann, K.M., et al.: Teaching machines to read and comprehend. In: NIPS, pp. 1693–1701 (2015)

Das, D., Schneider, N., Chen, D., Smith, N.A.: Probabilistic frame-semantic parsing. In: Human Language Technologies: The 2010 Conference of the North American Chapter of the Association for Computational Linguistics, vol. 40, pp. 948–956 (2010)

Das, D., Chen, D., Schneider, N., Smith, N.A.: Frame-semantic parsing. Comput. Linguist. **40**(1), 9–56 (2014)

Carreras, X., Litkowski, K.C., Stevenson, S.: Semantic role labeling: an introduction to the special issue. Comput. Linguist. **34**(2), 145–159 (2008)

Kalchbrenner, N., Grefenstette, E., Blunsom, P.: A Convolutional Neural Network for Modelling Sentences. eprint arxiv (2014)

Kim, Y.: Convolutional Neural Networks for Sentence Classification. eprint arxiv (2014)

Zeng, D.J., Liu, K., Lai, S.W., Zhou, G.Y., Zhao, J.: Relation classification via convolutional deep neural network. In: Proceedings of COLING, Dublin, Ireland (2014)

Chen, Y., Xu, L., Liu, K., Zeng, D., Zhao, J.: Event extraction via dynamic multi-pooling convolutional neural networks. In: The Meeting of the Association for Computational Linguistics, Beijing, China, pp. 167–176 (2015)

Dong, L., Wei, F., Zhou, M., Xu, K.: Question answering over freebase with multi-column convolutional neural networks. In: Meeting of the Association for Computational Linguistics and the International Joint Conference on Natural Language Processing, Beijing, China, pp. 260–269 (2015)

Severyn, A., Moschitti, A.: Learning to rank short text pairs with convolutional deep neural networks. In: The International ACM SIGIR Conference, pp. 373–382. ACM (2015)

Yin, W., Schütze, H.: MultiGranCNN: an architecture for general matching of text chunks on multiple levels of granularity. In: Meeting of the Association for Computational Linguistics and the International Joint Conference on Natural Language Processing, Beijing, China, pp. 63–73 (2015)

Gildea, D., Jurafsky, D.: Automatic labeling of semantic roles. Comput. Linguist. **28**(28), 245–288 (2002)

Carreras, X.: Introduction to the CoNLL-2004 shared task: semantic role labeling. In: Proceedings of 8th Conference on Natural Language Learning (CoNLL), Boston, MA, vol. 47, pp. 5–9 (2004)

Carreras, X., Màrquez, L.: Introduction to the CoNLL-2005 shared task: semantic role labeling. In: Proceedings of the 9th Conference on Computational Natural Language Learning (CoNLL), ANN Arbor, MI, pp. 152–164 (2005)

Fleischman, M., Kwon, N., Hovy, E.: Maximum entropy models for FrameNet classification. In: Conference on Empirical Methods in Natural Language Processing, Sapporo, pp. 49–56 (2003)

Erk, K.: Frame assignment as word sense disambiguation. In: proceedings of the 6th International Workshop on Computational Semantic (IWCS-6) (2005)

Johansson, R., Nugues, P.: LTH: semantic structure extraction using nonprojective dependency trees. In: Proceedings of SemEval, vol. 13, no. 4, pp. 227–230 (2007)

Bejan, C.A., Hathaway, C.: UTD-SRL: a pipeline architecture for extracting frame semantic structures. In: International Workshop on Semantic Evaluations, vol. 8764, pp. 460–463 (2007)

Hermann, K.M., Das, D., Weston, J., Ganchev, K.: Semantic frame identification with distributed word representations. In: Meeting of the Association for Computational Linguistics, vol. 1, pp. 1448–1458 (2014)

Bahdanau, D., Cho, K., Bengio, Y.: Neural machine translation by jointly learning to align and translate. arXiv: 1409.0473 (2014)

Collobert, R., Weston, J., Bottou, L., Karlen, M., Kavukcuoglu, K., Kuksa, P.: Natural language processing (almost) from scratch. J. Mach. Learn. Res. **12**(1), 2493–2537 (2011)

Yin, W.P., Schütze, H., Xiang, B., Zhou, B.W.: ABCNN: attention-based convolutional neural network for modeling sentence pairs. Action Editor: Brian Roark, vol. 4, pp. 259–272 (2016)

Collobert R., Weston J.: A unified architecture for natural language processing: deep neural networks with multitask learning. International Conference on Machine Learning, pp.160–167. ACM (2008)

Hinton, G. E., Srivastava, N., Krizhevsky, A., Sutskever, I., Salakhutdinov, R. R.: Improving neural networks by preventing co-adaptation of feature detectors. Comput. Sci. vol. 3(4), 212–223 (2012)

Zhang, Y., Wallace, B.: A sensitivity analysis of (and practitioners' guide to) convolutional neural networks for sentence classification. In: Proceedings of the 8th International Joint Conference on Natural Language Processing, Taipei, Taiwan, pp. 253–263 (2017)

Mikolov, T., Sutskever, I., Chen, K., Corrado, G., Dean, J.: Distributed representations of words and phrases and their compositionality. In: Advances in Neural Information Processing Systems, vol. 26, pp. 3111–3119 (2013)

Zhou, P., et al.: Attention-based bidirectional long short-term memory networks for relation classification. In: Meeting of the Association for Computational Linguistics, pp. 207–212 (2016)

Li, R., Liu, H.J., Li, S.H.: Chinese frame identification using T-CRF model. In: International Conference on Computational Linguistics, COLING 2010, pp. 674–682 (2010)

Zhao, H.Y., Li, R., Zhang, S., Zhang, L.W.: Chinese frame identification with deep neural network. J. Chin. Inf. Process. 30(6), 75–83 (2016)

Linked Document Classification by Network Representation Learning

Yue Zhang[1,2], Liying Zhang[1,2], and Yao Liu[1(✉)]

[1] Institute of Scientific and Technical Information of China, Beijing, China
{zhangyuejoslin, zhangliying}@pku.edu.cn,
liuy@istic.ac.cn
[2] School of Software and Microelectronics, Peking University, Beijing, China

Abstract. Network Representation Learning (NRL) can learn a latent space representation of each vertex in a topology network structure to reflect linked information. Recently, NRL algorithms have been applied to obtain document embedding in linked document network, such as citation websites. However, most existing document representation methods with NRL are unsupervised and they cannot combine NRL with a concrete task-specific NLP tasks. So in this paper, we propose a unified end-to-end hybrid Linked Document Classification (LDC) model which can capture semantic features and topological structure of documents to improve the performance of document classification. In addition, we investigate to use a more flexible strategy to capture structure similarity to improve the traditional rigid extraction of linked document topology structure. The experimental results suggest that our proposed model outperforms other document classification methods especially in the case of having less training sets.

Keywords: Document classification · NRL · Flexible random walk strategy

1 Introduction

Document classification is a very prevalent topic in the field of NLP, and there have been quite a lot of research results, such as the combination of SVM classifier and rule-based classifier (Prabowo and Thelwall 2009), the combination of Dependency Trees and CRF model (Nakagawa et al. 2010), and the ordinary BP neural network classification methods (Trappey et al. 2006). In general, the core of document classification is how to extract the key features of the text, and capture the mapping of features to categories.

The semantic features obtained by word2vec (Mikolov et al. 2013) and doc2vec (Le and Mikolov 2014) are commonly used in document classification, and those algorithms usually assume that documents are independent of each other. But in linked document networks, such as citation websites, documents are inherently connected (citation relationship), and this network structure information has been proven useful for machine learning and data mining tasks (Massa and Avesani 2007; Mei et al. 2009; Tang and Liu 2012). Therefore, in addition to the semantic features, the to-pology structure features among documents should also be considered in linked document classification. To join the linked features to documents, some researchers have

M. Sun et al. (Eds.): CCL 2018/NLP-NABD 2018, LNAI 11221, pp. 302–313, 2018.
https://doi.org/10.1007/978-3-030-01716-3_25

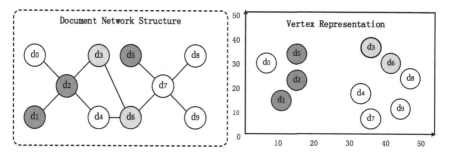

Fig. 1. Document network representation learning illustration by deepwalk

investigated to introduce the network representation learning algorithm into language models, such as LDE (Wang et al. 2016) and Tri-party (Pan et al. 2016). Figure 1 illustrates the document network representation learning by deepwalk (Perozzi et al. 2014). Deepwalk is one of the traditional algorithms utilized to capture the local structure information of the vertices in a network. In Fig. 1, each vertex denotes a paper, and the links are citation relationships. Different colors are the classification labels of documents, and the rest are unlabeled. Among them, d0 and d2 are connected, and then their vertex two-dimensional vector representations are close together even the label of d0 is unknown. However, deepwalk is not expressive enough to capture the diversity of connectivity patterns observed in networks because it can only obtain interconnected document information but lose identical structure information. For instance, as for d0 and d7, although they are not strictly connected, they share the same structural role in a hub document. So in theory, the distributed representation of d0 and d7 should also be close to each other, but according to the illustration, it's evident that deepwalk algorithm can't capture this kind of information.

Hence, in this paper, we study the novel problems of document classification based on network representation mainly from two specific aspects: (1) How to jointly learn document embedding and document topological structure, and apply the mixed feature into classification. (2) How to get more comprehensive and flexible document network embedding. Based on those problems, we propose a linked document framework for classification (LDC). The primary contribution of this paper is as follows:

(1) We provide a unified end-to-end model (LDC) to learn document semantic content and document topological structure jointly to improve the effectiveness of document classification.
(2) We get document network representations not only from the documents which are interconnected but also the documents which share similar topological structures. It improves the performance of document classification, especially in the case of small training sets.
(3) We evaluate our approach using DBLP[1]and Cite-Seer-M10[2] datasets. The results demonstrate the advantages of our method compared with 5 baselines.

---

[1] http://arnetminer.org/citation (V4 version is used)
[2] http://citeseerx.ist.psu.edu/

2 Related Work

The proposed model in this paper is based on network representation learning (network embedding). It aims to encode each vertex in the network as low-dimensional and dense vector, which can be easily and conveniently used as input in a machine learning model. Furthermore, the obtained vertex representation can be applied to common applications, such as visualization task, node classification, link prediction and community discovery.

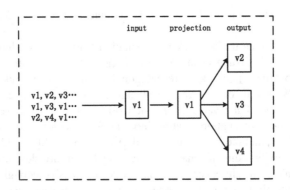

Fig. 2. Architecture of deepwalk algorithm

Deepwalk algorithm mentioned above is one of the most commonly used NRL algorithms. It introduces the skip-gram algorithm into NRL and employs word2vec model to embed all nodes into a continuous vector space. The architecture of deepwalk algorithm has been shown in Fig. 2. Each vertex in random walk such as $v_1 \rightarrow v_2 \rightarrow \cdots \rightarrow v_n$ can be considered as a word, and their random walk can be viewed as sentences. Then those sentences are input into word2vec model, which in turn yields the representation of each vertex.

Recently, some researchers have begun to introduce NRL models to NLP tasks. Yang proposed the text-associated deepwalk (TADW) model, taking into account the structure information and content information of the vertex, and combining the vertex text features into network representation learning under the framework of the matrix decomposition (Yang et al. 2015). Similar to TADW, such representation models based on network structure-content fusion learning include MFR (Li et al. 2015), Author2Vec (Genash et al. 2016), and etc. In addition to structure-content modeling, some algorithms even take document label information into consideration. Wang proposed the LDE model, employing distributional hypothesis idea for document embedding by combining link and label information with content simultaneously (Wang et al. 2016). Pan proposed the TriDNR model, Tri-party Deep Network Representation, exploited inter-node relationships, node-content correlation, and label-content correspondence in a network to learn representation for each vertex (Pan et al. 2016).

The similarity of those model is that they are unsupervised. Although LDE and TriDNR consider label information, they treat label as one of the features. So we put forward a combined model to incorporate task-specific supervision, which means to use label information supervise text and structure learning process simultaneously. The model improves the performance of text classification, at the same time, the intermediate training parameters of each vertex can better task-specific representation.

3 Model

3.1 Problem Statement

We first formally define the notation in linked document classification by network representation learning. Let $G = (V, E, D)$ denote a document network with texts and labels in each vertex, where $V = \{v_1, v_2 \cdots v_N\}$ is a set of connected N documents and v_i is the i-th vertex in the network. $e_{i,j} = (v_i, v_j) \in E$ represents the edge relationship between v_i and v_j, $D = \{d_1, d_2, \cdots d_N\}$ is the text information and $d_i = \{w_1, w_2, \cdots, w_n\}$ is the word representation of the i-th document. The primary task in this paper is to classify connected documents by learning the feature of the text and the topology structure jointly.

3.2 LDC

We propose a unified end-to-end hybrid model (LDC) to jointly capture the semantic and topology structure of linked document network. The overview of our model has been illustrated in Fig. 3. The overall model is divided into three major parts, which are semantic feature modeling, topological structure modeling and fusion modeling.

The inputs of the model are text content and linked information of documents. Among them, the word embedding is obtained from Glove which is a count-based model utilizing word co-occurrence matrix (Pennington et al. 2014), and the node embedding is trained by the node2vec algorithm (Grover et al. 2016).

We develop two deep neural network architectures, which are Convolution Neural Network (CNN) and Deep Neural Network (DNN). The input of word embedding is fed into CNN model to extract meaningful local semantic features in the text. The input of node embedding is fed into DNN model (two-layer) to get more abstract and smooth features in the document network structure.

After concating word and structure high-level features, the mixed representation is put into another fully four-layer connected deep neural network. The final softmax layer is to utilize classification label information to supervise the whole learning process. Through jointly learning the features of topology structure and document text, a better performance of text classification can be achieved.

3.3 Semantic Feature Modeling

There has been a variety of models to obtain document embedding from the word sequence. The traditional unsupervised algorithms, such as doc2vec (Micolov et al. 2013), are not able to consider classification label information during the training process. Some tasks (Pan et al. 2016) consider label after a classifier is trained with

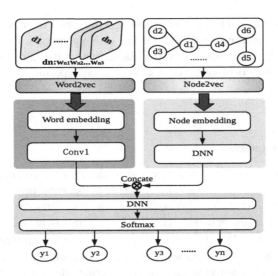

Fig. 3. LDC architecture. Conv: Convolution; DNN: Deep Neural Network.

document representations. However, experiment results prove that unsupervised text representation methods usually yield inferior results especially in particular machine learning tasks (Tang et al. 2015). In this section, we investigate different neural networks with label information for text modeling including CNN (Kalchbrenner et al. 2014), RNN (Cho et al. 2014), and Bidirectional RNN (Schuster and Paliwal 1997). The result is that CNN performs best as it can capture the local semantic dependency among words (Tu et al. 2017) by convolution and max-pooling layers. The input of CNN model is the word sequence representation of each document, and then CNN model gets document embedding though three layers, i.e. embedding layer, convolution layer and max-pooling layer.

Embedding Layer. The underlying assumption of word2vec is that "you shall know a word by the company it keeps," (Li J et al. 2016). In a word, a sound word representation should be used in predicting its nearby words. Word representation layer transfers the word sequence $D = (w_1, w_2, \ldots, w_n)$ in each document to corresponding word embedding. For instance, a sentence of length n (has been zero padded) can be represented as:

$$w_{1:n} = w_1 \oplus w_2 \cdots \oplus w_n \tag{1}$$

Where $\oplus$ is the concatenation operator. $w_{i:j}$ is the concatenation of words $w_i, w_{i+1}, \cdots w_{i+j}$.

Convolution Layer. After embedding layer, convolution layer can extract local features of word embedding of S. The convolution kernel is a sliding window of h words with a convolution matrix $K \in \mathbb{R}^{d \times (h \times d'')}$. Then:

$$Z_i = K \times S_{i:i+h-1} + b \tag{2}$$

Where $S_{i:i+h-1}$ is the concatenation of word embedding within the i-th window and b is the bias vector.

Global Max-Pooling Layer. The vector obtained after convolution layer is combined into a vector sequence $(x_0^i, \cdots, x_n^i)$. In order to capture the most important information related to tasks, we employ the global max-pooling and non-linear transformation as following:

$$d_i = \tanh\left(\max\left(x_0^i, \cdots, x_n^i\right)\right) \tag{3}$$

At last, we encode the text information of each document as $d^t = [d_1, \cdots, d_n]^T$. The document embedding trained by CNN model not only acquires the rich local features, but also embeds the label information into its own representation.

3.4 Topology Structure Feature Modeling

If the semantic feature of the text is not apparent, for instance, there are no same surrounding words between texts, then the topological structure feature should be used to indicate document's closeness in document network. As mentioned above, deepwalk is proposed as a method to learn network graphs based on skip-gram, and it employs the language model word2vec to learn a latent representation of each vertex. Mathematically, we simulate a random walk of fixed length l with a source node $t = c_0$ then the distribution of i-th node v_i in the walk is following:

$$P(c_i = x|c_{i-1} = v) = \begin{cases} \frac{\pi_{vx}}{Z} & \text{if } (v,x) \in E \\ 0 & \text{otherwise} \end{cases} \tag{4}$$

Where π_{vx} is the normalized transition probability between nodes v and x, and z is the normalizing constant. The assumption of deepwalk is that interconnected documents are more likely to be in the same category. However, this assumption is hidebound because documents could be far apart in the network but still have the same structural role, which also can indicate the same category. So we adjust the random walk strategy by utilizing document topology structure feature itself.

Rather than the rigid defining neighborhood for each vertex, the node2vec algorithm designs a flexible neighborhood strategy. It obtains a more comprehensive vertex representation mainly from two aspects, which are homophily and structural equivalence. In terms of homophily, vertices that are highly interconnected should be embedded close together. In contrast, in terms of structural equivalence, the documents sharing the same structural role should also be embedded close together even without interconnection. Among them, the homophily and structural equivalence are controlled by Breadth-first Sampling (BFS) and Depth-first Sampling (DFS) respectively. The concrete method is to multiple biases α to the weight of edges, which can be calculated through linked probability. Then π_{vx} in node2vec is:

$$\pi_{vx} = \alpha_{pq}(v, x) \cdot w_{vx} \tag{5}$$

Where w_{vx} is the weight of $E(v, x)$. Two parameters p and q are designed to adjust the random walk as follows:

$$\alpha_{pq}(v, x) = \begin{cases} \frac{1}{p} & \text{if } d_{vx} = 0 \\ 1 & \text{if } d_{vx} = 1 \\ \frac{1}{q} & \text{if } d_{vx} = 2 \end{cases} \tag{6}$$

Where d_{vx} is the shortest path between v and x. Assume the source vertex is t, and t moves to vertex x. If p > 1 (DFS), the next walk of x will traverse biasedly the vertices away from t, while if q > 1(DFS), the next random of x will traverse biasedly the vertices neighboring of t.

After getting the vertex representation, we did a comparative experiment. On one hand, we concate the node representation and document semantic representation directly; on the other hand, we feed the vertex representation into a deep neural network (two-layers) before concating with semantic representation. The experiment result shows that DNN performs better. The reason is that DNN can extract different levels of features and low-level features can be converted into a more abstract high-level representation through combination. In general, in section of topology structure feature modeling of documents, we not only use node2vec algorithm to get vertex representation, but also feed the representation into a DNN model to obtain a more abstract feature. The feature we get can be represented as $v^t = [v_1, \cdots, v_n]^T$.

3.5 Fusion

Given the semantic representation d^t and topology structure representation v^t, we can obtain the mix embedding as $u^t = d^t \otimes v^t$. Then u^t is fed into fully connected layers (four-layers). Finally, the final softmax layer, which contains probability distribution over the label, supervises the training process of document semantic embedding and structure embedding simultaneously. Let $Y \in R^{d \times L_c}$ be the label embedding matrix, where L_c is the number of unique labels. Then:

$$\max_{Y, U} \frac{1}{|y|} \sum_{i:y_i \in y} \log P(y_{u_i}|u_i) \tag{7}$$

Where U represents the documents with the combination of semantic and topology structure features. y_i is the label of the i-th document. And $P(y_{u_i}|u_i)$ is the probability that u_i's label is y_{u_i}, which is given as

$$P(y_{u_i}|d_i) = \frac{\exp\left(y_{u_i}^T u_i\right)}{\sum_{k=1}^{L_c} \exp\left(y_k^T u_i\right)} \tag{8}$$

The objective of LDC model is to minimize the following log-likelihood:

$$\zeta = \delta \sum_{i=1}^{N} \sum_{r \in R} \sum_{-t \leq j \leq t} \log P(v_{i+j}|v_i) + \mu \sum_{i=1}^{N} \sum_{-t \leq j \leq t} \log P(w_j|d_i) \tag{9}$$

The first term aims to learn vertex network structure information, and t is the window size, r is the random walks generated by the node2vec algorithm. The second term is to obtain the semantic information, and w_j is the j-th word in window size. δ and μ are the weights that balance network structure and text information.

4 Experiment Result

In order to investigate the effectiveness of LDC model, we conducted an experiment of document classification on two datasets. Both of them are paper citation networks.

4.1 Datasets

DBLP is a computer science bibliography website. Each paper may cite or be cited by other papers and form a citation website. We selected 60744 papers and labeled them into 4 categories, which are about the database, AI, computer vision and data mining (Table 1).

Cite-Seer-M10 is a subset of Cite-Seer-M10 data which consists of scientific publications. We selected 10310 papers and labeled them into 10 categories, which are about archaeology, financial economics, agriculture, biology, archeology, material science, industrial engineering, petroleum chemistry, physics and social science.

Table 1. Paper citation datasets

Datasets	DBLP	Cite-Seer-M10
#vertices	60744	10310
#edges	52890	77218
#group	4	10

4.2 Evaluation Metrics and Experiment Settings

We use the standard metrics of classification, Average Macro-F1, and Average Micro-F1. The larger the Macro-F1 and Micro-F1 are, the better the document representation is for the classification task. The default parameters for algorithms as follows: window size R = 5, dimension b = 300, training size p = 70%, epoch = 20, batch_size = 100, $\delta = 1$, $\mu = 1$. For fairness, we set the same dimension and training size for all algorithms. The hidden units of DNN in semantic feature modeling are 300, and the hidden units of DNN in Fusion are 300, 128, 64, and 32.

4.3   Experiment Performance Comparison

Table 2 shows the performance of different algorithms on two test datasets. The input of SVM, LSTM and CNN are the document embeddings obtained from d2v algorithm. The results show that the neural network methods perform better than traditional machine learning algorithms. d2v + dw represents the concatenation of the vector representation learned by doc2vec and deepwalk, and its representation is fed into a CNN model. d2v + dw has better results because it combines semantic and structural information at the same time. Furthermore, we utilize TriDNR algorithm to get each vertex representation and input it into an SVM classifier, the results show that it performs better than d2v + dw because the label information is considered. But TriDNR is still far from optimal, comparing with LDC algorithm. To verify the effect of different dimensions of representation on the classification effect, we verify dimension b increasing from 100 to 300, Fig. 4 shows the variation of Macro-F1 with different dimensions and there is a slight rise for LDC, which proves LDC is quite stable.

Table 2. Document classification comparison on datasets

Datasets	DBLP		Cite-Seer-M10	
Metric	Macro-F1	Micro-F1	Macro-F1	Micro-F1
SVM	0.635 ± 0.002	0.612 ± 0.003	0.655 ± 0.006	0.632 ± 0.005
LSTM	0.683 ± 0.006	0.681 ± 0.004	0.715 ± 0.004	0.695 ± 0.004
CNN	0.733 ± 0.004	0.721 ± 0.003	0.749 ± 0.008	0.752 ± 0.003
d2v + dw	0.757 ± 0.003	0.745 ± 0.002	0.771 ± 0.003	0.773 ± 0.002
TriDNR	0.782 ± 0.003	0.770 ± 0.004	0.785 ± 0.006	0.789 ± 0.004
LDC	$\mathbf{0.810 \pm 0.005}$	$\mathbf{0.813 \pm 0.006}$	$\mathbf{0.821 \pm 0.002}$	$\mathbf{0.819 \pm 0.005}$

At the same time, we vary the percentage of training sets p from 10% to 70% on DBLP datasets, and we change different network representation learning algorithms. Table 3 is the results. As for node2vec, the best exploration strategy (p = 0.25, q = 0.5) turns out to perform better than deepwalk (p = 1, q = 1) and LINE. Especially under the circumstance of 10% training sets. The reason is that when we have small training sets, the influence of feature of semantic would be little, which leads to the feature of topology structure is more apparent. When we lower the parameter of p and q, the feature of topological structure can be adequately captured from both macro and micro views.

4.4   Case

As we mentioned before, the intermediate training parameters of each vertex in LDC can be treated as task-specific representation. For the purpose of validating this conclusion, we visualize the DBLP document representation (4 categories, 100 dimensions) in 2D space in Fig. 5 by Google Embedding Projector (Smikov et al. 2016), which is a tool to visualize high-dimension data. The dimension reduction method we

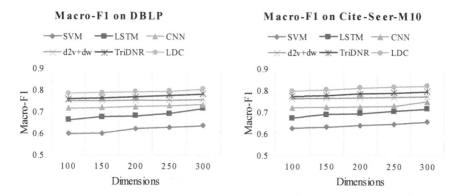

Fig. 4. Experiments with different dimensions

Table 3. Average macro-F1 score on DBLP datasets

%p	Deepwalk	Line	node2vec
10	64.3%	58.4%	71.1%
30	72.2%	66.2%	73.5%
50	77.3%	68.2%	79.2%
70	79.4%	70.2%	80.9%

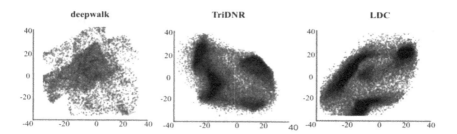

Fig. 5. Two-dimension visualization of DBLP

choose is custom linear projection. It can help discover useful direction in the dataset, such as the difference between a formal tone and an informal tone in a language generation model.

According to the results, the category boundaries of deepwalk are very blurred, which indicates that network feature alone is not apparent. TriDNR performs much better than deepwalk because it combines semantic, structural and label information. However, TriDNR is still slightly overlapping among four categories. The result of LDC is quite clear, which proves that our proposed supervised model can yield better task-specific linked document representation than other methods.

5 Conclusion

This paper presents a unified end-to-end LDC model, a new approach for document classification on citation networks. We not only propose a method learning text information and network topology structure jointly, but also improve the flexibility of random walk strategies by learning the document network structure feature itself. LDC gains significantly better performance than other document classification baseline methods. Same as other document classification methods, LDC can be used not only in classification, but also in sentiment analysis and information retrieval.

The future work has two main directions: (1) We will expand our investigation to heterogeneous network application in the language model. (2) We have proved the effectiveness of LDC on document embedding for classification, and we will investigate its performance on other specific tasks such as ranking or recommendation.

References

Blei, D.M., Ng, A.Y., Jordan, M.I.: Latent dirichlet allocation. J. Mach. Learn. Res. **3**, 993–1022 (2003)

Cai, D., Zhang, D., Zhai, C.: Topic modeling with network regularization. In: Proceedings of the 17th International Conference on World Wide Web, pp. 101–110. ACM, China (2008)

Cho, K., et al.: Learning phrase representations using RNN encoder-decoder for statistical machine translation. arXiv preprint arXiv:1406.1078 (2014)

Ganguly, S., Gupta, M., Varma, V., Pudi, V.: Author2Vec: learning author representations by combining content and link information. In: International Conference Companion on World Wide Web. International World Wide Web Conferences Steering Committee, pp. 49–50 (2016)

Grover, A., Leskovec, J.: node2vec: scalable feature learning for networks. In Proceedings of the 22nd ACM SIGKDD International Conference on Knowledge Discovery and Data Mining, pp. 855–864. ACM (2016)

Kalchbrenner, N., Grefenstette, E., Blunsom, P.: A convolutional neural network for modelling sentences. arXiv preprint arXiv:1404.2188 (2014)

Landauer, T.K., Foltz, P.W., Laham, D.: An introduction to latent semantic analysis. Discourse Process. **25**(2–3), 259–284 (1998)

Le, Q., Mikolov, T.: Distributed representations of sentences and documents. In: International Conference on Machine Learning, pp. 1188–1196 (2014)

Li, J., Ritter, A., Jurafsky, D.: Learning multi-faceted representations of individuals from heterogeneous evidence using neural networks. arXiv preprint arXiv:1510.05198 (2015)

Li, J., Zhu, J., Zhang, B.: Discriminative deep random walk for network classification. Meeting of the Association for Computational Linguistics, pp. 1004–1013 (2016)

Massa, P., Avesani, P.: Trust-aware recommender systems. In: Proceedings of the 2007 ACM Conference on Recommender systems, pp. 17–24. ACM (2007)

Mei, Q. Ma, H., Lyu, M.R., King, I.: Learning to recommend with trust and distrust relationships. In: RecSys, pp. 189–196. ACM (2009)

Mikolov, T., Chen, K., Corrado, G., Dean, J.: Efficient estimation of word representations in vector space. arXiv preprint arXiv:1301.3781 (2013)

Nakagawa, T., Inui, K., Kurohashi, S.: Dependency tree-based sentiment classification using CRFs with hidden variables. In: Human Language Technologies: The 2010 Annual Conference of the North American Chapter of the Association for Computational Linguistics, pp. 786–794. Association for Computational Linguistics (2010)

Pan, S., Wu, J., Zhu, X., Zhang, C., Wang, Y.: Tri-party deep network representation. In: International Joint Conference on Artificial Intelligence, pp. 1895–1901. AAAI Press (2016)

Perozzi, B., Al-Rfou, R., Skiena, S.: Deepwalk: online learning of social representations. In: Proceedings of the 20th ACM SIGKDD International Conference on Knowledge Discovery and Data Mining, pp. 701–710. ACM (2014)

Pennington, J., Socher, R., Manning, C.: Glove: global vectors for word representation. In: Proceedings of the 2014 Conference on Empirical Methods in Natural Language Processing. EMNLP, pp. 1532–1543 (2014)

Prabowo, R., Thelwall, M.: Sentiment analysis: a combined approach. J. Inf. **3**(2), 143–157 (2009)

Salton, G., Buckley, C.: Term-weighting approaches in automatic text retrieval. Inf. Process. Manag. **24**(5), 513–523 (1988)

Schuster, M., Paliwal, K.K.: Bidirectional recurrent neural networks. IEEE Trans. Signal Process. **45**(11), 2673–2681 (1997)

Smilkov, D., Thorat, N., Nicholson, C., Reif, E., Viégas, F.B., Wattenberg, M.: Embedding Projector: interactive visualization and interpretation of embeddings. arXiv preprint arXiv: 1611.05469 (2016)

Sun, X., Guo, J., Ding, X., Liu, T.: A general framework for content-enhanced network representation learning. arXiv preprint arXiv:1610.02906 (2016)

Tang, J., Liu, H.: Feature selection with linked data in social media. In: Proceedings of the 2012 SIAM International Conference on Data Mining, pp. 118–128. Society for Industrial and Applied Mathematics (2012)

Tang, J., Qu, M., Mei, Q.: PTE: predictive text embedding through large-scale heterogeneous text networks. In: Proceedings of the 21th ACM SIGKDD International Conference on Knowledge Discovery and Data Mining, pp. 1165–1174. ACM (2015)

Trappey, A.J., Hsu, F.C., Trappey, C.V., Lin, C.I.: Development of a patent document classification and search platform using a back-propagation network. Expert Syst. Appl. **31** (4), 755–765 (2006)

Tu, C., Liu, H., Liu, Z., Sun, M.: CANE: context-aware network embedding for relation modeling. In: Proceedings of the 55th Annual Meeting of the Association for Computational Linguistics, vol. 1, pp. 1722–1731 (2017)

Yang, C., Liu, Z., Zhao, D., Sun, M., Chang, E.Y.: Network representation learning with rich text information. In: International Conference on Artificial Intelligence, pp. 2111–2117. AAAI Press (2015)

Wang, S., Tang, J., Aggarwal, C., Liu, H.: Linked document embedding for classification. In: Proceedings of the 25th ACM International on Conference on Information and Knowledge Management, pp. 115–124. ACM (2016)

Zhang, D., Yin, J., Zhu, X., Zhang, C.: Network representation learning: a survey. arXiv preprint arXiv:1801.05852 (2017)

A Word Embedding Transfer Model
for Robust Text Categorization

Yiming Zhang[1](✉) ⓘ, Jing Wang[2] ⓘ, Weijian Deng[1] ⓘ, and Yaojie Lu[3] ⓘ

[1] Institute of Information Engineering, CAS, Beijing, China
zhangyiming@iie.ac.cn, dengwj16@gmail.com
[2] Beijing University of Posts and Telecommunications, Beijing, China
jingw@bupt.edu.cn
[3] Institute of Software, CAS, Beijing, China
yaojie.lu@outlook.com

Abstract. It is common to fine-tune pre-trained word embeddings in text categorization. However, we find that fine-tuning does not guarantee improvement across text categorization datasets, while could introduce considerable parameters to model. In this paper, we study new transfer methods to solve the problems above, and propose "Robustness of OOVs" to provide a perspective to reduce memory consumption further. The experimental results show that the proposed method is proved to be a good alternative to fine-tuning method on large dataset.

Keywords: Word embedding · Text categorization · Transfer learning

1 Introduction

For many natural language processing (NLP) tasks such as text categorization, the word-level features of text usually derive from pre-trained word embeddings [1]. The word embeddings are learned by unsupervised learning algorithms such as Skip-gram, cBoW [15], Glove [18], and FastText [2] on very large-scale corpus.

For text categorization task, it is common to fine-tune pre-trained word embeddings (noted as "fine-tuning method" in this paper) to learn task-specific word-level features. Previous work [8,24] report that this is effective on several datasets. And researchers usually attribute improvement to the learned task-specific word-level features [8,21].

However, we argue that fine-tuning word embeddings is not always a good choice. To avoid confusion, we refer the words appearing in training set as in-vocabulary words (IVs), and words not appearing in training set as out-of-vocabulary words (OOVs) in the rest of this section.

As shown in Fig. 1(a), we denote the pre-trained word embedding space as α. Since the vocabulary of training set is limited, the fine-tuning model can only transform the IVs into task-specific word embedding space (denoted as β) and OOVs stay in space α. With the increasing number of training steps, the distribution of words in space α will become increasingly different from the

© Springer Nature Switzerland AG 2018
M. Sun et al. (Eds.): CCL 2018/NLP-NABD 2018, LNAI 11221, pp. 314–323, 2018.
https://doi.org/10.1007/978-3-030-01716-3_26

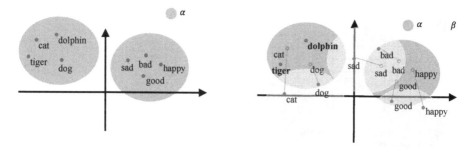

(a) Before fine-tuning word embedding.   (b) Intermediate state of word embedding space.

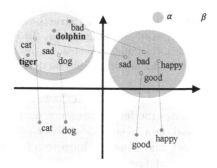

(c) Final state of the word embedding space.

Fig. 1. Illustration of a disadvantage of fine-tuning method. Blue areas represent pre-trained embedding space (denoted as space α). Orange areas represent generated task-specific embedding space (denoted as space β). From top to bottom: (a) the state before fine-tuning word embedding; (b) the beginning of fine-tuning where the updates on words generally occur within or between clusters; (c) the final state of fine-tuning where there are many overlapping regions among space α and space β, thus the meaning of words in α casts to the meaning of words in β. (Color figure online)

distribution of words in space β (the knowledge in space α is general while the knowledge in space β is task-specific). As a result, there would exist two embedding space for words in test set to choose from.

However, as shown in Fig. 1(c), there could exist overlapping regions among α and β. In these overlapping regions, since the model generally cannot distinguish multiple word embedding spaces (can only work on space β), the meaning of the OOVs casts to the meaning of IVs automatically. However, these casting operations are usually unreasonable. For example, as shown in Fig. 1(c), suppose that "tiger" and "dolphin" are OOVs and the given task is sentiment classification. Because "tiger" and "dolphin" are unseen for model, a word cluster representing negative sentiment information is likely generated nearby. As a result, the "tiger" and "dolphin" will carry negative information if they appear in testing set, which could greatly mislead the decision of sentiment polarity for the corresponding samples.

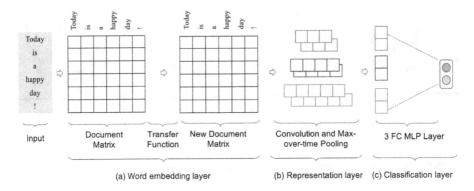

(a) Word embedding layer   (b) Representation layer (c) Classification layer

Fig. 2. The structure of our model. The model structure mainly consists of three parts (from input to output): (a) word embedding layer, (b) representation layer, and (c) classification layer. "FC" denotes Full-Connected. In this work, we focus on word embedding layer, and study the influence of transfer methods to model performance.

Furthermore, our experimental results on AGNews and Yelp Reviews demonstrate that fine-tuning degrades model performance dramatically. On the other hand, the fine-tuning method on a large-scale dataset could introduce considerable parameters to the model, and consume hard-to-afford memory resources.

To solve the above issues of fine-tuning method, we propose another two new transfer methods based on the fixing method, i.e., from the simple method, noted as "scaling method", to the complicated method "linear transformation method", noted as "lin-trans method". In addition, we also introduce "Robustness to OOVs" as a metric of memory consumption and we discuss it in detail in the Sect. 4.3.

2   Related Work

Distributed representation of words [1] which represent a word with a low-dimensional and dense embedding, alleviate the data sparsity problem and has been the basic methods in text categorization [5,6,14]. [7] and [8] first apply word-level convolutional neural network [12] to text categorization task and have achieved significant progress.

Recently, we have witnessed increasing efforts in utilizing the pre-trained word embedding [22], which introduce useful external knowledge to their model. [16] investigate the transferability of neural networks in NLP by Layer-by-Layer analysis, and found that word embeddings could be transferable to semantically different tasks.

Most of works [7,8] suggest that fine-tuning word embedding in the training stage can get a better effect. On the contrary, almost all of the works on machine reading comprehension task [10,19,23,25] are prone to fixing pre-trained word embeddings after transferred to MRC model, even [13] points out fine-tuning method can dramatically degrade model performance. Inspired by this, we report the disadvantages of fine-tuning method in Sect. 1, and propose two new transfer methods with the word embedding layer fixed.

3 Model

3.1 Preliminaries

Given a document $S = w_1 w_2 \ldots w_i \ldots w_l$, where l is the document length and w_i is the i-th token. Each token w_i is first transformed to v_i by looking up the $V * D$ word embedding table E, where V is vocab size and D is the feature dimension of each word vector. Then, S can be represented by the output matrix $S_{mat} = [v_1, v_2, \ldots, v_l]$. Given a "transfer function" $F_{tr}(v)$ to represented different transfer methods, each element v_i in S_{mat} is fed to $F(v)$ then S_{mat} can be represented by the new output matrix $S_{tr\_mat} = [v_{T1}, v_{T2}, \ldots, v_{Tl}]$, where $v_{Ti} = F_{tr}(v_i)$.

As introduced in Sect. 1, there are two choices for the "transfer function": scaling method and lin-trans method.

3.2 Scaling Method

Assume v is a D-dimensional vector, then in scaling method, the transfer function $F_{tr}(v)$ is defined as

$$F_{tr}(v) = v \otimes u \tag{1}$$

where $\otimes$ means element-wise multiplication and U is a trainable real vector, $u = [u_1, u_2, \ldots, u_D]$. Each element u_i in u denotes the scaling degree of the i-th element of input vector v.

3.3 Lin-Trans Method

Assume v is a D-dimensional vector, then in lin-trans method, the transfer function $F_{tr}(v)$ is defined as

$$F_{tr}(v) = U \cdot v + b \tag{2}$$

where U is a $D*D$ real matrix, b is bias vector $b = [b_1, b_2, \ldots, b_D]$, and "$\cdot$" denotes matrix multiplication. U and b are trainable parameters. After this transfer function, pre-trained embedding space is linearly transformed to a new embedding space.

Compare with Fine-Tuning Method. The proposed two transfer methods are all applied on the whole embedding feature space rather than directly fine-tuning word embedding layer. Therefore, the proposed two methods would not suffer from unaffordable memory problem on large-scale datasets.

3.4 Model Structure

As demonstrated in Fig. 2, the model structure mainly contains three parts: word embedding layer, representation layer and classification layer.

We adopt shallow-and-wide convolution structure [8,11] as our representation layer. We adopt MLP with two hidden layers as our classification layer.

The activation functions used in convolution layer and classification layer are all rectified linear unit (ReLU) [17]. And the softmax activation function is used in the output layer to normalize the output logits to get the predictive probabilities for all target labels. In addition, Dropout [20] is applied on the result of representation layer to improve the robustness of model.

3.5 Model Training

Given all of training instances and their labels from dataset D, the objective function $J(\theta)$ is as follows:

$$J(\theta) = \sum_{(x_i, y_i \in D)} \log p(y_i | x_i, \theta), \tag{3}$$

where θ is the parameters of model; x_i is a instance from dataset D and y_i is its label.

We maximize the log likelihood $J(\theta)$ through stochastic gradient descent over shuffled mini-batches method. We adopt Adam [9] as optimizer in this paper.

4   Experiment

4.1   Datasets

We select four available datasets, from a relative small scale (120K samples in training set) to a relative large scale (1400K samples in training set) on text categorization for experiment. They are AGNews, Yelp Reviews Polarity (noted as "Yelp2"), Yelp Reviews Full (noted as "Yelp5") and Yahoo Answers (noted as "Yahoo"). These four datasets introduced by previous work [26] are summarized in Table 1. To avoid confusion, the OOVs in this paper is referred to the words appear in training or testing set while not appear in pre-trained word embeddings. The vocab size in this paper refers to words in both training and testing set.

Table 1. Statistics of four text categorization datasets used in this paper.

Dataset	#Train	#Test	#Classes	Vocab size	#OOVs & proportion
AGNews	120K	7.6K	4	90549	8946 (9.8%)
Yelp2	560K	38K	2	291541	143708 (49.2%)
Yelp5	650K	50K	5	314526	161224 (51.2%)
Yahoo	1,400K	60K	10	1156111	848237 (73.3%)

4.2   Implementation Details

Before experiment, we tokenize all datasets using *WordpunktTokenizer* provided in NLTK and pad five tokens "<pad>" to each side of document. Public available *word2vec* [15] is used as pre-trained word embeddings in this paper. The out-of-vocabulary words are initialized randomly in reset settings. We fix random seed to 1 for all experiments to ensure fairness of comparisons.

In all experiments, we keep all hyper-parameters the same. We set batch size 32. We use Adam [9] as optimizer and set learning rate 0.001. The convolution layer has four filter groups with $2, 3, 4, 5$ sizes, respectively, and each filter group contains 1500 filters. We use Xavier [3] method to initialize the parameters of filters. The dropout rate is set to 0.5. Each hidden layer in MLP has 300 hidden units and initialized using [4]. In scaling method, the parameters in transfer function F_{tr} are randomly initialized with Gaussian distribution with a mean of 1.0 and a standard deviation of 0.1. In lin-trans method, the W in transfer function F_{tr} is initialized with a unit diagonal matrix and Gaussian noise with a mean of 0.0 and a standard deviation of 0.1 is added, and b in F_{tr} in initialized to 0.

4.3   Evaluation

The evaluation results of the four transfer methods are shown in Table 2. To test the effectiveness of our re-implemented model, we extra give a baseline (fine-tune.CNN) [11], which is consist of a embedding layer using fine-tuning method, a representation layer with the same structure as ours, and a linear classification layer.

Table 2. The accuracy of various methods on four text categorization datasets.

Method	AGNews	Yelp2	Yelp5	Yahoo
Fine-tune.CNN	92.2	95.9	64.9	73.0
Fixing	93.75	96.27	65.57	74.40
Fine-tuning	93.11	96.27	65.31	74.90
Scaling (ours)	**93.82**	96.35	**65.96**	**75.09**
Lin-trans (ours)	93.73	**96.42**	65.73	74.68

Fine-Tuning Method is Unaffordable on Large Dataset. There are more than 1,000,000 words in Yahoo. When using the Adam optimizer, the word embedding layer can not even be loaded into a single graphics card (e.g., Titan Xp with 12 GB GPU memory). As a result, the word embedding layer has to be loaded into CPU memory and updated by CPU, which greatly slows down the training speed. The training hours of different transfer methods (with the same training steps across methods for a dataset) are shown in Table 3. Noted that the fine-tuning method needs training for 208 h on Yahoo, 6X slower than fixing method.

Table 3. The training hours of different transfer methods

Method	AGNews	Yelp2	Yelp5	Yahoo
Fixing	0.4	9	9.7	33
Fine-tuning	0.5	9.5	10.5	208
Scaling	0.6	11	12	37
Lin-trans	1.5	13	14.5	39

Fine-Tuning Method Does Not Guarantee Improvement. Noted that fine-tuning method degrades the performance on AGNews, Yelp2 and Yelp5. We guess that the model overfits these not very large datasets due to the introduced considerable parameters by fine-tuning method. This phenomenon is not consistent with that in [8], we guess the reason is datasets in [8] are too small that model could gain significant improvement from task-specific word-level knowledge. Only if a dataset is relative large, the limited task-specific word-level knowledge could not easy to cancel out the bad effect of overfitting.

Scaling Method is a Good Solution on a Large-Scale Dataset. It is noted that the performance of scaling method is significant in relative large datasets such as Yelp5 (further improve 0.4% than fixing method) and Yahoo (further improve 0.69% than fixing method), even surpasses fine-tuning method on Yahoo. Noted that training with scaling method only need 37 h while fine-tuning method need 208 h on Yahoo. Besides, the lin-trans method is inferior to scaling method both in testing performance and training time. Therefore, scaling method is better in large dataset under the considerations above.

Robustness to Out-of-Vocabulary Words (OOVs). As summarized in Table 1, OOVs occupy a very large proportion in datasets (surpass 50% of vocabulary in a relative large dataset). It can further reduce the memory consumption by replacing OOVs to some appointed symbol because the size of embedding layer is $O(N)$ space complexity to vocab size N. The reduction of memory resources provides conditions for designing larger and more complex presentation and categorization layers. However, this reduction can only be safely conducted when the transfer method is robust enough to the replacement of OOVs.

We again conduct experiments with the reset settings, except for the replacement of all OOVs to "<unk>". This could leads to performance decrement because the information about the distinction of OOVs is discarded. Then we regard performance decrement (reduced accuracy compared to Table 2) as a simple metric of robustness to OOVs. We say a transfer method is more robust to OOVs if its performance decrement is smaller (if the decrement is less than 0, it means replacing OOVs could get extra improvement). The experimental results are shown in Table 4.

It is clear that fine-tuning method has the worst robustness to OOVs. Even on large dataset Yahoo, the performance of fine-tuning method will drop to a lower level than that of the fixing method. So it is unwise to replace OOVs when using

Table 4. Accuracy decrement of four transfer methods.

Method	AGNews	Yelp2	Yelp5	Yahoo
Fixing	0.16	0.01	0.24	0.11
Fine-tuning	0.26	0.05	**0.07**	0.94
Scaling	**−0.11**	**0.00**	0.15	**0.10**
Lin-trans	0.06	**0.00**	0.11	0.12

fine-tuning method. Conversely, replacing OOVs in proposed scaling method and lin-trans method only slightly degrades performance across all datasets in our experiments. Thus, the memory consumption could be further reduced by the proposed method.

Selection of Transfer Methods. Although in previous works [8], fine-tuning has been proved to be effective on very small datasets, e.g., MR, SST and TREC, we report that it is unwise to use fine-tuning method on larger datasets, e.g., from AGNews with 120K samples to Yahoo with 1400K samples, either because of poor testing performance or unaffordable memory consumption. In contrast, the proposed scaling method as an improved fixing method is proved to be effective and memory friendly on relative large datasets in this paper. Therefore, using fine-tuning method on very small datasets and using the scaling method on a larger dataset is recommendable.

5   Conclusion

In this work, we report fine-tuning word embedding could suffer from poor testing performance and unaffordable memory consumption problems on relatively large scale text categorization datasets. To alleviate those problems, we propose new transfer methods based on the fixing method, and introduce "Robustness of OOVs" to provide a perspective to further reduce memory consumption. The experimental results demonstrate that the proposed scaling method improves the accuracy of fixing method, while only introduces minor parameters and has the better robustness to OOVs. Thus, the scaling method is an effective and robust alternative for text categorization.

References

1. Bengio, Y., Ducharme, R., Vincent, P., Janvin, C.: A neural probabilistic language model. J. Mach. Learn. Res. **3**, 1137–1155 (2003)
2. Bojanowski, P., Grave, E., Joulin, A., Mikolov, T.: Enriching word vectors with subword information. Trans. Assoc. Comput. Linguist. **5**, 135–146 (2017)
3. Glorot, X., Bengio, Y.: Understanding the difficulty of training deep feedforward neural networks. In: Teh, Y.W., Titterington, M. (eds.) Proceedings of the Thirteenth International Conference on Artificial Intelligence and Statistics. Proceedings of Machine Learning Research, vol. 9, pp. 249–256. PMLR, Chia Laguna Resort, Sardinia, 13–15 May 2010. http://proceedings.mlr.press/v9/glorot10a.html

4. He, K., Zhang, X., Ren, S., Sun, J.: Delving deep into rectifiers: surpassing human-level performance on imagenet classification. CoRR abs/1502.01852 (2015). http://arxiv.org/abs/1502.01852

5. Ji, Y., Smith, N.A.: Neural discourse structure for text categorization. In: Proceedings of the 55th Annual Meeting of the Association for Computational Linguistics (Volume 1: Long Papers), pp. 996–1005. Association for Computational Linguistics, Vancouver, July 2017. http://aclweb.org/anthology/P17-1092

6. Johnson, R., Zhang, T.: Deep pyramid convolutional neural networks for text categorization. In: Proceedings of the 55th Annual Meeting of the Association for Computational Linguistics (Volume 1: Long Papers), pp. 562–570. Association for Computational Linguistics, Vancouver, July 2017. http://aclweb.org/anthology/P17-1052

7. Kalchbrenner, N., Grefenstette, E., Blunsom, P.: A convolutional neural network for modelling sentences. In: Proceedings of the 52nd Annual Meeting of the Association for Computational Linguistics (Volume 1: Long Papers), pp. 655–665. Association for Computational Linguistics, Baltimore, June 2014. http://www.aclweb.org/anthology/P14-1062

8. Kim, Y.: Convolutional neural networks for sentence classification. In: Proceedings of the 2014 Conference on Empirical Methods in Natural Language Processing (EMNLP), pp. 1746–1751. Association for Computational Linguistics, Doha, October 2014. http://www.aclweb.org/anthology/D14-1181

9. Kingma, D.P., Ba, J.: Adam: a method for stochastic optimization. CoRR abs/1412.6980 (2014). http://arxiv.org/abs/1412.6980

10. Krause, B., Murray, I., Renals, S., Lu, L.: Multiplicative LSTM for sequence modelling. arXiv: Neural and Evolutionary Computing (2017)

11. Le, H.T., Cerisara, C., Denis, A.: Do convolutional networks need to be deep for text classification (2017)

12. Lecun, Y., Bottou, L., Bengio, Y., Haffner, P.: Gradient-based learning applied to document recognition. Proc. IEEE **86**, 2278–2324 (1998)

13. Li, P., et al.: Dataset and neural recurrent sequence labeling model for open-domain factoid question answering. arXiv preprint arXiv:1607.06275 (2016)

14. Liu, P., Qiu, X., Huang, X.: Adversarial multi-task learning for text classification. In: Proceedings of the 55th Annual Meeting of the Association for Computational Linguistics (Volume 1: Long Papers), pp. 1–10. Association for Computational Linguistics, Vancouver, July 2017. http://aclweb.org/anthology/P17-1001

15. Mikolov, T., Chen, K., Corrado, G.S., Dean, J.: Efficient estimation of word representations in vector space. arXiv preprint arXiv:1301.3781 (2013)

16. Mou, L., et al.: How transferable are neural networks in NLP applications? In: Proceedings of the 2016 Conference on Empirical Methods in Natural Language Processing, pp. 479–489. Association for Computational Linguistics, Austin, November 2016. https://aclweb.org/anthology/D16-1046

17. Nair, V., Hinton, G.E.: Rectified linear units improve restricted Boltzmann machines. In: Proceedings of the 27th International Conference on Machine Learning, pp. 807–814 (2010)

18. Pennington, J., Socher, R., Manning, C.D.: Glove: global vectors for word representation. In: Proceedings of the 2014 Conference on Empirical Methods in Natural Language Processing (EMNLP), pp. 1532–1543 (2014)

19. Seo, M.J., Kembhavi, A., Farhadi, A., Hajishirzi, H.: Bidirectional attention flow for machine comprehension. In: International Conference on Learning Representations (2017)

20. Srivastava, N., Hinton, G.E., Krizhevsky, A., Sutskever, I., Salakhutdinov, R.: Dropout: a simple way to prevent neural networks from overfitting. J. Mach. Learn. Res. **15**(1), 1929–1958 (2014)
21. Tai, K.S., Socher, R., Manning, C.D.: Improved semantic representations from tree-structured long short-term memory networks, pp. 1556–1566, July 2015. http://www.aclweb.org/anthology/P15-1150
22. Wang, P., et al.: Semantic clustering and convolutional neural network for short text categorization, pp. 352–357 (2015)
23. Wang, W., Yang, N., Wei, F., Chang, B., Zhou, M.: Gated self-matching networks for reading comprehension and question answering. In: Proceedings of the 55th Annual Meeting of the Association for Computational Linguistics (Volume 1: Long Papers), vol. 1, pp. 189–198 (2017)
24. Yang, Z., Yang, D., Dyer, C., He, X., Smola, A.J., Hovy, E.H.: Hierarchical attention networks for document classification. In: Proceedings of the 2016 Conference of the North American Chapter of the Association for Computational Linguistics: Human Language Technologies, pp. 1480–1489 (2016)
25. Yu, A.W., et al.: QANet: combining local convolution with global self-attention for reading comprehension. arXiv preprint arXiv:1804.09541 (2018)
26. Zhang, X., Zhao, J., LeCun, Y.: Character-level convolutional networks for text classification. In: Proceedings of the 28th International Conference on Neural Information Processing Systems. NIPS 2015, vol. 1, pp. 649–657 (2015)

Review Headline Generation
with User Embedding

Tianshang Liu[1,2(✉)], Haoran Li[1,2], Junnan Zhu[1,2], Jiajun Zhang[1,2],
and Chengqing Zong[1,2,3]

[1] National Laboratory of Pattern Recognition, Institute of Automation,
Chinese Academy of Sciences, Beijing 100190, China
{tianshang.liu,haoran.li,junnan.zhu,jjzhang,cqzong}@nlpr.ia.ac.cn
[2] University of Chinese Academy of Sciences, Beijing 100049, China
[3] CAS Center for Excellence in Brain Science and Intelligence Technology,
Shanghai 200031, China

Abstract. In this paper, we conduct a review headline generation task
that produces a short headline from a review post by a user. We argue
that this task is more challenging than document summarization, because
the headlines generated by users vary from person to person. It not only
needs to effectively capture the preferences of the users who post the
reviews, but also requires to mine the emphasis of the users regarding the
review when they write the headlines. To this end, we propose to incorporate the user information as the prior knowledge into the encoder and
decoder for general sequence-to-sequence model. Specifically, we introduce user embedding for each user, and then we use these embeddings to
initialize the encoder and decoder, or as biases for decoder initialization.
We construct a review headline generation dataset, and the experiments
on this dataset demonstrate that our models significantly outperform
baseline models which do not consider user information.

Keywords: Review headline generation · User embedding
Sequence-to-sequence neural network

1 Introduction

Review headline generation is a task that aims to generate a headline for a
user review. A user review of an item generally provides user experience for this
item, including user satisfaction, and user preferences. Therefore, review headline
generation is very essential in helping potential users of an item to quickly get
the gist of the original review. In this work, we focus on generating a headline
for a single-review.

Review headline generation is a special case of document summarization
[15, 22, 25, 30]. Different from document summarization, headlines generated by
users are highly subjective. In other words, a review headline is not only correlated to the textual information of the review, but is also influenced by user

© Springer Nature Switzerland AG 2018
M. Sun et al. (Eds.): CCL 2018/NLP-NABD 2018, LNAI 11221, pp. 324–334, 2018.
https://doi.org/10.1007/978-3-030-01716-3_27

preference. However, we find that previous document summarization systems almost ignore user information. Thus, for review headline generation task, we believe that it is necessary to design a novel model that can capture the preference of the user and to mine the emphasis of the users regarding the review when they write the headlines.

To this end, we propose a method generating review headline with user embedding, which regards the user information as the prior knowledge. Specifically, for each user, we introduce a user embedding, and then we use user embedding to initialize the encoder and decoder, or as a bias for decoder initialization. To verify our idea, we construct a dataset for hotel review, and the experiments on this dataset demonstrate that our models significantly outperform baseline models which do not consider user information.

Our main contributions are as follows:

- We propose a single-review headline generation framework that incorporates user information into the attention-based sequence-to-sequence (seq2seq) framework.
- Experimental results show that using user embedding as the prior knowledge can significantly boost the performance of our model, which verifies our hypothesis on the essential role of user information for review headline generation task.

2   Related Work

Our work is mainly related to abstraction-based text summarization and natural language processing tasks using user information.

2.1   Abstractive Document Summarization

Abstractive document summarization generates a summary of the source document by building the semantic representation of a document. The generated summary may contain some words or sentences that not presented in the original document. The core of the abstractive document summarization lies in exactly representing the semantics of the original document, and then accurately attending the critical parts of the source document, and finally generating the gist of the source document. Comparing to the extraction-based summarization methods [2,10,28] that select important sentences or phrases from the source document, abstraction-based summarization [7,8,12,14,23,25,27] involves rewriting summary as a human-written summary usually does.

In recent years, researchers employ seq2seq framework to tackle the abstraction-based text summarization problem. [22] proposed a neural network based model with local attention modeling, which is trained on the Gigaword corpus, but combined with an additional log-linear extraction-based summarization model with hand-crafted features. [6] introduced a conditional recurrent neural network that acts as a decoder to generate the summary of an input sentence and at each time-step the decoder also takes a conditional input which is the output of an encoder module. [18] proposed to control the vocabulary size to improve the training efficiency. [23] augmented the seq2seq model by introducing soft copy mechanism for reproduction of information while retaining the ability to generate novel words. [30] extended the seq2seq framework and proposed a selective gate network to control the information flow from the encoder to decoder. [15] equipped the seq2seq oriented encoder-decoder model by adding a deep recurrent generative decoder which is used to learn the latent structure information implied in the target summaries.

Compared to those researches mentioned above, we argue that review headline generation is more challenging because users have their own preferences when writing the headlines for the reviews they post. Thus, we need to consider user information to generate a headline for the review.

2.2 User Information Used in Natural Language Processing Tasks

In other user-related natural language processing tasks, some researches have considered user information.

[21] and [27] used "neighbor" users' reviews to extend the traditional recommendation system. [4] proposed an approach to construct vectors to represent profiles of users and items under a unified framework to maximize word appearance likelihood. Then, the vectors were used for a recommendation task in which they predicted scores on unobserved user-item pairs without given text. [13] presented a weakly supervised approach to extract user attributes from user-generated text on Twitter which can provide the social connection information. [29] proposed two different neural networks to learn both user embeddings and text embeddings for scholarly Microblog recommendation. [3] presented a novel framework which used only the user id and user social contexts for gender prediction. Their key idea is to represent users in the embedding connection space by considering users' social contexts including family members, schoolmates, colleagues, and friends. [1] proposed to use user embeddings as lexical signals to recognize sarcasm. [26] proposed an embedding approach to learning user profiles, where users were embedded on a topical interest space, and then they directly utilized the user profiles for search personalization. [16] proposed mixture models which exploited user and item embedding in latent factor models for recommendations. [19] introduced an embedding model based on capsule network to model the 3-way (query, user, document) relationships for search personalization. In their model, each user was embedded as a vector in the same vector space as words.

In our task, headlines generated by users are different from person to person. However, we find that previous document summarization systems almost ignore user information. Thus, for review headline generation task, we believe that it is necessary to design a novel model that can take advantage of user information. To this end, we propose a method to generate review headline with user embedding, which regards the user information as the prior knowledge.

3 Background: Seq2Seq Model

In this section, we describe the basic seq2seq learning framework. Given a dataset of review-headline pairs, $\mathcal{D} = (x_i, y_i^*)_i^N$, the seq2seq model maximizes the conditional probability of a target sequence $y^* : p(y^*|x)$. Recurrent Neural Networks (RNNs) encoder [5,9] reads and converts a variable length input sequence x into a context representation c as follows:

$$h_t = f_{enc}(x_t, h_{t-1}) \tag{1}$$
$$c_t = f_c(h_t, \ldots, h_t) \tag{2}$$

where h_t is the hidden state at time-step t. c_t is a context vector generated from the sequence of the hidden states. f_{enc} and f_c are nonlinear activation functions.

The decoder generates word y_t given the context vector c_t and the previously generated words $\{y_1, \ldots, y_{t-1}\}$:

$$p(y_t|y_1, \ldots, y_{t-1}) = f_{dec}(y_{t-1}, s_t, c_t) \tag{3}$$

where s_t is the hidden state of the decoder and f_{dec} is a nonlinear activation function that computes the probability vector for output words at time-step t. The maximum likelihood (ML) framework tries to minimize negative log-likelihood loss of the parameters.

$$\mathcal{L}_{ML}(\mathcal{D}) = \sum_{(x,y^*)\in\mathcal{D}} -\log\ p(y^*|x) \tag{4}$$

4 Our Model

4.1 Overview

We begin by introducing the review headline generation task. The input of the task is a review that is post by a user, and the output is a headline for this review. The sentence encoder is a bidirectional LSTM [9] (BiLSTM). Our summary decoder is a uni-directional LSTM with an attention mechanism and a softmax layer over the target vocabulary to generate words. Specifically, beyond general seq2seq architecture, we explore the information of user who posts the review to navigate our model to generate a headline that is unique to that user. To this end, we design multiple strategies to incorporate user into our model, namely, user-specific encoder initialization, user-specific decoder initialization, user-bias decoder initialization, as depicted in Fig. 1. Furthermore, we combine above-mentioned strategies to initialize the encoder and decoder with the help of user information.

4.2 User-Specific Encoder Initialization

In the general attention-based seq2seq model, the initial hidden states of the encoder, $\overrightarrow{h}_0$ and $\overleftarrow{h}_{n+1}$, are initialized by the zero vectors. In this work, we

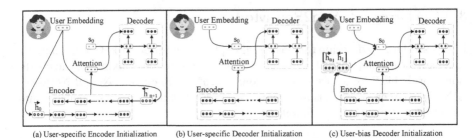

(a) User-specific Encoder Initialization   (b) User-specific Decoder Initialization   (c) User-bias Decoder Initialization

Fig. 1. Framework of our model. We introduce user embedding for each user, and then we use these embeddings to initialize the encoder (a), initialize the decoder (b), or as biases for decoder initialization (c).

propose a user-specific encoder initialization strategy which initializes the hidden states of the encoder by the user embedding. In details, we use two single-layer feed-forward neural networks to compute the initial hidden states of the forward and backward encoder, respectively, as follows:

$$\overrightarrow{h}_0 = \tanh(\mathbf{W}_p u + b_p) \tag{5}$$

$$\overleftarrow{h}_{n+1} = \tanh(\mathbf{W}_q u + b_q) \tag{6}$$

where $\mathbf{W}_p$, $\mathbf{W}_q$, b_p, and b_q are model parameters, and u denotes user embedding.

The idea behind this strategy is that different users have different preferences when they post the review, and we intend to explore the user preferences as the prior knowledge for the encoder.

4.3   User-Specific Decoder Initialization

Generally, the initial hidden state of the decoder, s_0, is initialized by concatenating forward and backward vectors for the last hidden states of encoder as follows:

$$s_0 = \tanh(\mathbf{W}_h[\overrightarrow{h}_n; \overleftarrow{h}_1] + b_h) \tag{7}$$

where $\mathbf{W}_h$ and b_h are model parameters.

In this work, to incorporate user information into the decoder, we propose a user-specific decoder initialization strategy which initializes the hidden state of the decoder by the user embedding. Specifically, we use a single-layer feed-forward neural networks to calculate the initial hidden state of the decoder instead of the one in Eq. 7, as follows:

$$s_0 = \tanh(\mathbf{W}_u u + b_u) \tag{8}$$

where $\mathbf{W}_u$ and b_u are model parameters.

The motivation for this strategy is that different users have different emphasis when they write the headlines for the review post by themselves, and we mine the emphasis regarding the whole review, and we argue that the user-specific emphasis can provide the prior knowledge for the decoder.

4.4   User-Bias Decoder Initialization

In this strategy, instead of initializing decoder only by the user information as in Eq. 8, we include the user embedding as an additional input to initialize the decoder hidden state as follows:

$$s_0 = \tanh(\mathbf{W}_v[\overrightarrow{h}_n; \overleftarrow{h}_1] + \mathbf{W}_w u + b_v) \tag{9}$$

where $\mathbf{W}_v$, $\mathbf{W}_w$, and b_v are model parameters.

In this way, the decoder initialization is related to both the source and user information. In other words, user-specific emphasis is used as a bias to the source representation.

4.5   Initializing the Encoder and Decoder with User Embedding

To jointly incorporate user information into the encoder and decoder, we combine (1) user-specific encoder initialization and user-specific decoder initialization; (2) user-specific encoder initialization and user-bias decoder initialization. We argue that these two strategies can simultaneously regard user information as prior knowledge for the encoder and decoder, which are more powerful than other strategies.

5   Experiments

5.1   Dataset

We construct the hotel review dataset by crawling data from TripAdvisor[1], which is a travel review website including review, headline, and the user who posts them. We finally get 315,396 user-review-summary triples. Statistics and division of the dataset are shown in Table 1.

Table 1. Dataset statistics.

#users	#reviews	#headlines	#reviews/user	#words/review	#words/headline
6,682	315,396	315,396	47.2	158	7.7
Dataset			Train	Valid	Test
#Review-Headline			305,396	5000	5000

5.2   Comparative Methods

We compare the following methods.

– **Lead** baseline uses the first sentence in the review as the headline.

---

[1] https://www.tripadvisor.com/.

- **ABS** [22] uses an attentive CNN encoder and a neural network language model decoder to summarize the sentence.
- **SEASS** [30] is a state-of-the-art sentence summarization systems, which employs a selective encoding model to control the information flow from the encoder to the decoder.
- **Seq2seq** model is a standard attention-based seq2seq model without using user information.
- **Seq2seq + USE** model is a seq2seq model with user-specific encoder initialization.
- **Seq2seq + USD** model is a seq2seq model with user-specific decoder initialization.
- **Seq2seq + UBD** model is a seq2seq model with user-bias decoder initialization.
- **Seq2seq + USE + USD** model is a seq2seq model with user-specific encoder initialization and user-specific decoder initialization.
- **Seq2seq + USE + UBD** model is a seq2seq model with user-specific encoder initialization and user-bias decoder initialization.

5.3 Experimental Settings

We initialize model parameters by uniform randomly with range $[-0.1, 0.1]$. We set the size of word embedding and user embedding to 300 and LSTM hidden state size to 512. We use the full source and target vocabularies collected from the training data, which have 119,602 and 18,520 words, respectively. We use dropout [24] with probability of 0.2 and gradient clipping [20] with range $[-5, 5]$. We set the initial learning rate for Adam [11] to 5×10^{-4}, $\alpha = 0.001$, $\beta_1 = 0.9$, $\beta_2 = 0.999$, and $\epsilon = 10^{-8}$. we use mini-batch size 64 by grid search. At training time, we test ROUGE-2 F1 score on the development set for every 2,000 batches, and we halve the learning rate if model performance drops. Our models typically converge within 20 epochs using an early stopping strategy. At test time, we use beam search with beam size 10 to generate the summary.

5.4 Evaluation Metric

We employ ROUGE [17] as our evaluation metric. ROUGE measures the quality of summary by computing overlapping textual units including unigram, bigram, longest common subsequence (LCS), and skip bigram, which is a standard evaluation metric for summarization tasks. In this work, we report F1 score for ROUGE-1 (unigram), ROUGE-2 (bi-gram), ROUGE-L (LCS), and ROUGE-SU4 (bigram that allows for a skip distance up to 4 words) in our experimental results.

Table 2. Main experimental results (%). Our seq2seq models with user information perform significantly better than baseline models by the 95% confidence interval in the ROUGE script.

Model	ROUGE-1	ROUGE-2	ROUGE-L	ROUGE-SU4
Lead	13.19	3.20	11.50	2.39
ABS [22]	15.64	4.83	14.87	3.71
SEASS [30]	18.41	5.33	17.41	4.19
Seq2seq	17.93	5.32	16.88	4.24
Seq2seq + USE	19.95	6.40	18.85	5.19
Seq2seq + USD	20.83	6.68	19.59	5.59
Seq2seq + UBD	20.25	6.38	19.00	5.37
Seq2seq + USE + USD	**20.83**	**6.79**	**19.61**	**5.61**
Seq2seq + USE + UBD	20.12	6.45	18.90	5.35

5.5 Experimental Results

Table 2 shows the results of the comparative methods and our proposed methods. The models in the first portion of Table 2 are widely compared baseline methods. The state-of-the-art sentence summarization systems **SEASS** performs better than other baselines and our **seq2seq** model. Our proposed models considering user information achieve better performances than other models, which proves that user information plays a significant role for review headline generation task. **User-specific decoder initialization** strategy is more effective than **user-specific encoder initialization**, and we argue that a better decoder initialization is more important because decoder initialization is directly related to the generated words in the headlines. **User-bias decoder initialization** strategy performs worse than **user-specific decoder initialization**, which may attribute to that the source representation dilutes the influence of user information when initializing the decoder. Combining **user-specific encoder and decoder initialization** leads to a further improvement, which achieves +2.90% ROUGE-1, +1.47% ROUGE-2, +2.73% ROUGE-L, and +1.37% ROUGE-SU4 improvements over the basic **seq2seq** model.

5.6 Case Study

In this section, we show several review-headline pairs post by a specific user. For the review-headline pair in the test set, we show comparisons of the reference headline and the headlines generated by **seq2seq model, seq2seq model with user-specific encoder and decoder initialization**. From Fig. 2, we can conclude that user have specific preferences when they post the reviews and write the headlines, and our model considering user information successfully captures these preferences while baseline model fails.

Review-headline pairs (in the training set) post by User-A:

Review 1: this is a solid 3 rating (out of 5) . *right across the street from the stadium and university* . *many restaurants only 1 minute walk* . very clean hotel (...) small selection but ok for the price of the room .

Headline 1: *good location to the stadium* . very low price . very nice staff

Review 2: they have an excellent gym . free weights , many machines , treadmills , etc . (...) *awesome location , 1 minute walk to bangla road and new shopping mall with movie theater* . *1 minute walk to 2 starbucks* (...) great value for a low price .

Headline 2: *excellent location* , gym , sauna , pool . delicious food at the restaurant

Review 3: this is a solid 5 star hotel . *location is the best you can get* . *1 minute walk to the beautiful modern shopping mall* . *2 minute walk to the outdoor shopping and food market* . the hotel is new , *very clean* , excellent security to remove any worries . (...) i will stay here every trip to udon thani because of the *excellent location* , low cost , *clean room* .

Headline 3: *great location* . *very clean* . very low price

Review 4: this was a very nice 3 star hotel . *great location across the street from the beach* . *5 minute walk to walking street* . *5 minute walk to the new mall* . (...) *room was perfect clean* .

Headline 4: *great location* , *clean hotel* , friendly staff , next to the beach

---

Review-headline pairs (in the test set) post by User-A:

Review : *very clean hotel* . rooms are like new . nice showers . beds are comfortable . *great location* . *just 2 minute walk to the bars* . *10 minute walk to the big modern shopping mall* . staff speaks english . very low price for a new hotel .

Headline written by User-A: *very clean* . *great location*

Headline generated by seq2seq model: new hotel, very low price

Headline generated by seq2seq + UE + UD model: *very clean rooms* . *great location*

Fig. 2. From the review-headline pairs in the training set, we can figure out that User-A concerns more about the location (in red and *italic*) and cleanliness (in green and *italic*) than other respects for the hotel. Our **seq2seq + UE + UD** model taking advantage of user information successfully predicts the preference for User-A, while **seq2seq** model fails. (Color figure online)

6   Conclusion

This paper addresses a review headline generation task, namely, how to produce a headline for a review post by a user. We prove that user information as the prior knowledge is vital for this task. Our proposed models explore the effectiveness of the user embedding to initialize the encoder and decoder, and the experiments on our constructed review headline generation corpus show our proposed framework significantly outperforms the baseline models which do not consider user information. We conclude that decoder initialization with only user embedding is a more valid strategy than others. When we simultaneously initialize the encoder and decoder using user embedding, the performance is further improved.

References

1. Amir, S., Wallace, B.C., Lyu, H., Carvalho, P., Silva, M.J.: Modelling context with user embeddings for sarcasm detection in social media. In: Proceedings of The 20th SIGNLL Conference on Computational Natural Language Learning, pp. 167–177 (2016)
2. Carenini, G., Cheung, J.C.K., Pauls, A.: Multi-document summarization of evaluative text. Comput. Intell. **29**, 545–576 (2013)
3. Chen, L., Qian, T., Zhu, P., You, Z.: Learning user embedding representation for gender prediction. In: 2016 IEEE 28th International Conference on Tools with Artificial Intelligence (ICTAI), pp. 263–269 (2016)
4. Chen, W., Zhang, Z., Li, Z., Zhang, M.: Distributed representations for building profiles of users and items from text reviews. In: Proceedings of the 26th International Conference on Computational Linguistics: Technical Papers, pp. 2143–2153 (2016)
5. Cho, K., van Merrienboer, B., Bahdanau, D., Bengio, Y.: On the properties of neural machine translation: encoder-decoder approaches. In: Proceedings of SSST-8, Eighth Workshop on Syntax, Semantics and Structure in Statistical Translation, pp. 103–111 (2014)
6. Chopra, S., Auli, M., Rush, A.M.: Abstractive sentence summarization with attentive recurrent neural networks. In: Proceedings of the 2016 Conference of the North American Chapter of the Association for Computational Linguistics: Human Language Technologies, pp. 93–98 (2016)
7. Gerani, S., Mehdad, Y., Carenini, G., Ng, R.T., Nejat, B.: Abstractive summarization of product reviews using discourse structure. In: Conference on Empirical Methods in Natural Language Processing, pp. 1602–1613 (2014)
8. Gu, J., Lu, Z., Li, H., Li, V.O.: Incorporating copying mechanism in sequence-to-sequence learning. In: Proceedings of the 54th Annual Meeting of the Association for Computational Linguistics, pp. 1631–1640 (2016)
9. Hochreiter, S., Schmidhuber, J.: Long short-term memory. Neural Comput. **9**, 1735–1780 (1997)
10. Hu, M., Liu, B.: Mining and summarizing customer reviews. In: Tenth ACM SIGKDD International Conference on Knowledge Discovery and Data Mining, Seattle, Washington, USA, August. pp. 168–177 (2004)
11. Kingma, D.P., Ba, J.: Adam: a method for stochastic optimization. arXiv preprint arXiv:1412.6980 (2014)
12. Li, H., Zhu, J., Zhang, J., Zong, C.: Ensure the correctness of the summary: incorporate entailment knowledge into abstractive sentence summarization. In: Proceedings of the 27th International Conference on Computational Linguistics (2018)
13. Li, J., Ritter, A., Hovy, E.: Weakly supervised user profile extraction from twitter. In: Proceedings of the 52nd Annual Meeting of the Association for Computational Linguistics, pp. 165–174 (2014)
14. Li, P., Lam, W., Bing, L., Guo, W., Li, H.: Cascaded attention based unsupervised information distillation for compressive summarization. In: Conference on Empirical Methods in Natural Language Processing, pp. 2081–2090 (2017)
15. Li, P., Lam, W., Bing, L., Wang, Z.: Deep recurrent generative decoder for abstractive text summarization. In: Proceedings of the 2017 Conference on Empirical Methods in Natural Language Processing, pp. 2091–2100 (2017)
16. Li, Z., Huang, J., Zhong, N.: Exploiting user and item embedding in latent factor models for recommendations. In: Proceedings of the International Conference on Web Intelligence, pp. 1241–1245 (2017)

17. Lin, C.Y.: Rouge: a package for automatic evaluation of summaries. In: Proceedings of the Workshop on Text Summarization Branches Out, pp. 74–81 (2004)
18. Nallapati, R., Zhou, B., dos Santos, C., Gulcehre, C., Xiang, B.: Abstractive text summarization using sequence-to-sequence RNNs and beyond. In: Proceedings of The 20th SIGNLL Conference on Computational Natural Language Learning, pp. 280–290 (2016)
19. Nguyen, D.Q., Vu, T., Nguyen, T.D., Phung, D.: A capsule network-based embedding model for search personalization. arXiv preprint arXiv:1804.04266 (2018)
20. Pascanu, R., Mikolov, T., Bengio, Y.: On the difficulty of training recurrent neural networks. In: International Conference on Machine Learning, pp. 1310–1318 (2013)
21. Poussevin, M., Guigue, V., Gallinari, P.: Extended recommendation framework: generating the text of a user review as a personalized summary. arXiv preprint arXiv:1412.5448 (2014)
22. Rush, A.M., Chopra, S., Weston, J.: A neural attention model for abstractive sentence summarization. In: Proceedings of the 2015 Conference on Empirical Methods in Natural Language Processing, pp. 379–389 (2015)
23. See, A., Liu, P.J., Manning, C.D.: Get to the point: summarization with pointer-generator networks. In: Proceedings of the 55th Annual Meeting of the Association for Computational Linguistics, pp. 1073–1083 (2017)
24. Srivastava, N., Hinton, G., Krizhevsky, A., Sutskever, I., Salakhutdinov, R.: Dropout: a simple way to prevent neural networks from overfitting. J. Mach. Learn. Res. **15**, 1929–1958 (2014)
25. Takase, S., Suzuki, J., Okazaki, N., Hirao, T., Nagata, M.: Neural headline generation on abstract meaning representation. In: Proceedings of the 2016 Conference on Empirical Methods in Natural Language Processing, pp. 1054–1059 (2016)
26. Vu, T., Nguyen, D.Q., Johnson, M., Song, D., Willis, A.: Search personalization with embeddings. In: Jose, J.M., et al. (eds.) ECIR 2017. LNCS, vol. 10193, pp. 598–604. Springer, Cham (2017). https://doi.org/10.1007/978-3-319-56608-5_54
27. Wang, L., Ling, W.: Neural network-based abstract generation for opinions and arguments. In: Proceedings of the 2016 Conference of the North American Chapter of the Association for Computational Linguistics: Human Language Technologies, pp. 47–57 (2016)
28. Xu, S., Yang, S., Lau, F.: Keyword extraction and headline generation using novel word features. In: Twenty-Fourth AAAI Conference on Artificial Intelligence, pp. 1461–1466 (2010)
29. Yu, Y., Wan, X., Zhou, X.: User embedding for scholarly microblog recommendation. In: Proceedings of the 54th Annual Meeting of the Association for Computational Linguistics, pp. 449–453 (2016)
30. Zhou, Q., Yang, N., Wei, F., Zhou, M.: Selective encoding for abstractive sentence summarization. In: Proceedings of the 55th Annual Meeting of the Association for Computational Linguistics, pp. 1095–1104 (2017)

Social Computing and Sentiment Analysis

A Joint Model for Sentiment Classification and Opinion Words Extraction

Dawei Cong[1], Jianhua Yuan[1], Yanyan Zhao[2(✉)], and Bing Qin[1]

[1] Research Center for Social Computing and Information Retrieval,
Harbin Institute of Technology, Harbin, China
{dwcong,jhyuan,bqin}@ir.hit.edu.cn
[2] Department of Media Technology and Art,
Harbin Institute of Technology, Harbin, China
yyzhao@ir.hit.edu.cn

Abstract. In recent years, mining opinions from customer reviews has been widely explored. Aspect-level sentiment analysis is a fine-grained subtask, which aims to detect the sentiment polarity towards a particular target in a sentence. While most previous works focus on sentiment polarity classification, opinion words towards the target are also very important for that they provide details about target and contribute to judging polarity. To this end, we propose a hierarchical network for jointly modeling aspect-level sentiment classification and word-level opinion words extraction. Our joint model acquires superior performance in opinion words extraction and achieves comparable results in sentiment polarity classification on two datasets from SemEval 2014.

Keywords: Aspect-level sentiment analysis
Opinion words extraction · Neural network · Attention mechanism

1 Introduction

Aspect-level sentiment analysis [9,13,15] has received much attention these years both in academic communities and industry. Given a sentence and a target, aspect-level sentiment analysis aims at inferring the sentiment polarity (i.e. positive, negative, neutral) towards the target in the sentence. Sentiment analysis at aspect-level is difficult because the polarity of distinct targets in a sentence may be different or even opposite. For the example in Fig. 1, the sentence expresses a positive sentiment towards aspcet "price" while a negative sentiment to "service". With the development of deep learning, various methods have utilized neural network models [2,7,12,22] to capture relevant information and learn semantic representation for classification automatically. In context of aspect-level sentiment analysis, Target-Dependent LSTM (TD-LSTM) and Target-Connection LSTM (TC-LSTM) [20] take more target information into consideration by modeling contexts surrounding the target string and incorporating a target connection component.

© Springer Nature Switzerland AG 2018
M. Sun et al. (Eds.): CCL 2018/NLP-NABD 2018, LNAI 11221, pp. 337–347, 2018.
https://doi.org/10.1007/978-3-030-01716-3_28

Fig. 1. An example of a review with two aspect terms which have different sentiments. The underlined word are opinion words and point to their corresponding targets.

Intuitively, corresponding opinion expression about an aspect plays a vital role in aspect-level sentiment polarity classification. In addition, opinion words provide more information rather than polarity about the aspect. For instance, the aspect "food" may be praised for its taste ("delicious") or its freshness ("fresh"). With opinion words extracted from the sentence, we will know the specific reasons why polarities towards certain aspects are positive, negative or neutral. However, researchers pay more attention to aspect-level sentiment classification but less attention to opinion words extraction. Recently, Bailin et al. [1] notice the importance of opinion words extraction and propose a model combining bidirectional long short-term memory networks (BiLSTM) [3] and conditional random field to capture both polarity and opinion information. Unfortunately, their result of opinion words extraction is not ideal.

To alleviate this, we propose a joint model to solve aspect-level sentiment classification and opinion words extraction. Specifically, we employ two BiLSTM layers to acquire sentiment information. The first LSTM extracts opinion words by their attention weight in the sentence, and the second LSTM captures the semantic information for sentiment classification. Moreover, position information is exploited in the second LSTM to reflect relation between aspect and each other words in sentence. We further optimize our model by adding constraints to loss function.

The main contributions of this work can be summarized as follows:

- We extract opinion words and classify sentiment polarity jointly. Since opinion words play an important part in sentiment classification, we employ attention mechanism to focus on opinion words by adding constraints in loss function.
- We use position information rather than syntactic parser to make connection between aspect and words in sentence, which is computationally efficient. We employ gate mechanism to capture semantic information for sentiment classification, which proves to be effective.
- Experimental results indicate that our approach outperforms several baselines, and we achieve remarkable improvement on opinion words extraction.

The rest of this paper is structured as follows: Sect. 2 discusses related works, Sect. 3 gives a detailed description of our proposed model for aspect-level sentiment classification and opinion words extraction. Section 4 compares several model experiments to prove the effectiveness of our proposed model, and Sect. 5 summarizes this work.

2   Related Work

2.1   Aspect-Level Sentiment Classification

In many NLP tasks, earlier approaches mainly include rule-based and traditional machine learning methods. Approaches to sentiment analysis formerly include lexicon-based methods [5,11,16,18] and SVM-based methods [6]. Those methods usually rely heavily on manual features, and models work only if the sets of manual features take effect. However, hand-craft features may be time and labor consuming. Neural Networks (NN) solve the problem by capturing semantic features automatically. Some classical models, such as Recursive Neural Network(RNN) [2,17] and LSTM [4] and Tree-LSTMs [19], are applied to sentiment analysis and prove to be useful. But RNN suffers vanishing gradient and exploding gradient, tree-LSTMs highly depends on the result of syntactic parser. To avoid those problems, LSTM is widely adopted and has shown superior performance. In consideration of target information, TD-LSTM and TC-LSTM [20] average the target words vectors to represent the semantics of target and led to better performance. Wang et al. [24] propose an Attention-based LSTM to explore the connection between an aspect and the content of a sentence. The attention mechanism concentrates on different parts of a sentence when different aspects are given as input. Gated neural network [25] is used to model the interaction between the target mention and its surrounding contexts. However, the above methods only focus on sentiment polarity classification.

2.2   Joint Sentiment Analysis Model

Several subtasks are defined to analyze sentiment at aspect-level, e.g., opinion targets extraction, aspect category detection, etc. Some approaches are proposed to solve the above tasks jointly. Li et al. [8] capture both opinion expressions and the polarity information jointly to extract opinions using sequence labeling by adding sentiment polarity. Zhao et al. [26] model aspect and opinion words jointly for extraction with a MaxEnt-LDA Hybrid. Mitchell et al. [10] extract the sentiment target with its sentiment polarity based on the assumption that surrounding context provides target's sentiment detection with enough information. Bailin et al. [1] propose a segmentation based model which can capture the structural dependencies between the target and the sentiment expressions with a linear-chain conditional random field layer. While this model achieves the state-of-the-art performance in sentiment classification, its opinion words extraction result is not good.

To our best knowledge, few research has studied the joint task of sentiment classification and opinion words, and no approach to sentiment classification and opinion words extraction jointly achieves an acceptable result.

3   Methodology

In this section, we introduce our attention-based network approach for aspect level sentiment classification and opinion words extraction. We first give the task definition. Afterwards, we introduce our approach to aspect-level sentiment classification and opinion words extraction.

Given a sentence $s = \{s_1, s_2, ..., s_i, s_{i+1}, ...s_j, ...s_p, s_{p+1}, ...s_q, ...s_n\}$ (actually p may be less than i, order depends on means of expression) consisting of n words and an aspect phrase $a = \{s_i, ...s_j\}$ appearing in sentence s, aspect level sentiment classification aims at predicting sentiment polarity of sentence s towards aspect a, while opinion words extraction detects words $o = \{s_p, ...s_q\}$ implying polarity occurring in sentence s. For example, given s, "The price is reasonable" and a, "price", the sentence express a positive sentiment towards "price" by using the opinion word "reasonable".

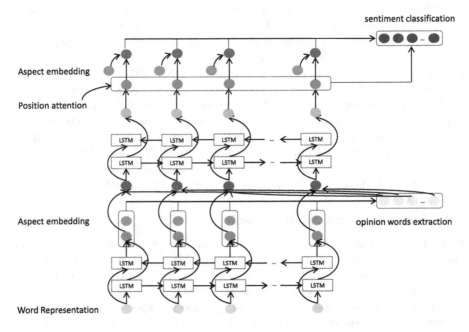

Fig. 2. Our proposed model for aspect-level sentiment classification and opinion words extraction.

Figure 2 presents an overview of our model for aspect level sentiment classification and opinion words extraction. Two BiLSTMs are applied to capture information at different levels and performs its own functions. Specifically speaking, the first BiLSTM acquires sentiment information by paying more attention to opinion words for extraction and feed it to the second BiLSTM. And the second BiLSTM learns semantic representation for sentiment polarity classification according to the output of the first BiLSTM and position information. Next, we will introduce our approach in detail.

3.1   Opinion Words Extraction

Representation of Aspects and Sentences. Word embedding is widely used for representing meanings of words. Formally, $W \in \mathbb{R}^{d_w \times |W|}$ is made up of all word embeddings, where d_w is dimension of word embedding, and $|W|$ is vocabulary size. To make full use of aspect information, we use aspect embedding for words in an aspect term to represent it just like word embedding. Aspect embedding vectors (unequal to word embedding vectors) of words in an aspect term are averaged and the result is regarded as representation of the aspect. Similar to word embedding, vector $v_{a_i} \in \mathbb{R}^{d_a}$ represents embedding of i-th word a_i in aspect a, where d_a is the dimension of aspect embedding. A is a set of all words in aspect terms. $A \in \mathbb{R}^{d_a \times |A|}$ contains all aspect embeddings. Representation of an aspect $a = \{a_1, a_2, ...a_m\}$ contains m words is computed as Eq. 1. We will treat the average result v_a as aspect embedding.

$$v_a = \frac{\sum\limits_{k=1}^{m} v_{a_k}}{m} \tag{1}$$

We then fed $\{w_1, w_2, ...w_n\}$ into a bidirectional LSTM to acquire sentence information, next we will introduce how our model incorporates aspect information.

Attention Mechanism for Opinion Words Extraction. As discussed above, more attention should be paid to opinion words about sentiment polarity in sentences. So an aspect-based attention mechanism is employed to compute relatedness of each word with aspect in sentence according to aspect and hidden state vectors of BiLSTM $H = \{h_1, h_2, ...h_n\}$, where n is length of the sentence. An activation function receives aspect information and sentence information to control the propagation of sentiment features. Two linear layers adjust dimensions of hidden and aspect vectors for combination. Finally, a softmax layer uses the vectors to predict attention weights of the input sentence. Let d_{h1} be dimension of hidden state vectors, $H \in \mathbb{R}^{2d_{h1} \times n}$. And v_a represents aspect embedding and $e_N \in \mathbb{R}^n$ is a vector of 1s. The calculation method of attention is as follows.

$$P = W_h H + b_h \tag{2}$$

$$Q = W_a(v_a \otimes e_n) + b_a \tag{3}$$

$$M = tanh(P \oplus Q) \tag{4}$$

$$\alpha = softmax(w^T M) \tag{5}$$

where $M \in \mathbb{R}^{(d_{h1}+d_a) \times n}$, $\alpha \in \mathbb{R}^n$, $W_h \in \mathbb{R}^{d_{h1} \times 2d_{h1}}$, $w \in \mathbb{R}^{d+d_a}$, $W_a \in \mathbb{R}^{d_{h1} \times d_{h1}}$, $b_h, b_a \in \mathbb{R}$, $v_a \otimes e_n$ means concatenates v_a for n times repeatedly and $\oplus$ means concatenation. Then we regard words with higher weights in attention α as opinion words. The selection rule is defined as follows.

$$o_i = \begin{cases} true, & if \quad there \ is \ a_j \ in \ a, \ a_i \ / \ a_j \ \geq \ \beta, \\ false, & otherwise. \end{cases} \tag{6}$$

where o_i stands for whether to select w_i for opinion words, β is a hyper parameter.

3.2 Sentiment Polarity Classification

Then we use the other BiLSTM to capture the semantic information for sentiment classification. After last step, we gain weight of every word in sentence and combine it with hidden vectors of the first LSTM in this step.

$$r_i = (\alpha_i \otimes e_{d_{h1}}) \odot h_i \tag{7}$$

where $\alpha_i \otimes e_{d_{h1}}$ means concatenates α_i for d_{h1} times repeatedly, $\odot$ means element-wise multiplication, $\{r_1, r_2, ...r_n\}$ is then fed into the second BiLSTM to learn semantic representation for sentiment polarity and output $\{q_1, q_2, ...q_n\}$. However, not every opinion word occurring in sentence contribute equally to sentiment polarity towards a particular aspect.

Position Information. Intuitively, a context word closer to the aspect should be more important than a farther one. Thus, we make use of position information based on this assumption. The details are described below.

$$p_i = n - l_i \tag{8}$$

$$\gamma = softmax(p) \tag{9}$$

$$g_i = \gamma_i \otimes e_{d_{h2}} \odot q_i \tag{10}$$

where l_i be the distance between aspect and w_i, d_{h2} is dimension of hidden state vectors, $p = \{p_1, p_2, ...p_n\}$ is a sequence of position information, q_i is i-th hidden vector from the second LSTM, $\gamma = \{\gamma_1, \gamma_2, ...\gamma_n\}$ is a sequence of position attention and g_i is semantic representation considering position information for i-th word in the sentence.

Gating Mechanism. The previous attention mechanism with aspect information is employed for opinion words extraction, which extracts opinion words well. In terms of sentiment polarity classification, sentiment-related words should be taken more attention to. We utilize a gating mechanism because it proves to be effective in LSTM for learning semantic information [25].

Here we apply ReLU-Tanh gate to filter information. Let β be attention for semantic of polarity and d_{h2} be dimension of hidden state vectors.

$$\beta_i = sum(relu(W_t q_i + V_t v_a + b_t)) \tag{11}$$

$$U = tanh(W_u(G\beta) + b_u) \tag{12}$$

where $W_t \in \mathbb{R}^{d_{h2} \times 2d_{h2}}$, $V_t \in \mathbb{R}^{d_{h2} \times d_{h2}}$, $b_t \in \mathbb{R}^{d_{h2}}$, $W_u \in \mathbb{R}^{d_{h2} \times 2d_{h2}}$, $b_u \in \mathbb{R}^{d_{h2}}$, $G = \{g_1, g_2, ...g_n\}$ and U is last representation of sentiment polarity. Then a softmax layer is employed to transform U to conditional probability distribution.

$$P = softmax(W_c U + b_c) \tag{13}$$

where $W_c \in \mathbb{R}^{|C| \times d_{h2}}$, $b_c \in \mathbb{R}^{|C|}$, C is the collection of sentiment polarity categories and P is conditional probability distribution of sentiment polarity.

3.3 Loss Function

The model is trained in a end-to-end fashion by minimizing the cross entropy error of sentiment classification and smooth l1 loss of opinion words extraction. And we use a hyper parameter η to balance loss of sentiment classification and opinion words extraction. Thus loss of whole model is computed as 14.

$$loss = \eta * loss_{polarity} + loss_{extraction} \tag{14}$$

Specially, loss function of sentiment classification $loss_{polarity}$ is defined as:

$$loss_{polarity} = -\sum_i \sum_j y_i^j log \hat{y}_i^j \tag{15}$$

where i, j is the index of sentence-aspect pair and class respectively, y_i^j is the probability of predicting y_i as category j by our model and $\hat{y}$ is 1 or 0, indicating whether the correct answer is j.

The calculation method of $loss_{extraction}$ is set as follows.

$$\hat{x}_i = \begin{cases} \frac{1}{z}, & if \ w_i \ is \ an \ opinion \ word, \\ 0, & otherwise. \end{cases} \tag{16}$$

$$loss_{extraction} = \frac{1}{n} \sum_i \sum_j \begin{cases} 0.5 * (x_i^j - \hat{x}_i^j)^2, & if \ |x_i^j - \hat{x}_i^j| < 1, \\ |x_i^j - \hat{x}_i^j| - 0.5, & otherwise. \end{cases} \tag{17}$$

where i, j is the index of sentence-aspect pair and class respectively, x is the attention weight produced by our system, z is number of opinion words in sentence.

4 Experiment

We describe experimental settings and report empirical results in this section.

Table 1. Statistics of the datasets from SemEval 2014.

Dataset	Pos.	Neg.	Neu.
Restaurant-Train	2164	807	637
Restaurant-Test	728	196	196
Laptop-Train	994	870	464
Laptop-Test	341	128	169

4.1 Dataset

We conduct experiment on datasets from SemEval 2014 task 4 [15] to verify the effectiveness of our approach. The datasets consist of customers reviews from two domains, namely restaurant and laptop. Each review contains aspects and corresponding polarities. Following previous work [21], we remove conflict category when preprocessing. Statistics of the datasets are given in Table 1. We also make use of the additional annotations for these two datasets from [23] which contain manually annotated labels for opinion words.

4.2   Experimental Setting

In our experiments, all word vectors are initialized by Glove [14]. The aspect vectors are initialized by sampling from a uniform distribution $U(-0.01, 0.01)$. The dimension of word vectors, aspect embedding and the size of hidden layer are 300. On account of the size of corpus, we set different number of layer and dropout for two datasets: 1 layer of first LSTM for opinion words extraction and 2 layers of second LSTM for sentiment classification in restaurant dataset, both 1 layer in laptop dataset. PyTorch[1] is used for implementing our neural network models. We optimize our models using Adam with initial learning rate of 1e−4, weight delay of 5e−5. The batch size for training is set to 15. Two hyper parameters β, η are 5 and 7 respectively. About one-sixth of training data in restaurant and one-seventh in laptop is left out as the validation set for tuning hyper parameters and selecting model.

4.3   Comparison with Other Methods

We describe empirical results on sentiment classification and opinion words extraction in this section. In terms of sentiment polarity classification, we compare with the following baseline methods on both datasets. We use the same Glove word vectors for fair comparison.

(1) **Majority** is a basic baseline method, which assigns the majority sentiment label in training set to each instance in the test set.
(2) We compare with three LSTM models [20]. In **LSTM**, a LSTM based recurrent model reads the start to the end of a sentence, and the last hidden vector is used as the sentence representation. **TDLSTM** takes aspect information into consideration by using two LSTM networks, a forward one and a backward one, towards the aspect. **TDLSTM+ATT** extends TDLSTM by incorporating attention mechanism over the hidden vectors.
(3) To the best of our knowledge, **SA-LSTM-P** [1] is the only existing model that supports sentiment classification and opinion words extraction jointly, which achieves state-of-the-art on aspect level sentiment classification.

Table 2. Classification accuracy (%) of different methods on laptop and restaurant datasets.

	Laptop	Restaurant
Majority	53.45	65.00
LSTM	66.45	74.28
TDLSTM	68.13	75.63
TDLSTM+ATT	66.24	74.31
SA-LSTM-P	75.1	81.6
Our method	67.29	77.07

---

[1] https://pytorch.org/.

Experimental results of sentiment polarity classification on review datasets from SemEval 2014 task 4 are given in Table 2.

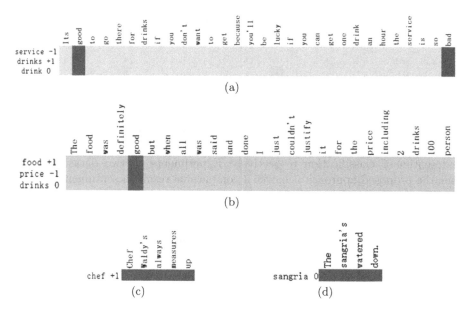

Fig. 3. Visualization of sentiment classification and opinion words results from our model. "+1", "0" and "−1" on the aspect indicate positive, negative and neutral sentiment respectively.

Table 3. Performance (%) of opinion words extraction at word level on laptop and restaurant datasets.

	Laptop			Restaurant		
	P	R	F	P	R	F
SA-LSTM-P	49.1	58.7	53.5	39.4	59.9	55.1
Our method	62.3	65.4	63.8	66.6	73.8	70.0

As shown in Table 3, our model achieves better performance in opinion words extraction. Although our sentiment classification accuracy is a little worsen than SA-LSTM-P, it significantly improves result of opinion words extraction.

4.4 Case Studies

To further show the advantages of our model, we visualize some examples from test data. As we can see from the first example in Fig. 3(a), the model successfully extracts opinion words in the sentence and distinguishes influence of different opinion words towards different aspects for sentiment polarity classification. Second example in Fig. 3(b) shows that when no explicit opinion word for aspect, model captures semantic information from context and classify correctly.

And we also analyze error. Model misunderstands latent sentiment occurring in sentences. It seems difficult to recognize the sentiment in sentences without obvious or usual expression in Figs. 3(c) and (d).

5   Conclusion

In this work, we propose a hierarchical model of reviews for joint aspect-level sentiment classification and word-level opinion extraction. The key idea is to learn semantic information from two level LSTMs for opinion word extraction and sentiment classification respectively. Experiments on two datasets verify that the proposed approach achieves a comparable result in sentiment polarity classification and makes a significant improvement in extraction at word level.

Acknowledgements. We thank the anonymous reviewers for their valuable suggestions. This work was supported by the National Natural Science Foundation of China (NSFC) via grant 61632011, 61772153 and 71490722.

References

1. Bailin, W., Lu, W.: Learning latent opinions for aspect-level sentiment classification (2018)
2. Dong, L., Wei, F., Tan, C., Tang, D., Zhou, M., Xu, K.: Adaptive recursive neural network for target-dependent Twitter sentiment classification. In: Proceedings of the 52nd Annual Meeting of the Association for Computational Linguistics (Volume 2: Short Papers), pp. 49–54. Association for Computational Linguistics, Baltimore, June 2014. http://www.aclweb.org/anthology/P14-2009
3. Graves, A., Mohamed, A.R., Hinton, G.: Speech recognition with deep recurrent neural networks. In: IEEE International Conference on Acoustics, Speech and Signal Processing, pp. 6645–6649 (2013)
4. Hochreiter, S., Schmidhuber, J.: Long short-term memory. Neural Comput. **9**(8), 1735–1780 (1997)
5. Kaji, N., Kitsuregawa, M.: Building lexicon for sentiment analysis from massive collection of HTML documents. In: Joint Conference on Empirical Methods in Natural Language Processing and Computational Natural Language Learning (2007)
6. Kiritchenko, S., Zhu, X., Cherry, C., Mohammad, S.: NRC-Canada-2014: detecting aspects and sentiment in customer reviews. In: International Workshop on Semantic Evaluation, pp. 437–442 (2014)
7. Lakkaraju, H., Socher, R., Manning, C.D.: Aspect specific sentiment analysis using hierarchical deep learning (2014)
8. Li, F., et al.: Structure-aware review mining and summarization. In: International Conference on Computational Linguistics, pp. 653–661 (2010)
9. Liu, B.: Sentiment Analysis and Opinion Mining. Morgan Claypool Publishers, San Rafael (2012)

10. Mitchell, M., Aguilar, J., Wilson, T., Durme, B.V.: Open domain targeted sentiment (2014)
11. Mohammad, S.M., Kiritchenko, S., Zhu, X.: NRC-Canada: building the state-of-the-art in sentiment analysis of tweets. In: Computer Science (2013)
12. Nguyen, T.H., Shirai, K.: PhraseRNN: phrase recursive neural network for aspect-based sentiment analysis. In: Conference on Empirical Methods in Natural Language Processing, pp. 2509–2514 (2015)
13. Pang, B., Lee, L.: Opinion mining and sentiment analysis. Found. Trends Inf. Retr. **2**(12), 1–135 (2008)
14. Pennington, J., Socher, R., Manning, C.: GloVe: global vectors for word representation. In: Conference on Empirical Methods in Natural Language Processing, pp. 1532–1543 (2014)
15. Pontiki, M., Galanis, D., Pavlopoulos, J., Papageorgiou, H., Androutsopoulos, I., Manandhar, S.: SemEval-2014 task 4: aspect based sentiment analysis. In: Proceedings of International Workshop on Semantic Evaluation at (SemEval 2014), pp. 27–35 (2014)
16. Prez-Rosas, V.: Learning sentiment lexicons in Spanish. In: Eighth International Conference on Language Resources and Evaluation (2013)
17. Qian, Q., Tian, B., Huang, M., Liu, Y., Zhu, X., Zhu, X.: Learning tag embeddings and tag-specific composition functions in recursive neural network. In: Meeting of the Association for Computational Linguistics and the International Joint Conference on Natural Language Processing, pp. 1365–1374 (2015)
18. Rao, D., Ravichandran, D.: Semi-supervised polarity lexicon induction. In: Eacl 2009, Conference of the European Chapter of the Association for Computational Linguistics, Proceedings of the Conference, 30 March–3 April 2009, Athens, Greece, pp. 675–682 (2009)
19. Tai, K.S., Socher, R., Manning, C.D.: Improved semantic representations from tree-structured long short-term memory networks. Comput. Sci. **5**(1), 36 (2015)
20. Tang, D., Qin, B., Feng, X., Liu, T.: Effective LSTMs for target-dependent sentiment classification. In: Computer Science (2015)
21. Tang, D., Qin, B., Liu, T.: Aspect level sentiment classification with deep memory network. In: Proceedings of the 2016 Conference on Empirical Methods in Natural Language Processing, pp. 214–224. Association for Computational Linguistics, Austin, November 2016. https://aclweb.org/anthology/D16-1021
22. Vo, D.T., Zhang, Y.: Deep learning for event-driven stock prediction. In: Proceedings of IJCAI, Buenos Aires, Argentina, August 2015
23. Wang, W., Pan, S.J., Dahlmeier, D., Xiao, X.: Recursive neural conditional random fields for aspect-based sentiment analysis. CoRR abs/1603.06679 (2016). http://arxiv.org/abs/1603.06679
24. Wang, Y., Huang, M., Zhu, X., Zhao, L.: Attention-based LSTM for aspect-level sentiment classification. In: Conference on Empirical Methods in Natural Language Processing, pp. 606–615 (2016)
25. Zhang, M., Zhang, Y., Vo, D.T.: Gated neural networks for targeted sentiment analysis (2016)
26. Zhao, W.X., Jiang, J., Yan, H., Li, X.: Jointly modeling aspects and opinions with a MaxEnt-LDA hybrid. In: Conference on Empirical Methods in Natural Language Processing, pp. 56–65 (2010)

Network Representation Learning Based on Community and Text Features

Yu Zhu[1], Zhonglin Ye[2], Haixing Zhao[1,2(✉)], and Ke Zhang[1]

[1] School of Computer, Qinghai Normal University,
Xining 810008, China
h.x.zhao@163.com
[2] School of Computer Science, Shaanxi Normal University,
Xi'an 710062, China

Abstract. Network representation learning (NRL) aims at building a low-dimensional vector for each vertex in a network, which is also increasingly recognized as an important aspect for network analysis. Some current NRL methods only focus on learning representations using the network structure. However, vertices in lots of networks may contain community information or text contents, which could be good for relevant evaluation tasks, such as vertex classification, link prediction and so on. Since it has been proved that DeepWalk is actually equivalent to matrix factorization, we propose community and text-enhanced DeepWalk (CTDW) based on the inductive matrix completion algorithm, which incorporates community features and text features of vertices into NRL under the framework of matrix factorization. In experiments, we evaluate the proposed CTDW compared with other state-of-the-art methods on vertex classification. The experimental results demonstrate that CTDW outperforms other baseline methods on three real-world datasets.

Keywords: Network representation learning · Community and text features
Inductive matrix completion

1 Introduction

With the constant development of the networks, modern society has entered an era of information explosion, and life is full of information. The relevance between the information forms all sorts of information networks, such as various social networks, citation networks between academic papers and so on. Recently, in order to extract useful information from massive network data, some researchers have already focused on network representation learning (NRL), which aims to build low-dimensional vectors for vertices and is applied to lots of machine learning tasks, such as vertex classification [1], recommendation system [2, 3], and link prediction [4]. In particular, NRL can alleviate the sparse issue caused by the conventional representation method based on the graph spectrum.

Recently, there are mainly two types of NRL methods, one of which only takes network structures as input to learn vertex representations without considering other information, and the other simultaneously considers both network structure and

© Springer Nature Switzerland AG 2018
M. Sun et al. (Eds.): CCL 2018/NLP-NABD 2018, LNAI 11221, pp. 348–360, 2018.
https://doi.org/10.1007/978-3-030-01716-3_29

auxiliary information including text information or community information of vertices. For example, based on a word representation model in NLP named as Skip-Gram [5], DeepWalk [6] learns vertex representations from random walk sequences in social networks. Shortly afterwards, various methods based on DeepWalk have been proposed for representation learning, such as Line [7], GraRep [8] and SDNE [9]. The above-mentioned methods are only based on network structures. Nevertheless, vertices in real-world networks usually contain sufficient community or text information, which may also be important to NRL. For example, one paper is more easily referred by other papers similar or same research fields or communities with it. In addition, there is lots of text information in each paper regarded as a vertex of the network. Inspired by this, some researchers incorporate texts or communities into training models to learn better representations, such as CENR [10], CNRL [11], TriDNR [12]. CENR utilizes a neural network method to jointly learn the inter-node and node-sentence network relationships. CNRL decomposes all contents into sentences, and then it applies Wavg, RNN and BiRNN to verify the feasibility and reliability of this algorithm. TriDNR utilizes two neural networks to learn the representations based on inter-node, node-word, and label-word network relationships.

Besides the above-mentioned methods, Inductive Matrix Completion (IMC) [13] also takes advantage of additional information to complete gene-disease matrix. IMC is actually a matrix reduction algorithm and utilizes two feature matrices to factorize the objective matrix. The result matrix obtained from the objective matrix factorization contains influence factors of the feature matrices. Inspired by the idea of IMC, we propose a novel NRL method to take network structure, community features and text features together into consideration, named as community and text -enhanced Deep-Walk (CTDW).

We test our method against several baselines on three real-world datasets. The vertex classification accuracy of our method outperforms the accuracies of other baselines when the ratio of training set ranges from 10% to 90%. Meanwhile, our method shows strong clustering abilities by node clustering visualizations on Citeseer, Cora and DBLP. In addition, we find that CTDW can learn better network representations than only network-based DeepWalk with the help of community and text information by case study.

2 Related Work

As a new technique, network representation learning (NRL) becomes more and more popular in network analysis. Some researchers' NRL methods mainly focus on representation learning based on network structures without taking other information into account. For example, Hofmann [14] first introduces the concept of network representation learning. Inspired by Skip-Gram, DeepWalk is proposed to learn the representations from network structures. Recently, there are some new methods to incorporate additional information into NRL. For example, NetPLSA [15] takes both network structures and text information into account for topic modeling. Max-margin DeepWalk (MMDW) [16] incorporates label information of vertices into NRL. Inspired by DeepWalk, node2vec [17] designs a biased random walk algorithm, which sufficiently utilizes different neighborhood information. The proposed models by Cao *et al.* [8] and Wang *et al.* [9] are applied to capture the vertices of the neighborhoods by incorporating global information into NRL. CENR, CNRL, TriDNR and M-NMF [18] incorporate text or community information into network representations.

The rest of this paper is organized as follows. Section 3 gives the formal definition of NRL and DeepWalk, and demonstrates that DeepWalk is equivalent to matrix factorization in fact. We put forward our method for NRL with community and text features in Sect. 4. The datasets and experimental results are introduced in Sect. 5. Section 6 concludes this paper.

3   DeepWalk Model

3.1   Formalization of NRL

Network representation learning is formalized as follows. Suppose that there is a network $G = (V, E)$, where V denotes the set of vertices and E denotes the set of edges. We want to build a low-dimensional representation vector $r_v \in R^k$ for each vertex v of G, where k is expected to be much smaller than $|V|$, which is the number of vertices of V.

3.2   DeepWalk

As a word representation method, Skip-Gram was introduced by DeepWalk into the study of social network to learn vertex representation from the network structure.

DeepWalk performs short random walks over the given network G to generate a sequence of vertices $S = \{v_1, v_2, \ldots, v_{|s|}\}$. We regard the vertices $v \in \{v_{i-t}, \ldots, v_{i+t}\} \setminus \{v_i\}$ as the context of the center vertex v_i, where t is the window size. The objective of DeepWalk is to maximize the average log likelihood of all vertex-context pairs in the random walk vertex sequence S

$$\frac{1}{|S|} \sum_{i=1}^{|S|} \sum_{-t \leq j \leq t, j \neq 0} \log p(v_{i+j}|v_i) \tag{1}$$

where $p(v_j|v_i)$ is defined by softmax function,

$$p(v_j|v_i) = \frac{\exp(c_{v_j}^T r_{v_i})}{\sum_{v \in V} \exp(c_v^T r_{v_i})}. \tag{2}$$

Here, r_{v_i} and c_{v_j} are the representation vectors of the center vertex v_i and its context vertex v_j, respectively. In other words, each vertex v has two representation vectors: r_v when v is a center vertex and c_v when v is a context vertex.

3.3   DeepWalk as Matrix Factorization

Fortunately, Yang *et al.* [19] has proved that given a network $G = (V, E)$, DeepWalk actually factorizes a matrix $M \in R^{|V| \times |V|}$, where each entry M_{ij} is logarithm of the average likelihood to perform a random walk from the vertex v_i to the vertex v_j in fixed steps. As shown in Fig. 1, the matrix M is factorized into the product of two low-dimensional matrices $W \in R^{k \times |V|}$ and $H \in R^{k \times |V|}$, where k $\ll |V|$. Note that we regard each column of the matrix W as a low-dimensional representation vector $r_v \in R^k$ for each vertex v.

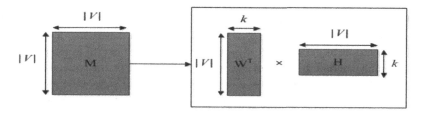

Fig. 1. DeepWalk as matrix factorization

4   Our Method

In this section, we first introduce inductive matrix completion, and then give a detailed introduction to community and text matrix. At last, we formulate our NRL method to take network structure, community features and text features together into consideration, under the framework of matrix factorization.

4.1   Inductive Matrix Completion (IMC)

Matrix is a common way to represent relationships among the vertices of the networks, whose inherent structure information can be figured out by means of the matrix analysis. In order to make good use of abundant information of the networks, researchers resort to inductive matrix completion, which can make these features participate in representation learning by incorporating two feature matrices into the objective function, if there are additional features in the items of matrix $M \in R^{b \times d}$. Assume that two feature matrices are $X \in R^{f_x \times b}$ and $Y \in R^{f_y \times d}$. We want to solve matrices $W \in R^{k \times f_x}$ and $H \in R^{k \times f_y}$ to minimize square loss function, where $k \ll \{b, d\}$.

$$\min_{W,H} \sum_{(i,j) \in \Omega} (M_{ij} - (X^T W^T HY)_{ij})^2 + \frac{\lambda}{2} (\|W\|_F^2 + \|H\|_F^2) \qquad (3)$$

Where Ω and λ are an observation set of matrix M and a harmonic factor to balance two components respectively. In addition, $\|\cdot\|_F$ is Frobenius norm of the matrix.

Note that some researchers originally put forward IMC to complete gene-disease matrix with gene and disease features. Although the goal of IMC differs from that of our method, we can utilize the idea of IMC, which takes the two aforementioned feature matrices as the auxiliary parameters to factorize the target matrix M. In this paper, we take community and text matrices as auxiliary matrices to learn better network representations.

4.2 Community and Text Matrix

The text in the dataset is converted to a matrix, which is factorized by Singular Value Decomposition (SVD) to obtain a text matrix $T \in R^{k \times |V|}$, where k $\ll |V|$. The graph in the dataset is converted to adjacency matrix $A \in R^{|V| \times |V|}$, and then SVD is adopted to factorize the matrix $M = (A + A^2)/2 \in R^{|V| \times |V|}$ to obtain a matrix $B \in R^{n \times |V|}$, where n $\ll |V|$. By means of the matrix B, we call K-means algorithm [20] to return a vector IDX containing community labels of each point in the matrix B. The community matrix $P\_C \in R^{|V| \times |V|}$ is obtained from the vector IDX, and then we use SVD algorithm to factorize the matrix P_C to obtain a community matrix $C \in R^{k \times |V|}$. In fact, the community and text matrix $C\_T \in R^{2k \times |V|}$ is a concatenation of the community matrix C and the text matrix T.

4.3 Community and Text-Enhanced DeepWalk (CTDW)

Given a network $G = (V, E)$ and its corresponding community and text matrix $C\_T \in R^{2k \times |V|}$, we propose community and text-enhanced DeepWalk (CTDW) to learn representations of each vertex $v \in V$ from both network structure and community and text features. Since Yang *et al.* [19] has found a tradeoff between speed and accuracy in their method by factorizing the matrix $M = (A + A^2)/2$ from the derivation of MF-style DeepWalk, we also factorize the same matrix in our method:

$$M = (A + A^2)/2 \tag{4}$$

As shown in Fig. 2, the matrix M is factorized into the product of four matrices $E \in R^{|V| \times |V|}$, $W \in R^{2k \times |V|}$, $H \in R^{2k \times 2k}$, $C\_T \in R^{2k \times |V|}$ by using IMC algorithm, where E is an identity matrix, W and H are both target matrices and C_T is an additional community and text matrix.

Here, we factorize M instead of $\log M$ for computational efficiency. The reason is that $\log M$ has much more non-zero entries than M, and the complexity of matrix factorization with square loss [3] is proportional to the number of non-zero elements of the matrix M. Our task is to solve matrices $W \in R^{2k \times |V|}$ and $H \in R^{2k \times 2k}$ to minimize

$$\min_{W,H} \left\| M - EW^T HC\_T \right\|_F^2 + \frac{\lambda}{2} \left(\|W\|_F^2 + \|H\|_F^2 \right) \tag{5}$$

A direct method for representation learning is to independently train community or text features and network structure, and then we concatenate community or text

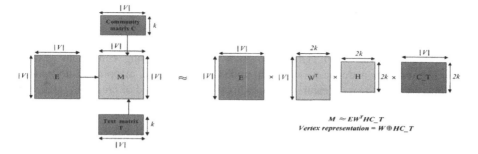

Fig. 2. CTDW framework

representations along with network structure representations. However, this method leads to a loss of joint learning between network structure and community or text information. Therefore, similar with inductive matrix completion, CTDW incorporates community and text features to learn network representations. As shown in Fig. 2, we factorize the matrix M with the help of the community matrix C and the text matrix T to obtain a target matrix W, which contains influence factors of the community and text matrix. The vertex representation is equivalent to a concatenation of W and HC_T, that is to say, vertex representation $= W \oplus HC\_T$. Since both W and HC_T obtained from CTDW can be regarded as low-dimensional representations of vertices, we build a unified 4 k-dimensional matrix for network representations.

5 Experiments

5.1 Datasets

Citeseer. Citeseer[1] contains 3312 publications from six classes and 4732 links, which are citation relationships between the documents. Each document is described by a binary vector of 3703 dimensions.

Cora. Cora[2] contains 2708 machine learning papers from seven classes and 5429 links, which are citation relationships between the documents. Each document is described by a binary vector of 1433 dimensions indicating the presence of the corresponding word.

DBLP. DBLP[3] is a bibliographic network composed of authors and papers containing 3119 nodes from four classes and 39516 links.

---

[1] http://citeseerx.ist.psu.edu/.

[2] https://people.cs.umass.edu/mccallum/data.html.

[3] http://arnetminer.org/citation.

5.2   Baseline Methods

Structure-Based Method

node2vec. node2vec is an algorithmic framework for learning continuous feature representations for nodes in networks. In node2vec, we learn a mapping of nodes to a 200-dimensional space of features that maximizes the likelihood of preserving network neighborhoods of nodes.

LINE. LINE is proposed to learn network representations for large scale networks, which takes both 1-order and 2-order proximity into account, and the concatenation of these two representations is used as the final embedding. Same as DeepWalk, the dimension of representation vectors is 200.

DeepWalk. DeepWalk is a popular network structure-only representation learning method, which learns network representations by using the Skip-Gram model. We set parameters as follows, walk length $\gamma = 80$ and window size $t = 10$, representation dimension $k = 200$.

MFDW. MFDW is the abbreviated form of DeepWalk. MFDW factorizes the target matrix $M = (A + A^2)/2$, where A is the adjacency matrix of the network and then it uses the matrix $W \in R^{200 \times |V|}$ to train classifiers.

Content-Based Method

Text. We take the text matrix $T \in R^{100 \times |V|}$ as 100-dimensional representation. The method is content-only baseline.

Community. We take the community matrix $C \in R^{100 \times |V|}$ as 100-dimensional representation. The method is also content-only baseline.

Community + Text. We can simply concatenate the vectors from both community features and text features into a 200-dimensional vector for network representations.

Combined Method

MV + Community + Text. We use SVD algorithm to factorize the matrix $M = (A + A^2)/2$ to obtain 100-dimensional vectors MV and then simply concatenate the vectors from the matrix, community features and text features into a 300-dimensional vector for network representations.

CT@E. CT@E is a variant of CTDW, based on IMC algorithm and its two auxiliary feature matrices are the matrix $P_C \in R^{|V| \times |V|}$ and the text matrix T to replace the identity matrix and the community and text matrix C_T in CTDW respectively. Its representation dimension is 200.

MMDW. As MFDW, MMDW also factorizes the matrix $M = (A + A^2)/2$ and uses the matrix $W \in R^{200 \times |V|}$ to train classifiers. MMDW utilizes the max-margin approach to optimize matrix W, and thus the learnt representation possesses a discriminative ability.

TADW. TADW [19] incorporates text features of vertices into network representation learning under the framework of matrix factorization, and also factorizes the same matrix $M = (A + A^2)/2$ as MFDW and MMDW. It uses the concatenation matrix $W \oplus HT \in R^{200 \times |V|}$ to train classifiers.

5.3 Classifiers and Experiment Setup

We conduct our experiments on three real-world network datasets. We adopt the classification tasks to verify the feasibility of our method. For all three datasets, we reduce the dimension of vectors to 100 via SVD decomposition of the related matrix, and obtain community and text matrix $C\_T \in R^{200 \times |V|}$ by concatenating the community matrix C and the text matrix T. We also take the community and text matrix C_T as a content-only baseline. To evaluate our method, we randomly select a portion of documents as training set, and the rest are testing set. We take representation vectors of vertices as features to train classifiers, and calculate the accuracy of vertex classifications based on different training ratios, which range from 10% to 90%. Note that, the dimension of representation vectors from CTDW is $4k$. Our experiment is repeated for 10 times to record the average classification accuracy.

5.4 Experimental Results and Analysis

The classification experimental results for three datasets are shown in Tables 1, 2 and 3. CTDW consistently outperforms the baseline methods on different datasets, which shows the feasibility of our method.

Table 1. Accuracy (%) of vertex classification on Citeseer

Training ratio	10%	20%	30%	40%	50%	60%	70%	80%	90%
node2vec	54.38	57.29	58.64	59.53	59.63	59.88	60.43	61.36	62.42
LINE	39.82	46.83	49.02	50.65	53.77	54.20	53.87	54.67	53.82
DeepWalk	49.09	55.96	60.65	63.97	65.42	67.49	66.80	66.82	63.91
MFDW	47.95	54.75	57.18	58.47	59.86	60.24	61.53	60.64	61.60
Text	57.76	66.02	69.70	70.12	70.54	70.58	70.76	70.34	70.41
Community	63.00	66.70	67.69	67.83	68.08	68.45	68.07	67.93	68.07
Community+Text	66.63	69.91	70.72	71.22	71.38	72.05	72.16	71.82	71.36
MV+Community+Text	68.29	70.54	71.34	71.90	71.47	72.56	72.52	72.39	72.51
CT@E	68.76	69.07	69.12	69.39	69.49	69.47	69.68	69.22	69.75
MMDW	55.49	60.70	63.66	65.27	66.02	69.14	69.34	69.47	69.72
TADW	70.20	71.23	73.17	73.45	74.02	74.06	75.48	76.74	75.12
CTDW	70.41	72.39	73.49	74.48	74.43	75.17	75.99	75.95	76.23

Tables 1, 2 and 3 show classification accuracies on Citeseer, Cora and DBLP datasets. From the three tables, we have following observations:

(1) CTDW consistently outperforms all the other baselines on all three datasets. For example, CTDW outperforms the best baseline, i.e. TADW, by about 1% for the

Table 2. Accuracy (%) of vertex classification on Cora

Training ratio	10%	20%	30%	40%	50%	60%	70%	80%	90%
node2vec	76.30	79.26	80.43	80.70	81.13	81.26	82.18	81.63	82.81
LINE	65.13	70.17	72.20	72.92	73.45	75.67	75.25	76.78	79.34
DeepWalk	68.51	73.73	76.87	78.64	81.35	82.47	84.31	85.58	85.61
MFDW	66.38	75.52	78.78	80.54	82.09	81.93	82.62	81.57	83.81
Text	57.70	67.26	70.10	71.05	71.47	71.74	72.04	73.17	74.02
Community	52.08	59.60	61.99	62.67	63.21	63.49	63.69	63.75	64.85
Community+Text	65.07	69.36	71.40	72.28	72.59	73.09	73.17	74.99	75.52
MV+Community+Text	75.18	79.69	81.46	81.90	82.48	83.67	83.70	84.16	84.33
CT@E	65.48	66.16	66.58	66.52	66.88	66.85	67.05	66.63	66.57
MMDW	73.61	79.99	80.43	81.92	83.76	84.97	86.39	86.70	87.45
TADW	80.09	80.70	83.47	84.94	85.37	85.87	85.94	85.26	86.01
CTDW	81.99	82.25	84.62	85.42	86.24	85.91	86.83	86.43	87.12

Table 3. Accuracy (%) of vertex classification on DBLP

Training ratio	10%	20%	30%	40%	50%	60%	70%	80%	90%
node2vec	82.71	83.66	84.07	84.51	84.18	84.71	85.28	84.99	84.69
LINE	79.13	79.81	80.41	81.22	82.95	83.39	83.04	84.74	83.85
DeepWalk	81.53	81.61	83.07	83.78	83.76	84.32	84.42	85.13	84.17
MFDW	75.17	81.44	84.03	84.29	84.99	85.01	85.24	85.44	86.04
Text	60.69	68.15	70.62	72.76	73.79	73.46	74.37	74.16	74.85
Community	58.61	64.50	66.01	66.85	67.31	67.55	68.24	67.72	67.56
Community+Text	65.35	71.09	73.31	74.97	75.28	75.78	76.88	75.35	76.97
MV+Community+Text	76.22	80.10	81.78	82.75	82.94	83.76	84.53	84.93	85.16
CT@E	69.26	69.92	70.04	70.11	70.04	70.36	69.96	70.11	70.37
MMDW	79.70	82.05	84.23	84.84	83.45	85.42	84.96	85.78	84.49
TADW	79.09	81.43	82.42	82.94	83.50	84.40	84.91	85.26	85.72
CTDW	81.61	82.52	83.69	83.99	85.22	85.21	85.95	86.37	86.98

three datasets. Meanwhile, CTDW outperforms the remaining baselines more or less to some extent. These experiments demonstrate that CTDW is effective and robust.

(2) From the above tables, we find that content-based methods, i.e. Text, Community as well as Community and Text have good performances on the vertex classification. Meanwhile, a simple concatenation from Community and Text has better performance than separate content, i.e. Community or Text.

(3) Simple concatenation of representation vectors from the network structure, i.e. the matrix $M = (A + A^2)/2$ along with the community features and text features yields better improvements of classification accuracy than the content-based methods on all three datasets, showing the importance of both network structure and contents. But the performance of this kind of simple concatenation cannot do better than that of CTDW.

From these observations we find that CTDW generates high-quality representations, by incorporating community and text features into inductive matrix completion (IMC). Moreover, CTDW is not task-specific and the representations can be conveniently used for different tasks, such as link prediction, similarity computation and so on. The classification accuracy of CTDW is also competitive with several recent collective classification algorithms, although we don't perform specific optimization for the tasks.

5.5 Parameter Sensitivity

CTDW has two hyper-parameters: dimension k and weight of regularization term λ. We fix training ratio to 60% and test classification accuracies with different k tand λ.

We let k vary from 60 to 140 and λ vary from 0.1 to 1 for Citeseer, Cora and DBLP datasets. Figure 3 shows the variation of classification accuracies with different k and λ. The accuracies vary within 2%, 2.4% and 1.8% for fixed k on Citeseer, Cora and DBLP respectively. Therefore, CTDW can keep stable when k and λ vary within a reasonable range.

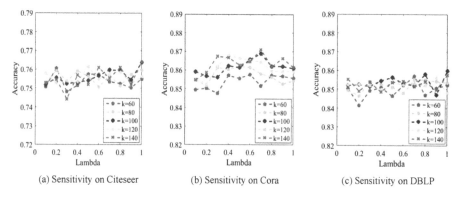

(a) Sensitivity on Citeseer   (b) Sensitivity on Cora   (c) Sensitivity on DBLP

Fig. 3. Parameter sensitivity

5.6 Visualizations

In our research, we propose CTDW to learn network representations on Citeseer, Cora and DBLP datasets. To demonstrate whether the representations generated from CTDW show the discriminative classification ability or not, we randomly select three categories of networks, and each category contains 150 nodes. Figure 4 shows node clustering visualizations on Citeseer, Cora and DBLP.

As shown in Fig. 4, we find that CTDW learns efficient representations with better clustering and representation ability. The representations on Citeseer and Cora datasets show strong clustering abilities, and the boundaries between the categories are clear and discriminative. However, the representation on DBLP dataset shows a relatively weaker clustering ability than Citeseer and Cora datasets. This reason is that there are more network links and more paths among communities in DBLP network, which

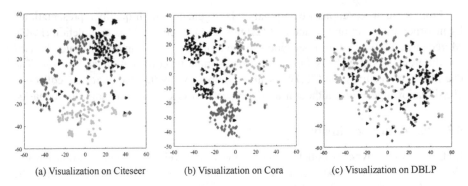

(a) Visualization on Citeseer   (b) Visualization on Cora   (c) Visualization on DBLP

Fig. 4. Clustering visualizations

Table 4. Five nearest documents found by DeepWalk and CTDW

Title	Cosine similarity	Label
5 nearest documents by DeepWalk		
A bootstrapping approach to unsupervised detection of cue phrase variants	0.7266	Artificial Intelligence
Towards the automatic identification of adjectival scales clustering adjectives according to meaning	0.7137	Artificial Intelligence
Similarity-based estimation of word cooccurrence probabilities	0.7002	Artificial Intelligence
Statistical sense disambiguation with relatively small corpora using dictionary definitions	0.6999	Artificial Intelligence
The distributional inclusion hypotheses and lexical entailment	0.6908	Artificial Intelligence
5 nearest documents by CTDW		
Dimension induced clustering	0.8246	Artificial Intelligence
Multiclass spectral clustering	0.8080	Artificial Intelligence
Top-down induction of clustering trees	0.7978	Artificial Intelligence
Refining initial points for k-means clustering	0.7829	Artificial Intelligence
Generative model-based clustering of directional data	0.7634	Artificial Intelligence

leads to closer spatial distances among the vectors obtained from CTDW than Citeseer and Cora networks. In a word, the results of visualization demonstrate effectiveness of our method.

5.7   Case Study

Network representation learning aims to build a low-dimensional vector for each vertex in a network. To verify the performance of CTDW, we conduct an experiment on DBLP dataset. The document title is "distributional clustering of English words", whose class label is "Artificial Intelligence". As shown in Table 4, using representations generated by DeepWalk and CTDW, we find 5nearest documents of the above-mentioned selected document ranked by cosine similarity.

We find that all these documents are cited by the document "distributional clustering of English words" or some of these documents cite the document "distributional clustering of English words". Since DeepWalk learns representations only based on network structures, 5 nearest documents by Deepwalk hardly contain relevant words with the selected document "distributional clustering of English words", while 5 nearest documents by CTDW totally contain relevant words with the selected document "distributional clustering of English words", such as "clustering". This indicates that CTDW can learn better network representations with the help of community and text features than only network-based DeepWalk.

6 Conclusion

In this paper, we propose community and text-enhanced DeepWalk (CTDW), which is a novel and discriminative network representation method to take network structure, community features and text features together into consideration based on inductive matrix completion. We conduct experiments with the tasks of vertex classification on three real-world datasets (Citeseer, Cora and DBLP). The experimental results show that CTDW is an effective and robust network representation method compared to other baseline methods. Meanwhile, the visualization results of learnt representations generated by CTDW demonstrate stronger discrimination ability. CTDW provides a normalized framework for joint learning with different types of resources via inductive matrix completion instead of simple concatenation towards these resources. For future work, we will extend our method to representation learning of large-scale networks. Meanwhile, we will explore some new technologies of matrix factorization, such as max-margin matrix factorization and matrix co-factorization.

Acknowledgements. The work is supported by the National Natural Science Foundation of China under grant 11661069, grant 61663041 and grant 61763041, the Program for Changjiang Scholars and Innovative Research Team in Universities under grant IRT_15R40, the Key Laboratory of Tibetan Intelligent Information Processing and Machine Translation in Qinghai Normal University, the Research Funds for Chunhui Program of Ministry of Education of China under grant Z2014022, the Natural Science Foundation of Qinghai Province under grant 2013-Z-Y17 and grant 2014-ZJ-721, the Fundamental Research Funds for the Central Universities under grant 2017TS045.

References

1. Tsoumakas, G., Katakis, I.: Multi-label classification: an overview. Int. J. Data Warehous. Min. **3**(3), 1–13 (2007)
2. Tu, C., Liu, Z., Sun, M.: Inferring correspondences from multiple sources for microblog user tags. In: Huang, H., Liu, T., Zhang, H.-P., Tang, J. (eds.) SMP 2014. CCIS, vol. 489, pp. 1–12. Springer, Heidelberg (2014). https://doi.org/10.1007/978-3-662-45558-6_1

3. Yu, H.F., Jain, P., Kar, P., et al.: Large-scale multi-label learning with missing labels. In: Proceedings of ICML, pp. 593–601 (2014)
4. Liben-Nowell, D., Kleinberg, J.: The link prediction problem for social networks. J. Assoc. Inf. Sci. Technol. **58**(7), 1019–1031 (2007)
5. Mikolov, T., Sutskever, I., Chen, K., et al.: Distributed representations of words and phrases and their compositionality. In: Advances in Neural Information Processing Systems, pp. 3111–3119 (2013)
6. Perozzi, B., Al-Rfou, R., Skiena, S.: DeepWalk: online learning of social representations. In: ACM SIGKDD International Conference on Knowledge Discovery and Data Mining, pp. 701–710 (2014)
7. Tang, J., Qu, M., Wang, M.Z., et al.: Line: large-scale information network embedding. In: Proceedings of WWW, pp. 1067–1077 (2015)
8. Cao, S.S., Lu, W., Xu, Q.K.: GraRep: learning graph representations with global structural information. In: Conference on Information and Knowledge Management, pp. 891–900 (2015)
9. Wang, D.X., Cui, P., Zhu, W.W.: Structural deep network embedding. In: The ACM SIGKDD International Conference, pp. 1225–1234 (2016)
10. Sun, X.F., Guo, J., Ding, X., et al.: A general framework for content-enhanced network representation learning. arXiv:1610.02906 (2016)
11. Tu, C.C., Wang, H., Zeng, X.K., et al.: Community-enhanced network representation learning for network analysis. arXiv:1611.06645 (2016)
12. Pan, S.R., Wu, J., Zhu, X.Q., et al.: Tri-party deep network representation. In: Proceedings of IJCAI 2016, pp. 1895–1901 (2016)
13. Natarajan, N., Dhillon, I.S.: Inductive matrix completion for predicting gene-disease associations. Bioinformatics **30**(12), 60–68 (2014)
14. Hofmann, T.: Probabilistic latent semantic indexing. In: Proceedings of ACM, pp. 50–57 (2000)
15. Mei, Q.Z., Cai, D., Zhang, D., et al.: Topic modeling with network regularization. In: Proceedings of WWW, pp. 101–110 (2008)
16. Tu, C.C., Zhang W.C., Liu, Z.Y., et al.: Max-margin DeepWalk: discriminative learning of network representation. In: Proceedings of IJCAI 2016 (2016)
17. Grover, A., Leskovec, J.: node2vec: scalable feature learning for networks. In: ACM SIGKDD International Conference on Knowledge Discovery and Data Mining, pp. 855–864 (2016)
18. Wang, X., Cui, P., Wang, J., et al.: Community preserving network embedding. In: AAAI Conference on Artificial Intelligence 2017 (2017)
19. Yang, C., Liu, Z.Y., Zhao, D.L., et al.: Network representation learning with rich text information. In: Proceedings of the Twenty-Fourth International Joint Conference on Artificial Intelligence 2015 (2015)
20. Krishna, K., Narasimha Murty, M.: Genetic K-means algorithm. IEEE Trans. Syst. Man Cybern.-Part B: Cybern. **29**(3), 433–439 (1999)

NLP Applications

Learning to Detect Verbose Expressions
in Spoken Texts

Qingbin Liu[1,2]([✉]), Shizhu He[1], Kang Liu[1], Shengping Liu[3], and Jun Zhao[1,2]

[1] National Laboratory of Pattern Recognition, Institute of Automation,
Chinese Academy of Sciences, Beijing 100190, China
{qingbin.liu,shizhu.he,kliu,jzhao}@nlpr.ia.ac.cn
[2] University of Chinese Academy of Sciences, Beijing 100049, China
[3] Beijing Unisound Information Technology, Beijing 100028, China
liushengping@unisound.com

Abstract. The analysis and understanding of spoken texts is an important task in artificial intelligence and natural language processing. However, there are many verbose expressions (such as mantras, nonsense, modal particle, etc.) in spoken texts, which brings great challenges to subsequent tasks. This paper devote to detect verbose expressions in spoken texts. Considering the correlation of verbose words/characters in spoken texts, we adapt sequence models to detect them with an end-to-end manner. Moreover, we propose a model with the long-short term memory (LSTM) and modified restrict attention (MRA) mechanism which are able to utilize the mutual influence between long-distance and local words in sentences. In addition, we propose a compare mechanism to model the repetitive verbose expressions. The experimental result shows that compared with the rule-based and direct classification methods, our proposed model increases F1 measure by 54.08% and 18.91%.

Keywords: Spoken texts · Verbose expressions · Text transformation
Modified restricted attention mechanism · Compare mechanism

1 Introduction

Spoken language understanding and processing are important tasks in artificial intelligence (AI) and natural language processing (NLP) [4,7,11,16,20]. In addition, the processing of spoken texts is very important for subsequent tasks such as generation tasks [6,10,13]. There are many verbose expressions in the spoken texts such as mantras, nonsense words, repetitions like 'this this (这个这个)', and modal particle like 'Ah (啊)' as shown in Fig. 1 that bring great challenges in spoken language processing. In the practical spoken systems such as reservation system, spoken context need to be converted into texts by speech recognition. The errors in the speech recognition also aggravate the above problems.

In this paper, we propose the detection task to detect the verbose expressions such as 'ah (啊)', 'this (这个)' in Fig. 1. Deleting these expressions will get normal

M. Sun et al. (Eds.): CCL 2018/NLP-NABD 2018, LNAI 11221, pp. 363–375, 2018.
https://doi.org/10.1007/978-3-030-01716-3_30

Fig. 1. The spoken texts and normal texts

texts. As far as we know, there is little work in the task, especially for Chinese. Then we construct a dataset based on the interview texts and manually annotate the verbose expressions. This dataset drives on the task. The direct methods in the task are the rule-based or direct classification methods. Rule-based methods usually count the frequency of verbose expressions and use rules to detect the expressions. The direct classification detects the verbose expressions based on word embedding. However, these methods ignore the relationship between different words. Actually, in a sentence, different expressions can affect each other. For example, some pronouns are not verbose words as sentence components, but as mantras are verbose words. In addition, the above methods cannot directly detect the repetitive verbose expressions like 'this this (这个这个)' in texts.

Recently, the recurrent neural networks (RNN) and its variant (LSTM) [8], have been applied extensively in many tasks. LSTM can obtain long-distance information in a sentence. Moreover, attention mechanism has been introduced to get "soft" correlated information to many tasks [3,12]. Even in the machine translation field, just using the attention mechanism can get the best performance in some languages [14,19]. The LSTM combined with a conditional random field (CRF) achieved best performance in many sequence-labeling tasks [9,18].

Although the above approaches can improve performance in the task, they still suffer from three problems: (1) Chinese word segmentation is inaccurate for spoken texts. Therefore, we need to incorporate proper word information in character-vector level. (2) The global attention mechanisms extract many irrelevant information in character level and degrade performance. (3) We need an explicit compare mechanism to detect the repetitive verbose characters such as 'this this (这个这个)' in Fig. 1. We propose a new model with LSTM and MRA to address these problems. MRA utilizes multiply restricted mask matrixes to extract local relevant information in Chinese-character level. Compared with global attention mechanisms, our proposed attention mechanism reduces a lot of irrelevant information. Furthermore, a gate is used to filter irrelevant information between different mask matrixes. We also propose the compare mechanism in our model to explicit recognize the repetitive cases.

Our main contributions are as follows:

1. We propose a new task which devotes to detect verbose expressions in spoken texts. It is very useful in understanding and processing spoken texts. As far as we know, there is little work in the task, especially for Chinese.

2. We propose a model with LSTM and the modified restricted attention mechanism to extract more accurate information for verbose expressions.
3. We constructed and published a dataset based on the interview dialogue, and we believe that it promotes the research progress of the task.
4. The experimental result shows that our proposed method can increase 54.08% and 18.91% F1 measure compared with the rule-based and direct classification methods.

Table 1. Examples of spoken texts from the annotated dataset.

Sentences	Verbose types
{那么¹}我们近些年以来{啊²}，利用这个宝贵的文化资源，打造了三个文化传承的[品牌 + 平台³]。 {Then ¹} in recent years, we have {ah ²} used this precious cultural resources, which has created three cultural inherited [brand + platform ³].	1, Needless conjunction. 2, Modal particle. 3, Replacement.
像我们罗五的{这个¹}荞丝糖{的话²}，它一年的收入就要接近[上{这个¹} + 上³]百万。 Such as {this¹} Shao-Silk sugar {uh²}, its annual income is [more {this¹} + more³] than one million dollars a year.	1, Needless pronoun. 2, Meaningless word. 3, Simple repetition.

2   Task Definition

2.1   The Task

The task is to detect the verbose characters in the spoken texts. Deleting these verbose characters will generate a more fluent text that preserves the original meaning.

Formally, we represent each sentence of the interview text as (S, Y), where $S = (s_1, ..., s_i, s_n)$ is a sentence with a length n and the labels $y_j \in Y$ indicates the verbose characters in the sentence. The task is to estimate a conditional probability $P(y_j|S)$ from the dataset. The normal sentences can be obtained from the prediction result.

2.2   Data

The dataset is constructed based on 207 Chinese interview texts. Each interview contains two participants: a presenter and an interviewee. The presenter will introduce the main topic firstly and ask questions to the interviewee. The interviewee answers the questions. Usually, the answer is a long paragraph with many verbose characters. We manually annotated the verbose characters as shown in Table 1. We classify the verbose characters into two main categories. The first category includes modal particle like 'ah', needless conjunction like 'then', pronoun like 'this' and meaningless characters like 'uh' in Table 1. They are marked with curly brackets. Another category includes the simple repetition like 'more more', replacement like 'brand platform' in Table 1. The second category is marked with square brackets and the plus sign. The characters before the plus sign is the verbose characters. Removing all labeled verbose characters will not affect the original meaning and fluency of the sentence.

These 207 texts are directly converted by speech recognition and contains many verbose expressions. Under the premise of following the original meaning, we require the labeling person to mark the verbose expressions as much as possible.

2.3 Challenges

The interview text contains various verbose characters. In dialog, people usually have special habits of speech that cause the verbose characters. For example, some people like to say specific modal characters such as 'ah', 'um'. In addition, people may realize that they have said something wrong and will correct it immediately. Those wrong characters are also converted into text by speech recognition. These all caused the diversification of verbose characters. Another challenge is that characters in different contexts may belong to different labels. For example, some pronouns like 'this' are not verbose characters when used as a sentence component, but as a mantra is verbose characters.

3   Method

In this section, we will firstly introduce the overall architecture of our model in Subsect. 3.1. Then, we will introduce the modified restrict attention (MRA) in detail.

3.1 Model Overview

We propose a neural networks with the MRA and compare mechanism to predict the probability distribution $P(y_j|S)$. Figure 2 shows our model's architecture. Due to the poor performance of the Chinese word segmentation on this dataset, the model is based on Chinese character units. The context layer extracts on the long-distance relevant information and gets the context focused (CF) representation. The MRA Layers with the normalize layers in the left part of Fig. 2 can gather relevant local focused (LF) information into the Chinese characters' representation. We augment the local information with these densely connected layers. The other layers in the right part of the model in Fig. 2 compose the compare mechanism. The compare mechanism takes the CF and LF representation as inputs and generates the local focused information behind the characters (BLF) to obtain the rear information.

Context layer is to incorporate long-distance information into the characters' vector. The habits of speech can be modeled in long-distance information. We utilize LSTM in bi-direction to encode each character.

$$\overrightarrow{h_i} = \overrightarrow{LSTM}(\overrightarrow{h}_{i-1}, s_i) \qquad i = 1, ..., N \qquad (1)$$

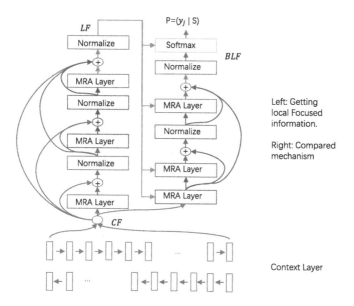

Fig. 2. The proposed model's architecture. CF: Context-Focused representation. LF: Local-Focused representation. BLF: Behind the Local-Focused representation.

$$\overleftarrow{h}_i = \overleftarrow{LSTM}(\overleftarrow{h}_{i+1}, s_i) \qquad i = N, ..., 1 \qquad (2)$$

After encoding, we concatenate the forward and reverse vectors together to represent the contextual information. It will be transmitted to next layers with the original characters' vector.

The MRA layer combined with the normalize layer in the left part of Fig. 2 can gather relevant local information. We will introduce the MRA Layer in next subsection. The output of the MRA Layer is densely connected with the output of the context layer and front normalize Layers. Compared with residual connection, the densely connected can focus more on the own information. The connection function F is additive operation.

$$LF_i = F(LF_1, ..., LF_{i-1}, CF) \qquad i = 1, ..., L \qquad (3)$$

The normalize layer ensures that the data does not become too large during the additive operation. This normalize layer also can accelerate model training. We can think of a MRA Layer, the additive operation and a normalize layer as a block. We use three blocks to encode the local information. Compared with other global attention mechanisms, the restrict attention focuses on the local information by the densely connect and restrict mask matrixes. The global attention usually gather too many contextual information which masking the original information in Chinese characters' vectors. Because the meaningful characters is much more than the verbose characters, the global attention on the long sentence may always predict the non-verbose label for all units. The output LF of the three blocks is transmitted to the compare mechanism.

The compare mechanism in the right part of Fig. 2 takes the CF and LF as inputs. The principle of the compare mechanism is to obtain the local focused information BLF which behind the characters. As shown in Table 1, the simple repetition or replacement characters are very relevant to the characters behind. The compare mechanism gets the rear information with different mask matrixes in the MRA layers from CF according to LF. The front information and other global information is masked. Thought different matrixes, the front and rear local information are respectively obtained from different layers. Then, we use a linear layer and softmax to predict the probability based the information.

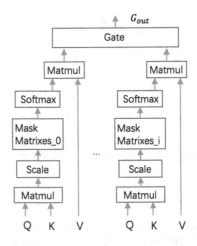

Fig. 3. The architecture of the MRA.

3.2 Modified Restrict Attention

Figure 3 shows the architecture of the MRA layer. It is based on the multi-head attention proposed by [19]. Q, K and V represent query, keys and values vector [19]. In the self-attention, Q, K and V was obtained from applying different mapping matrixes to the same sentence. In the mutual attention, Q comes from one sentence and K/V comes from another. $\sqrt{d_k}$ is the scaling factor [19]. Compared with multi-head attention, the most difference is the mask matrixes and gate in our MRA. The computation of the a single restrict attention A_i is as follow.

$$A_i = softmax(M_i(\frac{Q * K^T}{\sqrt{d_k}})) * V \tag{4}$$

The mask matrixes M_i replace the attention weight of the needless information with a small number like $(-2^{32}+1)$. Then, through softmax, it will become a very small weight. One mask matrix will be applied to many heads in multi-head attention. As shown in Table 2, the simply repetition exists in Utterance$_1$. We can easily identify this repetition in the fourth case of LF. The other characters will be masked by the matrixes. In Utterance$_2$, we may need the first cases in LF and BLF together to identify the verbose characters. Other information besides LF and BLF is mask.

The length of focused characters is a hyper-parameter and we set 4 in our model. Therefore, there are four cases of each Chinese character as shown in Table 2. The length should not be too large because we mainly focus on the local information. As we can see, there are only one or two valuable local information in LF and BLF. We add a gate for automatic learning to reduce the impact of irrelevant cases. We connect all the A_i and the input sentence representation S_e together to estimate a weight for each cases. S_e is the input sentence representation that will be converted to K/V as mentioned above. r is a tunable parameter that is to sharpen the weight. We tried r in [3, 5, 7] and set $r = 5$ in our model. The output of the gate pass through a feed forward layer and a normalize layer to as the MRA layer's output.

$$G_w = (softmax(W * [A_1, ..., A_i, S_e]))^r \qquad i = 1, ..., 4 \qquad (5)$$

Table 2. Examples of different focused information. x: difficult to express in English.

Utterance$_1$:	通过举办[泳 + 泳]博会，您觉得给城市面貌带来了哪些变化？ By holding the [Swimming + Swimming] Fair, what are the changes that you feel to our city?
LF:	1):过举办泳 (By holding the Swimming) 2):举办泳泳 (holding the Swimming Swimming) 3):办泳泳博 (x Swimming Swimming x) 4):泳泳博会 (the Swimming Swimming Fair)
BLF:	1):泳博会， (the Swimming Fair，) 2):博会，您 (Swimming Fair, what) 3):会，您觉 (Fair, what are) 4):，您觉得 (, what are the)
Utterance$_2$:	未来有没有[一些方向 + 新的方向]？ Are there [some directions + new directions] in the future?
LF:	1):一些方向 (some directions) 2):些方向新 (x directions new) 3):方向新的 (directions new) 4):向新的方 (x new x)
BLF:	1):新的方向 (new directions) 2):的方向？ (x directions ?) 3):方向？ (directions ?) 4):向？ (x ?)

4 Experiment

4.1 Dataset and Implementation Details

The dataset contains 207 spoken texts. We randomly cut the training set, development set and test set in a ratio of (8:1:1) and truncate all the sentences longer than 100 characters. The meaningful characters are labeled 0 and the verbose characters are labeled 1. We mainly focus on the performance on label 1. The label 0 characters account for 80.94% of the total Chinese characters. The label 1 only account for 19.06%. In our experiment, the character-embedding dimension is 128. The character embedding is pre-trained in Wiki corpus and fine-tuned in the train set. We use the Adam Optimizer with a fix learning rate 0.002. The model is trained on NVIDIA GTX 1080Ti GPU with the batch size of 128. We use F1 measure to evaluate all the models.

Table 3. Performance of baselines and our models. Bold data: Best data

Models	Label 0			Label 1			Avg.
	P	R	F1	P	R	F1	F1
Rule-based	91.71	49.47	64.27	31.35	**83.76**	45.62	54.95
Direct classification	94.09	87.45	90.65	50.96	70.38	59.11	74.88
Our model$_1$	**96.19**	88.00	**92.45**	56.83	80.41	66.59	79.52
Our model$_2$	94.49	90.17	92.28	62.59	75.79	68.56	80.42
Our model$_3$	94.69	90.09	92.33	62.61	76.68	68.93	80.63
Our model	93.34	**91.19**	92.25	**67.28**	73.56	**70.29**	**81.27**

4.2   Compared with Baseline Models and Ablation Study

Experimental results of baselines and our models are listed in Table 3. Rule-based method counts the verbose words frequency and uses some rules to predict the label for sentences. The rule-based method can get highest recall in label 1 but the lowest precision. Therefore, their F1 measure 45.62 is quite low. The direct classification uses two-layer Convolutional neural network (CNN) to recognize verbose characters. It can utilize the character embedding to obtain useful features. However, due to the lack of contextual information, it also achieves lower performance. Our model$_1$ is the model only with the densely connected MRA. Its F1 measure is 66.59 and 92.45. It prove the effectiveness of our attention mechanism to extract the local information. Our model$_2$ is the densely connected MRA with the compare mechanism. When the compare mechanism is added to our model, its performance is further improved with the F1 measure is 68.56 and 92.28. Model the front and rare local-focused information can detect semantic duplication well. Our model$_3$ only contains the context layer and achieves better performance for extracting long-distance information. Our model with all the layers achieves the best performance with F1 measure is 70.29 and 92.25. It proved that combining long-distance and local information is important for improving the performance. Compared with the baselines, our proposed model increased F1 measure of label 1 by 54.08% and 18.91%.

4.3 Compared with Different Attention Models and CRF Models

We compare our model with many different global attention models. The attention models in Table 4 are bi-direction LSTM with self-attention. The difference between these models is the computation methods of the self-attention. Q and K in self-attention are from the same sentence. Att_0 directly concatenate the Q and K vectors together with a linear mapping to get the attention information. Att_1 firstly mapping the Q and K with different mapping matrixes and computes the point-wise multiplication. Att_2 only computes the point-wise multiplication between vectors of Q and K. The point-wise multiplication divides the scaling factor $\sqrt{d_z}$ [19]. We also tried the BI-LSTM together with cosine similarity attention and multi-head attention [19] but they do not work. Gathering too much global contextual information makes the two model only predict the label 0, because label 0 account for 80.94% in all characters.

As we can see, every other kind of attention mechanisms achieves lower performance compared with our model. The more complex the other attention mechanisms is, the more performance is lost. The most complex attention mechanisms are the cosine similarity attention and multi-head attention that makes the model only predict label 0 for all characters. The Att_0, Att_1 and Att_2 are in a simpler order. The Att_0 only achieves 56.83 F1 value which is 19.15% lower than our model and is lower than CNN baseline. The most simple attention mechanism, the Att_2, only affects a little performance with F1 value is 66.47. All the above attention mechanisms prove that adding global information has no benefit to our model.

Table 4. Performance of different attention mechanisms and CRF models. Bold data: Best data.

Settings	Label 0			Label 1			Avg.
	P	R	F1	P	R	F1	F1
Bi-LSTM+Att_0	89.19	87.69	88.44	55.08	58.69	56.83	72.64
Bi-LSTM+Att_1	93.72	88.55	91.06	56.52	71.49	63.13	77.10
Bi-LSTM+Att_2	**96.75**	88.66	**92.53**	55.59	**82.65**	66.47	79.50
Bi-LSTM+CRF	94.71	90.05	92.32	62.45	76.67	68.83	80.58
Bi-LSTM+CRF+Att_0	92.06	84.70	88.23	40.32	58.59	47.77	68.00
Bi-LSTM+CRF+Att_1	95.25	87.66	91.3	51.89	75.28	61.44	73.37
Bi-LSTM+CRF+Att_2	93.23	89.87	91.52	62.30	71.95	66.78	79.15
Our model	93.34	**91.19**	92.25	**67.28**	73.56	**70.29**	**81.27**

The LSTM combined with a conditional random field (CRF) has achieved impressive performance in many sequence-labeling tasks [9,18]. In Table 4, the best performance of CRF models is 68.83. Therefore, the CRF has little influence on the performance. It proved that the verbose expressions are very diverse and have no obvious transfer relationship. Compared with CRF models, our model also achieved the best performance with the F1 measure is 70.29. It proves that our model could integrate local-focused information and long-distance information well to achieve better performance.

Table 5. Examples of the error prediction by our model.

Utterance	Original sentences	Gold truth	Predicted sentences
U_1	所以呢要形成这三家啊联动的这么一个一个机制，这么一种氛围。 Therefore, ah, it is necessary to form such a a linkage mechanism and such an atmosphere of the three sides, ah.	要形成三家联动的一个机制，一种氛围。 It is necessary to form a linkage mechanism and an atmosphere of the three sides.	要形成三家联动的机制，氛围。 It is necessary to form a linkage mechanism and atmosphere of the three sides.
U_2	他有一个恒定的规律。 He has a constant rule.	他有一个恒定的规律。 He has a constant rule.	他有恒定的规律。 He has a constant rule.
U_3	啊他他有这方面的考虑。 Ah he he has a consideration in this aspect.	他有这方面的考虑。 He has a consideration in this aspect.	他他有这方面的考虑。 He he has a consideration in this aspect.
U_4	而且我们是通过问题导向进行创新。 Moreover, we are innovating through problem orientation.	而且我们是通过问题导向进行创新。 Moreover, we are innovating through problem orientation.	通过问题导向进行创新。 Innovating through problem orientation.

4.4   Qualitative Analysis

Table 5 shows many error generated by our model. U_1 and U_2 is caused by inconsistencies annotations in the dataset. We can see that, the predicted sentences U_1 and U_2 also are normalized sentences. In English language, U_2 even has the same target translation. The inconsistent annotations in the dataset account for many error predictions. Therefore, our proposed model will performance better without the inconsistent annotations. U_3 is an error prediction of simple repetition. The compare mechanism does not capture this pattern. U_4 is a misrecognition of meaningful conjunction and noun because these conjunction and noun are meaningless in many other cases. Our proposed model does not properly understand contextual information in the sentence.

5   Related Work

The task most similar to ours is text normalization. It's very useful in many texts such as cell phone messages [2,4,5], social media texts [1,7,11,20,21] and broadcast transcription [1]. In cell phone messages normalization, they mainly focus on translate the brief and colloquial words into standard forms. In [2], they treated this as a Machine translation tasks. They achieved good performance using the phrase-based statistical model. In [5], they used an unsupervised model and achieved good performance. [4] proposed a method to utilize the rule-based and machine translation approaches to achieve better performance. However, we only need to delete the verbose characters without translation in the spoken texts.

In the field of social media texts, there are also a lot of text normalization work. [15] utilized many unsupervised features to choose candidate words and used graph-based approach to normalize sentences. [21] weighted different unsupervised features and employed a new training algorithm to search in the large space. In [17], they utilized distributed representations of words to gather the contextual relevant information into vectors and get good performance on a Twitter dataset. In social media texts, they mainly focus on using candidate words to replace the colloquial words. However, we focus on the verbose and noise expressions in spoken texts. The spoken texts have no abbreviated words and have many different cases.

The dataset proposed in [20], is similar to ours. They normalize the chat texts on the Internet chat texts. They proposed a phonetic mapping model to map the chat terms to a standard word via phonetic transcription [20]. They mainly solve the dynamic problem in the Internet chat. [16] used RNN to do text normalization. However, they also mainly focus on the transformation between different words' forms.

In a short word, we focus on the verbose and noise characters in the spoken texts. Meanwhile, deleting these characters will normalize the sentence and keep the original meaning of the sentence.

6   Conclusion

In this work, we propose a new task to detect verbose characters in spoken tests and construct a new dataset. The dataset drives the task of transforming texts with verbose and noise characters into normalized texts. We propose an attention mechanism that use the different mask matrixes and a gate to get relevant local focused information. We also propose a compare mechanism to leverage the front and rear local focused information. Experimental results on the dataset show that our proposed model performances better than many other models and achieves the state-of-the-art performance. In future work, we want to increase data filtering method in our model to reduce the influence of inconsistent annotations and apply our model in different tasks.

Acknowledgments. The research work is supported by the National Key Research and Development Program of China under Grant No.2017YFB1002101, the Natural Science Foundation of China (No.61533018 and No.61702512), and the independent research project of National Laboratory of Pattern Recognition.

References

1. Adda-Decker, M., Adda, G., Lamel, L.: Investigating text normalization and pronunciation variants for German broadcast transcription. In: Sixth International Conference on Spoken Language Processing (2000)
2. Aw, A., Zhang, M., Xiao, J., Su, J.: A phrase-based statistical model for SMS text normalization. In: Proceedings of the COLING/ACL on Main Conference Poster Sessions, pp. 33–40. Association for Computational Linguistics (2006)
3. Bahdanau, D., Cho, K., Bengio, Y.: Neural machine translation by jointly learning to align and translate. CoRR abs/1409.0473 (2014)
4. Beaufort, R., Roekhaut, S., Cougnon, L.A., Fairon, C.: A hybrid rule/model-based finite-state framework for normalizing SMS messages. In: Proceedings of the 48th Annual Meeting of the Association for Computational Linguistics, pp. 770–779. Association for Computational Linguistics, Uppsala, July 2010
5. Cook, P., Stevenson, S.: An unsupervised model for text message normalization. In: Proceedings of the Workshop on Computational Approaches to Linguistic Creativity, pp. 71–78. Association for Computational Linguistics (2009)
6. Dong, L., Mallinson, J., Reddy, S., Lapata, M.: Learning to paraphrase for question answering. In: Proceedings of the 2017 Conference on Empirical Methods in Natural Language Processing, pp. 875–886. Association for Computational Linguistics, Copenhagen, September 2017
7. Hassan, H., Menezes, A.: Social text normalization using contextual graph random walks. In: Proceedings of the 51st Annual Meeting of the Association for Computational Linguistics (Volume 1: Long Papers), pp. 1577–1586. Association for Computational Linguistics, Sofia, August 2013
8. Hochreiter, S., Schmidhuber, J.: Long short-term memory. Neural Comput. **9**(8), 1735–1780 (1997)
9. Huang, Z., Xu, W., Yu, K.: Bidirectional LSTM-CRF models for sequence tagging. CoRR abs/1508.01991 (2015)
10. Khashabi, D., Khot, T., Sabharwal, A., Roth, D.: Learning what is essential in questions. In: Proceedings of the 21st Conference on Computational Natural Language Learning (CoNLL 2017), pp. 80–89. Association for Computational Linguistics, Vancouver, August 2017
11. Liu, F., Weng, F., Wang, B., Liu, Y.: Insertion, deletion, or substitution? Normalizing text messages without pre-categorization nor supervision. In: Proceedings of the 49th Annual Meeting of the Association for Computational Linguistics: Human Language Technologies, pp. 71–76 (2011)
12. Liu, Y., Sun, C., Lin, L., Wang, X.: Learning natural language inference using bidirectional LSTM model and inner-attention. CoRR abs/1605.09090 (2016)
13. Qin, K., Wang, L., Kim, J.: Joint modeling of content and discourse relations in dialogues. In: Proceedings of the 55th Annual Meeting of the Association for Computational Linguistics (Volume 1: Long Papers), pp. 974–984. Association for Computational Linguistics, Vancouver, July 2017
14. Shaw, P., Uszkoreit, J., Vaswani, A.: Self-attention with relative position representations. CoRR abs/1803.02155 (2018)
15. Sonmez, C., Ozgur, A.: A graph-based approach for contextual text normalization. In: Proceedings of the 2014 Conference on Empirical Methods in Natural Language Processing (EMNLP), pp. 313–324. Association for Computational Linguistics, Doha, October 2014

16. Sproat, R., Jaitly, N.: RNN approaches to text normalization: A challenge. CoRR abs/1611.00068 (2016)
17. Sridhar, V.K.R.: Unsupervised text normalization using distributed representations of words and phrases. In: NAACL HLT 2015, pp. 8–16 (2015)
18. Sun, W., Sui, Z., Wang, M., Wang, X.: Chinese semantic role labeling with shallow parsing. In: Proceedings of the 2009 Conference on Empirical Methods in Natural Language Processing: Volume 3. EMNLP 2009, vol. 3, pp. 1475–1483. Association for Computational Linguistics, Stroudsburg (2009)
19. Vaswani, A., et al.: Attention is all you need. CoRR abs/1706.03762 (2017)
20. Xia, Y., Wong, K.F., Li, W.: A phonetic-based approach to Chinese chat text normalization. In: Proceedings of the 21st International Conference on Computational Linguistics and 44th Annual Meeting of the Association for Computational Linguistics, pp. 993–1000. Association for Computational Linguistics, Sydney, July 2006
21. Yang, Y., Eisenstein, J.: A log-linear model for unsupervised text normalization. In: Proceedings of the 2013 Conference on Empirical Methods in Natural Language Processing, pp. 61–72. Association for Computational Linguistics, Seattle, October 2013

Medical Knowledge Attention Enhanced Neural Model for Named Entity Recognition in Chinese EMR

Zhichang Zhang[(✉)], Yu Zhang, and Tong Zhou

College of Computer Science and Engineering, Northwest Normal University, Lanzhou, China
zzc@nwnu.edu.cn, zhangyu.wuyi@foxmail.com,
1083993141@qq.com

Abstract. Named entity recognition (NER) in Chinese electronic medical records (EMRs) has become an important task of clinical natural language processing (NLP). However, limited studies have been performed on the clinical NER study in Chinese EMRs. Furthermore, when end-to-end neural network models have improved clinical NER performance, medical knowledge dictionaries such as various disease association dictionaries, which provide rich information of medical entities and relations among them, are rarely utilized in NER model. In this study, we investigate the problem of NER in Chinese EMRs and propose a clinical neural network NER model enhanced with medical knowledge attention by combining the entity mention information contained in external medical knowledge bases with EMR context together. Experimental results on the manually labeled dataset demonstrated that the proposed method can achieve better performance than the previous methods in most cases.

Keywords: Chinese electronic medical record · Named entity recognition
Deep learning · Knowledge attention

1 Introduction

Named entity recognition (NER), which aims to identify boundaries and types of entities in text, has been one of the well-established and extensively investigated tasks in natural language processing (NLP), and is an essential component for a large number of NLP applications such as entity linking (Chabchoub *et al.* 2016), relation extraction (Liu *et al.* 2014), question answering (Cao *et al.* 2011) and knowledge base population (Carlson *et al.* 2010).

The electronic medical record (EMR), sometimes called electronic health record (EHR) or electronic patient record (EPR), is one of the most important types of clinical data and often contain valuable and detailed patient information for many clinical applications. Clinical NER in EMR text is therefore a fundamental task in medical NLP and has been extensively studied (Ye *et al.* 2011). Nevertheless, most previous studies on clinical NER have primarily focused on EMR in English text. With the rapid growth of clinical NLP applications in China, NER from Chinese clinical text has also become a hot research topic for biomedical informatics or NLP researchers.

© Springer Nature Switzerland AG 2018
M. Sun et al. (Eds.): CCL 2018/NLP-NABD 2018, LNAI 11221, pp. 376–385, 2018.
https://doi.org/10.1007/978-3-030-01716-3_31

Most previous research on NER have been dominated by applying traditional machine learning (ML) based models, which require a set of informative features that are well engineered and carefully selected. However, feature engineering is very time-consuming and costly, and resulting feature sets are both domain and model-specific.

In the past few years, the advent of deep neural networks with the capability of automatic feature engineering has leveraged the development of NER models, and this kind of models has also been studied on clinical NER (Le *et al.* 2018). While deep neural network approaches for clinical NER have achieved better performance compared to traditional models, many existing domain knowledge bases have rarely been utilized or combined in these deep models. Medical knowledge bases (MKBs), however, contain a large amount of clinical entity name and definition or description of them. These entity names can be applied as reference lexicon for clinical NER, meanwhile their definition and description provide rich context and entity relationship information which is still helpful for entity recognition.

In view of the above, this paper proposes a novel model for clinical NER in Chinese EMRs. The model trains character-level embedding representation of words using Convolutional Neural Network (CNN), and combines them with pretrained character embedding vectors obtained from large-scale background training corpus, then sends the combined vectors to a deep neural network called BILSTM-CRF to train entity recognition model. To enhance the representation and distinguish ability of words and their contexts, we integrate the medical knowledge attention (MKA) learned from entity names and their definition or descriptions in MKBs. The experimental results on the labeled Chinese EMR evaluation corpus show that the model achieved the best performance without any artificial features, and the F-values is 92.03%.

The remainder of this paper is composed as follows. In Sect. 2 we summarize the related work about clinical or medical NER. In Sect. 3 we present our medical knowledge attention enhanced neural network NER model in Chinese EMRs. In Sect. 4 we show the experimental results on the test data and give some analysis. Finally, we summarize our work and outline some ideas for future research.

2   Related Work

NER is typically treated as a sequence labeling problem and many researchers applied ML-based methods to learn named entity tagging decisions from annotated texts. Those techniques utilized for clinical NER are Support Vector Machines (SVM), structural SVM (SSVM), Conditional Random Fields (CRF), Maximum Entropy (ME). (Wang *et al.* 2009) applied CRF, SVM and ME to recognize symptoms and pathogenesis in ancient Chinese medical records and showed that CRF achieved a better performance. (Wang *et al.* 2012) conducted a preliminary study on symptom name recognition in clinical notes of traditional Chinese medicine. (Xu *et al.* 2014) proposed a joint model that integrates segmentation and NER simultaneously to improve the performance of both tasks in Chinese discharge summaries. (Lei *et al.* 2014) systematically evaluated the effects of different features and ML algorithms on NER in Chinese clinical text.

In recent years, unsupervised learned word embeddings have been seen tremendous success in numerous NLP tasks, including clinical or medical NER. (Tang *et al.* 2014) used CRF model and supplement artificial features with word embeddings to identify biological entity, achieved good performance on BioCreative II GM and JNLPBA corpus. (Chang *et al.* 2015) also utilized word embeddings in their CRF-based medical NER model and obtained performance improvement on JNLPBA corpora.

Deep neural network (DNN) architecture is also widely used in NER task. (Yao *et al.* 2015) trained a multilayer neural network model for biological entity recognition with word embeddings generated on unlabeled biological texts. (Li *et al.* 2016) used the Bi-directional Long Short Term Memory Network (BLSTM) method to achieve an 88.6% and 72.76% F-value on the Biocreative II GM and JNLPBA corpus respectively. (Dong *et al.* 2017) used the BLSTM model to identify named entity in Chinese electronic medical records. (Liu *et al.* 2017) used the BLSTM-CRF with features to identify clinical notes, achieve an 89.98% F-value on the 2016 N-GRID.

Although clinical or medical NER in English text has been extensively studied and many kinds of traditional ML-based and DNN-based models have been proposed, there is limited work on clinical NER in Chinese EMRs using word embeddings or deep learning methods. Furthermore, the application of MKBs and their effectiveness on clinical NER should also be carefully studied and analyzed.

3   The Proposed Method

In this paper, we propose a neural network architecture combining BI-LSTM-CNN-CRF network with Medical Knowledge-Attention that will learn the shared semantics between medical record texts and the mentioned entities in the MKBs. The architecture of our proposed model is shown in Fig. 1. After querying pretrained character embedding tables, the input sentence s will be transformed respectively to the corresponding sequences of pretrained character embeddings and random generated character embedding matrixes for every word. Then a CNN is used to form the character level representation and a bidirectional LSTM is used to encode the sentence representation after concatenating the pretained character embeddings and char-level representation of the sentence. Afterwards, we treat the entity information from MKBs as a query guidance and integrate them with the original sentence representation using a multimodal fusion gate and a filtering gate. At last, a CRF layer is used to decode.

3.1   Feature Extractor

Character-Level Representation with CNN. As described in Fig. 2, we firstly train character embeddings from a large unlabeled Chinese EMR corpus, then CNN is used to generate sentence character-level representation from the character embedding matrix sequence to alleviate rare character problems and capture helpful morphological information like special characters in EMRs. Since the length of sentences is not consistent, a placeholder (padding) is added to the left and right side of character embeddings matrix to make the length of every sentence character-level representation vector matrix sequence equal.

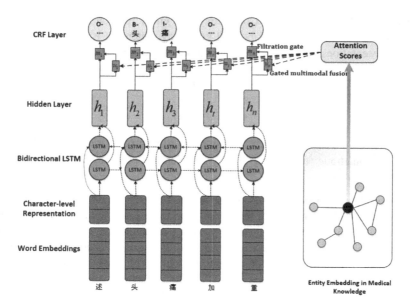

Fig. 1. The framework of NER in Chinese EMRs enhanced with medical knowledge attention

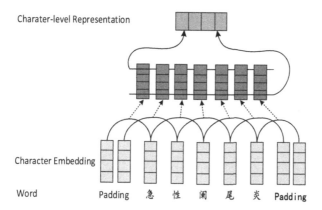

Fig. 2. Character-level representation of a sentence by CNN.

Sentence Encoding with Bidirectional LSTM. Bidirectional LSTM encodes the sentence twice from the start to the end and from the end to the start. Thus, at each time state t, we can obtain two representations $\overrightarrow{h_t}$ and $\overleftarrow{h_t}$ of sentence. Two representations are concatenated to form the final encoding representation of the sentence:

$$h_t = \left[\overrightarrow{h_t}, \overleftarrow{h_t} \right] \tag{1}$$

Medical Knowledge Attention. Concerning rich entity relation and definition information containing in MKBs, the medical knowledge attention is applied to integrate entity representations learned from external knowledge bases as query vector for encoding. We use a medical dictionary to encode entity information and entity relation information into attention scores as entity embeddins.

$$a_t = f(eW_A h_t) \tag{2}$$

Where e is the embedding for entity, and W_A is a bi-linear parameter matrix. We simply choose the quadratic function $f(x) = x^2$, which is positive definite and easily differentiate.

Gated Multimodal Fusion. Based on the output of LSTM and attention scoring, we design a gated multimodal fusion (GMF) method to fuse the features from EMR text context and external knowledge dictionary. When predicting the entity tag of a word, the GMF trades off how much new information of the network is considering from the query vector with the EMR text containing the word. The GMF is defined as:

$$h_{a_t} = \tanh(W_{a_t} a_t + b_{a_t}) \tag{3}$$

$$h_{h_t} = \tanh(W_{h_t} h_t + b_{h_t}) \tag{4}$$

$$g_t = \sigma(W_{g_t}(h_{a_t} \oplus h_{h_t})) \tag{5}$$

$$m_t = g_t h_{a_t} + (1 - g_t)h_{h_t} \tag{6}$$

Where W_{a_t}, W_{h_t}, W_{g_t} are parameters, h_{h_t} and h_{a_t} are the new sentence vector and new query vector respectively, after transformation by single layer perceptron. $\oplus$ is the concatenating operation, σ is the logistic sigmoid activation, g_t is the gate applied to the new query vector h_{h_t}, and m_t is the multimodal fused feature from the new medical knowledge feature and the new textual feature.

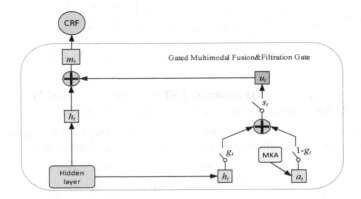

Fig. 3. The architecture of gated multimodal fusion and filtering gate.

Filtering Gate. The architecture of gated multimodal fusion and filtering gate are shown in Fig. 3. When decoding the combination of the word feature at position t and its corresponding multimodal fusion feature m_t, the impact and necessity of the MKB feature for different POS of word is different. We therefore use a filtering gate to combine different features. The filtering gate s_t is a scalar in the range of [0, 1] and its value depend on how much the multimodal fusion feature is helpful to label the entity tag of the word. s_t and the input feature to the decoder $\hat{m}_t$ are defined as follows:

$$s_t = \sigma(W_{s_t,h_t}h_t \oplus (W_{m_t,s_t}m_t + b_{m_t,s_t})) \tag{7}$$

$$u_t = s_t(\tanh(W_{m_t}m_t + b_{m_t})) \tag{8}$$

$$\hat{m}_t = W_{\hat{m}t}(h_t \oplus u_t) \tag{9}$$

where W_{m_t,s_t}, W_{s_t,h_t}, W_{m_t}, $W_{\hat{m}t}$ are parameters, h_t is the hidden state of bidirectional LSTM at time t, u_t is the reserved multimodal features after the filtering gate filter out noise, and $\oplus$ is the concatenating operation.

3.2 CRF Tagging Layer

Formally, we use $X = \{x_1, \cdots, x_n\}$ to represent a generic input sequence where x_i is the input vector of the ith word, and $y = \{y_1, \cdots, y_n\}$ denotes the set of possible entity tag sequences for X where y_i represent a generic tag for x_i. We use the maximum conditional likelihood estimation for CRF training. The logarithm of likelihood is given by:

$$L(p(y/X)) = \sum_i \log p(y/X) \tag{10}$$

In the decoding phrase, we predict the output sequence that obtains the maximum score given by:

$$y^* = \arg\max_{y' \in Y} p(y/X) \tag{11}$$

4   Experiments

In this section, we evaluate our method on a manually annotated dataset. Following (Nadeau and Sekine 2007), we use Precision, Recall, and F1 to evaluate the performance of the models.

4.1   Data Set

We use our own manually annotated corpus as evaluation dataset, which consists of 800 de-identified EMR texts from different clinical departments of a grade-A hospital of second class in Gansu Province. Five entity types, including symptom, disease, laboratory test, body parts and medicine are labeled. The number of every entity type in dataset is shown in the Table 1.

Table 1. Number statistics of different entity type in the evaluation dataset.

Statistics	Train	Test	Total
Symptom (症状)	823	412	1235
Disease (疾病)	1013	506	1519
Laboratory test (检查)	637	315	952
Body parts (身体部位)	761	380	1141
Medicine (药物)	547	273	820
Total	3781	1886	5667

4.2 Embedding

We use Google's Word2Vec to train character embeddings on our 30 thousand unlabeled Chinese EMR texts which is from a grade-A hospital of second class in Gansu Province. Random generated character embeddings are initialized with uniform samples from $\left[-\sqrt{\frac{3}{dim}}, \sqrt{\frac{3}{dim}}\right]$, where we set dim = 30.

4.3 Parameter Setting

Table 2 gives the chosen hyper-parameters for all experiments. We tune the hyper-parameters on the development set by random search. We try to share as many hyper-parameters as possible in experiments.

Table 2. Parameter setting.

Parameter	Value
Character embedding size	50
Word embedding size	100
Learning size	0.014
Decay rate	0.05
Dropout	0.5
Batch size	10
LSTM state size	200
LSTM initial state	0
CNN window size	3
CNN number of filters	50

4.4 Experimental Results

We carry out the experiments to compare the performance of the following models.

CNN+BILSTM+CRF: It was proposed by (Ma and Hovy 2016) and is a truly end-to-end system. The model was reported to have achieved the best 91.21% F1 score on the CoNLL 2003 test set. This model will be used as the baseline.

CNN+BILSTM+MKA+CRF: In this model, the fusion gate and the filtering gate are not used, and the medical knowledge attention score is simply added to the output value of LSTM as the query vector.

CBMFiC: This model is formed after leaving the fusion gate from the architecture described in Fig. 1. It directly concatenates the features from different modalities. At each time step, we use a filtering gate to filter out the noise introduced by the attention score, then concatenate the external multimodal feature of the word with the text feature as CRF input.

CBMFuC: This model is formed after only leaving the filtering gate from the architecture described in Fig. 1. After obtaining the fused features, both Medical Knowledge-Attention and text representation, we concatenate the fusion feature (query vector) with the text feature at this time step, then the CRF is applied for learning.

CBMFuFiC: This model is totally corresponding to the architecture in Fig. 1, and contains all components investigated in this study including.

Table 3. The performance of different models on the evaluation dataset.

Model	Precision (%)	Recall (%)	F1 (%)
CNN+BILSTM+CRF(Baseline)	88.88	88.31	88.59
CNN+BILSTM+MKA+CRF	90.21	88.72	89.46
CBMFiC	90.49	89.65	90.07
CBMFuC	91.27	**92.03**	91.65
CBMFuFiC	**92.19**	91.87	**92.03**

4.5 Effect of Dropout

Table 4 compares the experiment performance when the dropout layer is or not used. All other hyper parameters remain the same as in Table 2. It demonstrates the effectiveness of the dropout in reducing overfitting.

Table 4. Results with and without dropout

	Train	Dev	Test
No	99.63	93.74	90.21
Yes	**99.19**	94.87	**92.03**

4.6 Discussion

The experiment results of different models on our manually annotated datasets are shown in Table 3. Compared with the baseline model, all other models have improved performance and it shows that the updated neural network model is better than the traditional deep neural network model only composed of CNN, BILSTM and CRF on the clinical NER task.

The performance of the CBMC model is better than baseline. This result shows that, the rich information of entities and their relations from MKBs is useful for clinical NER in Chinese EMR. CBMFuC model is slightly better than CBMC model and shows that it is surely helpful for the clinical NER task in Chinese EMRs to fuse the features from EMR text context with external knowledge dictionary and utilize gated multimodal fusion (GMF) is help for our model. The supplement of external information in MKBs sometimes causes noise to the model. We therefore use a filtering gate to weight and combine different features. As shown by our experimental results, CBMFiC model is also slightly better than CBMC model. Therefore, the filtering gate is helpful to improve the performance of our model.

Due to their sublanguage characteristic of Chinese EMRs, the expression of clinical named entity is different from those in general text. Using the entity information in the MKB as the classification query vector can lead the decoder to focus on the entity itself. So we combine text features and MKB features together with a multimodal fusion gate as query vector. We then set up a filtering gate to filter out useless feature vectors. The experiment result shows that our model CBMFuFiC, which integrates CNN, BILSTM, medical knowledge attention, gated multimodal fusion, filtering gate, and CRF, achieves the best F1 score on our labeled evaluation corpus.

5 Conclusion

In this work, we proposed a medical knowledge-attention enhanced neural clinical entity recognition model, which makes use of the external MKBs in the way of attention mechanism. A gated multimodal fusion module is introduced to decide how much MKB features are fused into the query vector at each time step. We further introduced a filtering gate module to adaptively adjust how much multimodal information can be considered at each time step. We built a manually annotated Chinese EMR dataset for evaluation. The experiment results on the evaluation dataset show that our proposed approach improved the clinical NER performance obviously compared to the baseline models.

In the future, we will explore a fine-grained clinical entity recognition model for Chinese EMR and to extract entity relation in Chinese EMRs.

Acknowledgements. We would like to thank the anonymous reviewers for their valuable comments. The research work is supported by the National Natural Science Foundation of China (No. 61762081, No. 61662067) and the Key Research and Development Project of Gansu Province (No. 17YF1GA016).

References

Cao, Y.-G., Liu, F., Simpson, P., Antieau, L., Bennett, A.: AskHERMES: an online question answering system for complex clinical questions. J. Biomed. Inform. **44**(2), 277–288 (2011)

Carlson, A., Betteridge, J., Wang, R.C., et al.: Coupled semi-supervised learning for information extraction. DBLP, pp. 101–110 (2010)

Chabchoub, M., Gagnon, M., Zouaq, A.: Collective disambiguation and semantic annotation for entity linking and typing. In: Sack, H., Dietze, S., Tordai, A., Lange, C. (eds.) SemWebEval 2016. CCIS, vol. 641, pp. 33–47. Springer, Cham (2016). https://doi.org/10.1007/978-3-319-46565-4_3

Chang, F.-X., Guo, J., Xu, W.-R., Chung, S.-R.: Application of word embeddings in biomedical named entity recognition tasks. J. Digit. Inf. Manag. **13**(5), 321–327 (2015)

Dong, X., Chowdhury, S., Qian, L., et al.: Transfer bi-directional LSTM RNN for named entity recognition in Chinese electronic medical records. In: The Proceedings of International Conference on E-Health Networking, Applications and Services, pp. 1–4. IEEE (2017)

Le, H.-Q., Nguyen, T., Vu, S., Dang, T.-H.: D3NER: biomedical named entity recognition using CRF-biLSTM improved with fine-tuned embeddings of various linguistic information. Bioinformatics (2018). https://doi.org/10.1093/bioinformatics/bty356

Lei, J., Tang, B., Lu, X., Gao, K., Jiang, M., Xu, H.: A comprehensive study of named entity recognition in Chinese clinical text. J. Am. Med. Inform. Assoc. **21**(5), 808–814 (2014)

Li, L., Jin, L., Jiang, Y., Huang, D.: Recognizing biomedical named entities based on the sentence vector/twin word embeddings conditioned bidirectional LSTM. In: Sun, M., Huang, X., Lin, H., Liu, Z., Liu, Y. (eds.) CCL/NLP-NABD-2016. LNCS (LNAI), vol. 10035, pp. 165–176. Springer, Cham (2016). https://doi.org/10.1007/978-3-319-47674-2_15

Liu, Y., Liu, K., Xu, L.-H. Zhao, J.: Exploring fine-grained entity type constraints for distantly supervised relation extraction. In: Proceedings of COLING 2014, Dublin, Ireland, 23–29 August (2014)

Liu, Z., Tang, B., Wang, X., et al.: De-identification of clinical notes via recurrent neural network and conditional random field. J. Biomed. Inform. **75S**, S34 (2017)

Ma, X., Hovy, E.: End-to-end sequence labeling via bi-directional LSTM-CNNs-CRF (2016). https://arxiv.org/pdf/1603.01354

Nadeau, D., Sekine, S.: A survey of named entity recognition and classification. Lingvisticae Investig. **30**(1), 3–26 (2007)

Tang, B.-Z., Cao, H., Wang, X.-L., Chen, Q.-C., Xu, H.: Evaluating word representation features in biomedical named entity recognition tasks. Biomed Res. Int. **2014**, 6 (2014). https://doi.org/10.1155/2014/240403. Article ID 240403

Wang, S., Li, S., Chen, T.: Recognition of Chinese medicine named entity based on condition random field. J Xiamen Univ. (Nat. Sci.) **48**, 349–364 (2009)

Wang, Y., Liu, Y., Yu, Z., et al.: A preliminary work on symptom name recognition from free-text clinical records of traditional Chinese medicine using conditional random fields and reasonable features. In: Proceedings of the 2012 Workshop on Biomedical Natural Language Processing, Stroudsburg, PA, USA, pp. 223–30 (2012)

Xu, Y., Wang, Y., Liu, T., et al.: Joint segmentation and named entity recognition using dual decomposition in Chinese discharge summaries. J. Am. Med. Inform. Assoc. **21**, e84–e92 (2014)

Yao, L., Liu, H., Liu, Y., et al.: Biomedical named entity recognition based on deep neutral network. Int. J. Hybrid Inf. Technol. **8**, 279–288 (2015)

Ye, F., Chen, Y.Y., Zhou, G.G., et al.: Intelligent recognition of named entity in electronic medical records. Chin. J. Biomed. Eng. **30**(2), 256–262 (2011)

Coherence-Based Automated Essay Scoring Using Self-attention

Xia Li[1,2(✉)], Minping Chen[2], Jianyun Nie[3], Zhenxing Liu[2],
Ziheng Feng[2], and Yingdan Cai[2]

[1] Key Laboratory of Language Engineering and Computing,
Guangdong University of Foreign Studies, Guangzhou, China
shelly_lx@126.com
[2] School of Information Science and Technology/School of Cyber Security,
Guangdong University of Foreign Studies, Guangzhou, China
minpingchen@126.com, liuzhenxingw@126.com,
zihengfeng@126.com, ldchoy@126.com
[3] Department of Computer Science and Operations Research,
University of Montreal, Montreal, Canada
nie@iro.umontreal.ca

Abstract. Automated essay scoring aims to score an essay automatically without any human assistance. Traditional methods heavily rely on manual feature engineering, making it expensive to extract the features. Some recent studies used neural-network-based scoring models to avoid feature engineering. Most of them used CNN or RNN to learn the representation of the essay. Although these models can cope with relationships between words within a short distance, they are limited in capturing long-distance relationships across sentences. In particular, it is difficult to assess the coherence of the essay, which is an essential criterion in essay scoring. In this paper, we use self-attention to capture useful long-distance relationships between words so as to estimate a coherence score. We tested our model on two datasets (ASAP and a new non-native speaker dataset). In both cases, our model outperforms the existing state-of-the-art models.

Keywords: Self-attention · Automated essay scoring · Neural networks

1 Introduction

Traditional Automated Essay Scoring (AES) methods are based on manually determined features, which require much manual work. In contrast, the recent neural-network-based models [1–5] can automatically extract features to avoid feature engineering work. It turns out that neural-network-based methods can achieve better performance than the traditional methods and human raters [2, 4].

Previous neural network models for AES use word embedding of essay's words as input and use CNN and LSTM to capture the content information and local relationship between the words within sentences. However, the captured relationships remain local (within sentences), and no global relationships among the words across sentences and paragraphs can be obtained.

© Springer Nature Switzerland AG 2018
M. Sun et al. (Eds.): CCL 2018/NLP-NABD 2018, LNAI 11221, pp. 386–397, 2018.
https://doi.org/10.1007/978-3-030-01716-3_32

In human essay scoring, we observe that global relationships play an important role. When rating an essay, human raters not only consider the goodness of an essay's content, but also pay much attention to the structure of the essay. In particular, the *coherence* between different parts of the essay is an important rating criterion. A good essay should contain related parts with strongly connected words, while a bad essay may contain parts that are unrelated.

For example, the following text fragments (a) and (b) are all grammatically correct:

(a) *I was born in* **Glasgow**. **Glasgow** *is the largest city in Scotland.*
(b) *I was born in* **Glasgow**. *It is very nice in Scotland.*

However, the first fragment reads better than the second one because sentences in the first fragment are more connected and it is more coherent. A higher coherence implies that the sentences read more smoothly [6, 7]. A human rater would rate the first segment higher than the second one according to the coherence criterion. For a whole essay, the principle is the same. We expect a good essay to be coherent between different parts.

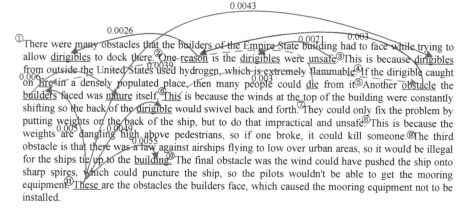

Fig. 1. Coherences learned in a 4-score essay from prompt 6. In this figure, we show that the coherences of the essay are important for judging of good essay for human raters. We can see that the word "reason" in sentence ② has a high relationship with the word "dirigibles" in sentence ①, which represent the causal relationship between sentence ① and ②. We can also see that the word "These" in sentence ⑪ has four high relationships with "dirigibles" in sentence ③, "builders" in sentence ⑤, "dirigible" in sentence ⑥ and "building" in sentence ⑨. These relationships can help and improve the essay's readability which gives the ground of high score for human raters.

Most previous studies on AES did not take into account the coherence aspect. Tay et al. [5] is one of the few exceptions. It incorporated the relationship between two LSTM hidden unit outputs as neural coherence features. These relationships were used as auxiliary neural coherence features to predict essay scores. However, these neural coherence features were captured only between words within windows of fixed size.

The relationships among words are thus limited to local relationships. The model was unable to capture the relationships between distant words and thus to estimate the global coherence.

In general, it is known that LSTM and CNN have difficulty to capture long-distance dependencies. To solve the problem, recent work [8] used self-attention instead, which is shown to be a mechanism capable of capturing long-distance relationships between words in a sequence. Inspired by this work, we propose to use self-attention to learn the relationships between words in the whole essay. The words in relation can be from the same sentence or from different sentences. Figure 1 shows some of the relationships recognized between words by self-attention[1]. We use different colored lines with arrow to indicate the relationships of different words. The values beside the lines represent the attention weights of the relationship between the two words. We can see that self-attention mechanism can learn these explicit relationships between words at long distance. In this figure, we can also notice that the connected words are also semantically related (based on their contents).

Based on these observations, our intuition used in this paper is that an essay with strong connections between words has a high coherence, and thus should be rated high. This coherence criterion is combined with the usual criterion on content. It may not be possible to simply sum up all the connections as a coherence measure. However, the notion of coherence may be much more complex. Therefore, we use a more sophisticated mechanism to aggregate the connections into a rating score. We observe that there is a natural sequence between the connections among words: a connection usually connects a word with some previous words in the sentence or in other sentences. Based on this observation, we use a LSTM layer stacked on the output of self-attention to learn the whole score of the essay's coherence based on the recognized connections.

The architecture of our model is illustrated in Fig. 2. Multi-head self-attention is first applied to the words (embeddings) and their positions are used to recognize the connections between words. Each head is intended to capture one type of relationship. Then LSTM is used to aggregate the outputs of multiple heads into a rating score.

An important difference between this approach and the previous ones using LSTM is that our LSTM works on relations between words rather than their representations (contents). Therefore, our LSTM will also account for the connections among words, in addition to what the words represent, which is the common focus of previous LSTM-based approaches. The aggregated output of the LSTM will then encode the global coherence of the essay.

To our knowledge, no prior work has investigated using coherence based on self-attention for AES. We will show in our experiments that our model outperforms the state-of-the-art methods. The main contributions of our paper include:

(1) We propose a model based on self-attention mechanism to capture the relationships between different parts of the essay, and we show that the method is appropriate for AES.

---

[1] The essay is from prompt 6 of ASAP dataset - https://www.kaggle.com/c/asap-aes/data. We only show some of the strong relationships for clarity.

(2) We tested our approach on two sets of essays, one from native English speakers and another from English learners. In both cases, we show that our model outperforms the existing approaches.

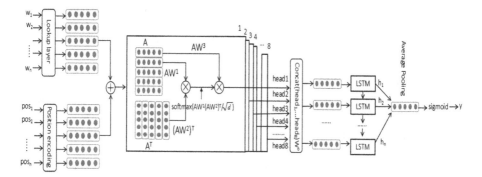

Fig. 2. Structure of our model.

2 Coherence-Based Scoring Model

Different from previous work, we use multi-head self-attention to learn the relationships between different words in the essay and obtain a relationship-based essay representation. Then, we use LSTM to learn the essay's overall score based on the output of multi-head self-attention. In this section, we will describe our model in detail.

2.1 Words and Positions Encoding

Look-Up Layer. First, the words of an essay are inputted into the look-up layer to obtain a dense representation. We use the length of the longest essay in each topic as the length of all the essays in the topic and use the padding operation when the length of other essays is shorter. We use the Stanford's open source 50-dimensional GloVe word embedding [9] as our word representations. The word embedding will be fine-tuned during training.

Position Encoding. Position encoding intends to mark the position of each word in the essay, so that a word can be connected to the word around it. Following [8], we encode the absolute position information of the words to obtain a position embedding with the same dimension as the word embedding (i.e. 50). This embedding is obtained by Eqs. (1) and (2), where *pos* is the position of the word, *i* is the *i*-th dimension in the position embedding, and *d* is the dimension of embedding.

$$PE(pos, 2i) = sin\left(\frac{pos}{10000^{\frac{2i}{d}}}\right) \quad (1)$$

$$PE(pos, 2i+1) = cos\left(\frac{pos}{10000^{\frac{2i}{d}}}\right) \tag{2}$$

PE defines a sinusoidal function, which could allow the model to attend relative positions [8]. Through this operation, each word position is represented by a vector with values between 0–1 and has a relation with the word's position. We add the word embedding and the position embedding to form the input matrix to self-attention in the same way as in [8]. This produces the input matrix A used for self-attention networks.

2.2 Getting Relationships with Self-attention

After obtaining a dense representation of each word and position, we use self-attention to capture the relationships between each pair of words across the essay. In our model, we use 8 heads of self-attention that work in parallel. A head of self-attention is intended to capture one type of relationship.

Vaswani et al. [8] used self-attention for machine translation. Self-attention is based on key-value network, in which key memories (K) encode the information to calculate the attention for a given input (Q), while the value memories (V) encode the corresponding context information and the distribution for the output (next translation word). Vaswani et al. [8] defined the following scaled-dot-product attention for their purpose:

$$Attention(Q, K, V) = softmax\left(\frac{QK^T}{\sqrt{d}}\right)V \tag{3}$$

where QK^T determines where the attention should be paid to, $\frac{1}{\sqrt{d}}$ is a scaling factor (d is the dimension of the embedding) to counteract the fact that the softmax function is pushed into regions where it has extremely small gradients when d is large. Similar mechanism has been used for question answering – to determine the best answer for a given question (Q).

In our case, the situation is different: we intend to estimate the connections between words in the same essay. This can be basically estimated through AA^T, which tells us how different words are connected. The value network is also based on the same representation space as A. However, instead of using A, we use 3 different linear projections of A as in [8]: AW^1, AW^2 and AW^3. Attention is thus determined by Eq. (4).

$$Attention(AW^1, AW^2, AW^3) = softmax\left(\frac{AW^1(AW^2)^T}{\sqrt{d}}\right)AW^3 \tag{4}$$

We also use 8 heads of self-attention in parallel. The 8 heads are concatenated at the end, which is projection with another weight matrix W^0. The multi-head attention mechanism is then defined as follows:

$$head_i = Attention\left(AW_i^1, AW_i^2, AW_i^3\right) \tag{5}$$

$$MultiHead(A) = Concat(head_1, \ldots, head_h)W^0 \tag{6}$$

where h is the number of attention heads (i.e. 8) and W^0, W^1, W^2, W^3 are trainable parameters.

2.3 Estimating Global Score with LSTM

The output of multi-head attention should be used according to the task. For machine translation, it is used to help the decoding, i.e. determining the next translation word. In our task, we want to obtain a global rating for an essay. As we noticed, the connections between words in an essay are sequential: a word is usually connected to the words before it. Therefore, we use LSTM [10, 11] to cope with the sequential nature of the connections. LSTMs have been generally used to compose a document (text) representation in most previous models [2–4]. The main difference of our case is that we deal with relationship-based representation.

Assuming that an essay consists of m words $w_1, w_2, \ldots, w_m$. After the operation of self-attention, the output is m vectors $w_1', w_2', \ldots, w_m'$, each vector w_i' representing the relationship encoding between the word w_i and other words $w_j (j = 1, 2, \ldots, m)$. We consider these representations as a sequence and use LSTM to produce the internal states for the sequence at each timestep t: $(h_1, h_2, \ldots, h_t, \ldots, h_n)$.

Then, we use an average pooling to get the final representation of essay S. Here average pooling is used because we believe that all the connections between different sentences and different words are equally important. Finally, the final essay score is obtained through a fully-connected layer with a nonlinear activation function. The activation function is showed in Eq. (7).

$$\hat{y} = sigmoid(WS + b) \tag{7}$$

Where W is the weight matrix, b is the bias, and $\hat{y}$ is a predicted score. As in previous work, we use the Mean Square Error (MSE) as a loss function, as in Eq. (8), where y is the essay's score rated by human and $\hat{y}$ is the score predicted by the model.

$$MSE(y, \hat{y}) = \frac{1}{N} \sum_{i=1}^{N} (y_i - \hat{y}_i)^2 \tag{8}$$

3 Experiments

3.1 Datasets

In this paper, we use two datasets in our experiments. The first dataset is ASAP which is widely used in the existing work. The ASAP contains 12,978 essays in eight different prompts (essay topics) which are written by students from grades 7–10. Another dataset was collected from the College Entrance English Examination (CEEE dataset) of Guangdong province in China. CEEE contains 3,958 essays written by Chinese English-learners who are senior students. There is only one prompt in CEEE, which asks to write an essay based on four given pictures. Some detailed description of these two datasets is showed in Table 1. Following previous studies [3, 4], we split each prompt into 60% training data, 20% development data and 20% testing data. We use 5-fold cross-validation in our evaluations.

Table 1. Details of ASAP and CEEE datasets.

Data	Prompt	#Essay	Avg len.	Score range	Score median
ASAP	1	1783	350	2–12	8
	2	1800	350	0–6	3
	3	1726	150	0–3	1
	4	1772	150	0–3	1
	5	1805	150	0–4	2
	6	1800	150	0–4	2
	7	1569	250	0–30	16
	8	723	650	0–60	30
CEEE	1	3958	145	0–25	13

3.2   Evaluation Metric

We use Quadratic Weighted Kappa (QWK) as the evaluation metric in our experiments. This metric is widely used in many previous studies. It is defined in Eq. (9).

$$k = 1 - \frac{\sum W_{ij} O_{ij}}{\sum W_{ij} E_{ij}} \tag{9}$$

Where O_{ij} is the number of essays that receive a rating i by the human rater and a rating j by the AES system, and the matrix E is the outer product of vectors of human ratings and system ratings. Matrix E needs to be normalized such that the sum of elements in E and the sum of elements in O are the same. The quadratic-weight matrix W_{ij} is defined in Eq. (10), where i and j are the human rating and the system rating respectively, and N is the number of the essays.

$$W_{ij} = \frac{(i-j)^2}{(N-1)^2} \tag{10}$$

Following Taghipour and Ng [3], Dong and Zhang [4], we performed one-tailed t-test to determine the statistical significance of improvements.

3.3 Experiment Setup

To make our results comparable, we use the same preprocessing as in Taghipour and Ng, Dong and Zhang [3, 4]: NLTK is used to tokenize each essay, all the words are lowercased, and the score is normalized within the range of [0, 1]. During the model evaluation phase, the score is converted back to an integer within the original score range to facilitate the calculation of the QWK value. We select 4000 words with the highest frequency from the training data as the vocabulary and treat all other words as unknown words. The hyper-parameters of our model are showed in Table 2.

We use RMSprop [12] as our optimizer and the initial learning rate is set to 0.001. The model is trained for 50 epochs in each prompt and evaluation is performed after each training epoch. We retain the model that produces the best performance on the development set as the final model to be used on testing data.

Table 2. Parameters of our model.

Layer	Parameters	Value
Look-up	Word embedding dim	50
Self-attention	Number of heads	8
	Size per head	16
	Number of layers	1
LSTM	Hidden units	100
Dropout	Dropout rate	0.5
Others	Optimization	RMSprop
	Epoch	50
	Batch size	10
	Initial learning rate	0.001

3.4 Experimental Results and Discussion

In this section, we will introduce the baselines and present the results of our model on the two datasets.

Baselines. We use three state-of-the-art models as our baselines in the experiments:

- **LSTM-MoT** model (Taghipour and Ng [3]): The model inputs all words of the essay into the LSTM model and uses the average of all hidden layer outputs of the LSTM model as the final representation of the essay.
- **LSTM-CNN-attention** model (Dong and Zhang [4]): The LSTM-CNN-attention model [4] uses the soft-attention mechanism [13, 14] to learn the n-grams weights of the sentences and then inputs these sentence representations to a LSTM layer. The model uses soft-attention to all the hidden outputs of the LSTM layer to obtain the final representation of the essay.
- **SKIPFLOW-LSTM** model (Tay et al. [5]): The SKIPFLOW-LSTM model adopts a tensor layer to model the relationship between each pair of hidden unit output of the LSTM, which aims to capture the textual coherence.

Our model described in Sect. 2 is named **Self-attention-LSTM**. In addition, we also tested the **Self-attention-MoT** model. Instead of stacking an LSTM operation on the relationship representations, the model simply performs an average pooling and uses the average pooled result as the final representation of the essay.

Results on ASAP Dataset. The experimental results are showed in Table 3. As we can see, our model has achieved better performance than other state-of-the-art methods, except for prompt 4. The average QWK of our Self-attention-LSTM model on ASAP data is 3.0% higher than LSTM-MoT and 1.2% higher than LSTM-CNN-attention and SKIPFLOW-LSTM. These differences are statistically significant. On prompt 4, the performance of our model is close to that of LSTM-CNN-attention.

Table 3. QWK results on Kaggle data set. * means statistical significance.

Models	Prompts								
	1	2	3	4	5	6	7	8	Avg.
LSTM-MoT	0.775	0.687	0.683	0.795	0.818	0.813	0.805	0.594	0.746
LSTM-CNN-attention	0.822	0.682	0.672	**0.814**	0.803	0.811	0.801	0.705	0.764
SKIPFLOW-LSTM	0.832	0.684	0.695	0.788	0.815	0.810	0.800	0.697	0.764
Self-attention-MoT	0.830	0.683	0.667	0.796	0.810	0.812	0.801	0.703	**0.763**
Self-attention-LSTM	**0.834**	**0.692**	**0.700**	0.811	**0.819**	**0.822**	**0.816**	**0.713**	**0.776***

We also do the experiments on ASAP dataset using only self-attention without LSTM layer (Self-attention-MoT), in which the relationship-based representation are directly fed into an average pooling layer to get the final representation of the essay. In this case the average QWK on ASAP dataset 0.763 is slightly lower than the best baseline methods, which indicates that self-attention can learn useful relationship-based representation of the essay to some extend, despite the simplicity of the model. This difference with Self-attention-LSTM, although not very large, reflects the gain brought by LSTM to aggregate the sequential information about the relationships between words.

Results on CEEE Dataset. The CEEE dataset is different from ASAP in which the essays are written by English learners. We expect that sentences are less fluent in CEEE than in ASAP. The results are showed in Table 4. We can see that Self-attention-LSTM method still achieves the highest QWK value 0.731, which is 2.2% higher than LSTM-MoT and 0.6% higher than LSTM-CNN-attention. The differences are again statistically significant with $p < 0.05$. This result confirms that our model can be used for scoring different types of essays, whether by native English speakers or by English learners.

Discussion. The experiments demonstrate that the model we propose can produce superior performance than the existing state-of-the-art models. The key difference between our model and previous models lies in the use of relationships between words across the essay so that scoring relies on coherence. Compared to the previous models that capture only local relationships between words (mainly within sentences),

Table 4. QWK results on CEEE data set. SKIPFLOW-LSTM (tensor) has not been tested on this dataset, so the result is missing. * means statistical significance.

Models	CEEE
LSTM-MoT	0.709
LSTM-CNN-attention	0.725
SKIPFLOW-LSTM (tensor)	–
Self-attention-MoT	0.729
Self-attention-LSTM	**0.731***

the self-attention mechanism is capable of detecting relationships between words at any positions in the essay. Such global relationships can better reflect an essay's coherence. Our experimental results have confirmed that the coherence criterion is an important factor in essay scoring.

4   Related Work

Many of traditional AES models are devoted to the design of features. These works including supervised methods based on machine learning algorithms [15–20] and unsupervised methods based on ranking methods [21, 22]. All these previous methods are based on handcrafted features, which require significant amount of manual works.

In recent year, neural networks have been used in AES task [1–5]. Compared to the traditional approaches, neural approaches do not need any handcrafted features. Alikaniotis et al. [1] train a score-specific word embeddings (SSWEs) to represent words. They use a two-layer bidirectional LSTM and take the last hidden state as the final representation of essays. Taghipour and Ng [3] input the essay's words into a layer of LSTM, taking the average of all hidden states of LSTM as the essay's representation. Dong and Zhang [2] proposed a hierarchical CNN model for sentence-level and text-level representation by processing text into sentences. In their subsequent work [4], they used the soft-attention mechanism to learn the n-grams weights in sentences and then use these sentence representations as input to a LSTM layer to obtain the final representation of the quality of essays. Zhao et al. [23] use memory network model for AES task. The model predicts a score for an ungraded essay by computing the relevance between the ungraded essay and each selected essay as grading criteria specified in memory.

These previous works use the word embedding as input and can be seemed as content-based representations for the next LSTM layer. But relationships of words and sentences from different parts of the essay are important for evaluating the quality of an essay.

Tay et al. [5] propose a SKIPFLOW-LSTM model for AES task, they adopt a tensor layer to model the relationship between each two LSTM hidden unit outputs as neural coherence features to represent and approximate textual coherence. Although SKIPFLOW-LSTM model can capture the relationship between snapshots of essay, these snapshots are continued words of sequence with a fixed size. Their model still can't capture the relationships between different parts of the essay, for example, the first sentence and the last sentence.

Inspired by works using self-attention on different tasks [8, 24–29], we propose to use self-attention to learn the relationships between words from different parts of the essay. And we use a LSTM layer to capture the overall coherence of the essay based on these relationship-level representations.

5 Conclusion

In this paper, we propose an automated essay scoring model based on relationship-representations via self-attention and LSTM framework. Firstly, we add word embeddings of the essay and position encoding matrices as first input to self-attention layer. And then, we use self-attention mechanism to learn the relationship between words from different parts of the essay. Before self-attention, we use content-based representation as input and obtain the relationship-based representation of the essay as output. In the end, we use a LSTM layer to capture the overall coherence of the essay through averaging all the hidden unit output of LSTM as the final representation of the quality of essays. The results on Kaggle data set and CEEE data set show that our model outperforms the current state-of-the-art model.

Acknowledgement. This work is supported by the National Science Foundation of China (61402119) and Special Funds for the Cultivation of Guangdong College Students' Scientific and Technological Innovation. ("Climbing Program" Special Funds.)

References

1. Alikaniotis, D., Yannakoudakis, H., Rei, M.: Automatic text scoring using neural networks. arXiv preprint arXiv:1606.04289 (2016)
2. Dong, F., Zhang, Y.: Automatic features for essay scoring. In: Proceedings of the Conference on Empirical Methods in Natural Language Processing (EMNLP), pp. 968–974 (2016)
3. Taghipour, K., Ng, H.T.: A neural approach to automated essay scoring. In: Proceedings of the 2016 Conference on Empirical Methods in Natural Language Processing (EMNLP), pp. 1882–1891 (2016)
4. Dong, F., Zhang, Y., Yang, J.: Attention-based recurrent convolutional neural network for automatic essay scoring. In: Proceedings of the 21st Conference on Computational Natural Language Learning (CoNLL), pp. 153–162 (2017)
5. Tay, Y., Phan, M., Tuan, L., Hui, S.: SkipFlow: incorporating neural coherence features for end-to-end automatic text scoring. arXiv preprint arXiv:1711.04981 (2017)
6. Halliday, M.A.K., Hasan, R.: Cohesion in English. Longman, London (1976)
7. McNamara, D.S., Kintsch, W.: Learning from texts: effects of prior knowledge and text coherence. Discourse Process. 22(3), 247–288 (1996)
8. Vaswani, A., et al.: Attention is all you need. In: Neural Information Processing Systems (NIPS), pp. 6000–6100 (2017)
9. Pennington, J., Socher, R., Manning, C.: Glove: global vectors for word representation. In: Proceedings of the 2014 Conference on Empirical Methods in Natural Language Processing (EMNLP), pp. 1532–1543 (2014)

10. Hochreiter, S., Schmidhuber, J.: Long short-term memory. Neural Comput. **9**(8), 1735–1780 (1997)
11. Pascanu, R., Mikolov, T., Bengio, Y.: On the difficulty of training recurrent neural networks. In: Proceedings of International Conference on International Conference on Machine Learning (ICML), pp. 1310–1318 (2013)
12. Dauphin, Y.N., Vries, H.D, Bengio, Y.: Equilibrated adaptive learning rates for non-convex optimization. In: Proceedings of International Conference on Neural Information Processing Systems (NIPS), pp. 1504–1512 (2015)
13. Xu, K., et al.: Show, attend and tell: neural image caption generation with visual attention. In: Proceedings of the 32nd International Conference on Machine Learning (ICML), pp. 77–81 (2015)
14. Li, J., Luong, M.T., Jurafsky, D.: A hierarchical neural autoencoder for paragraphs and documents. arXiv preprint arXiv:1506.01057 (2015)
15. Page, E.B.: Computer grading of student prose, using modern concepts and software. J. Exp. Educ. **62**(2), 127–142 (1994)
16. Landauer, T.K., Foltz, P.W., Laham, D.: An introduction to latent semantic analysis. Discourse Process. **25**(2–3), 259–284 (1998)
17. Foltz, P.W., Laham D., Landauer T.K.: Automated essay scoring: applications to educational technology. In: Proceedings of EdMedia, pp. 40–64 (1999)
18. Larkey, L.S.: Automatic essay grading using text categorization techniques. In: Proceedings of the 21st Annual International ACM SIGIR Conference on Research and Development in Information Retrieval, pp. 90–95 (1998)
19. Rudner, L.M.: Automated essay scoring using Bayes' theorem. Nat. Counc. Measur. Educ. Orleans La **1**(2), 3–21 (2002)
20. Attali, Y., Burstein, J.: Automated essay scoring with e-rater R V. 2.0. ETS Research Report Series, pp. 1–21 (2004)
21. Phandi, P., Chai, K.M.A., Ng, H.T.: Flexible domain adaptation for automated essay scoring using correlated linear regression. In: Proceedings of the 2015 Conference on Empirical Methods in Natural Language Processing (EMNLP), pp. 431–439 (2015)
22. Yannakoudakis, H., Medlock, B., Medloc, B.: A new dataset and method for automatically grading ESOL texts In: Proceedings of the 49th Meeting of the Association for Computational Linguistics (ACL), pp. 180–189 (2011)
23. Zhao, S., Zhang, Y., Xiong, X., Botelho, A., Heffernan, N.: A memory-augmented neural model for automated grading. In: Proceedings of the Fourth ACM Conference on Learning at Scale (L@S), pp. 189–192 (2017)
24. Cheng, J., Dong, L., Lapata, M.: Long short-term memory-networks for machine reading. arXiv preprint arXiv:1601.06733 (2016)
25. Parikh, A.P., Täckström, O., Das, D., Uszkoreit, J.: A decomposable attention model for natural language inference. In: Proceedings of the 2016 Conference on Empirical Methods in Natural Language Processing (EMNLP), pp. 2249–2255 (2016)
26. Lin, Z., et al.: A structured self-attentive sentence embedding. arXiv preprint arXiv:1703.03130 (2017)
27. Shen, T., Zhou, T., Long, G., Jiang, J., Pan, S., Zhang, C.: DiSAN: directional self-attention network for RNN/CNN-free language understanding. arXiv preprint arXiv:1709.04696 (2017)
28. Tan, Z., Wang, M., Xie, J., Chen, Y., Shi, X.: Deep semantic role labeling with self-attention. arXiv preprint arXiv:1712.01586 (2017)
29. Paulus, R., Xiong, C., Socher, R.: A deep reinforced model for abstractive summarization. arXiv preprint arXiv:1705.04304 (2017)

Trigger Words Detection by Integrating Attention Mechanism into Bi-LSTM Neural Network—A Case Study in PubMED-Wide Trigger Words Detection for Pancreatic Cancer

Kaiyin Zhou[1,2], Xinzhi Yao[1], Shuguang Wang[1], Jin-Dong Kim[3],
Kevin Bretonnel Cohen[4], Ruiying Chen[1], Yuxing Wang[1,2],
and Jingbo Xia[1,2(✉)]

[1] College of Informatics, Huazhong Agricultural University, Wuhan, China
xiajingbo.math@gmail.com, xjb@mail.hzau.edu.cn
[2] Hubei Key Laboratory of Agricultural Bioinformatics, Wuhan, China
[3] Database Center for Life Science (DBCLS),
Research Organization of Information and Systems (ROIS), Tokyo, Japan
[4] School of Medicine, University of Colorado Denver,
Anschutz Medical Campus, Aurora, USA

Abstract. A Bi-LSTM based encode/decode mechanism for named entity recognition was studied in this research. In the proposed mechanism, Bi-LSTM was used for encoding, an Attention method was used in the intermediate layers, and an unidirectional LSTM was used as decoder layer. By using element wise product to modify the conventional decoder layers, the proposed model achieved better F-score, compared with other three baseline LSTM-based models. For the purpose of algorithm application, a case study of causal gene discovery in terms of disease pathway enrichment was designed. In addition, the causal gene discovery rate of our proposed method was compared with another baseline methods. The result showed that trigger genes detection effectively increase the performance of a text mining system for causal gene discovery.

Keywords: Natural language processing · LSTM
Encoder/decoder model · Trigger words

1 Introduction

Named Entity Recognition (NER) is to detect mentions that we concerned from text [1], and NER is generally the first stage for complex natural language processing tasks (NLP). In tradition, most sequence labeling models were linear statistical models. However, these models usually heavily depended on specific feature engineering and high-quality labeled data. In recent years, the development of deep learning has broke this limitation dramatically. Performance of

© Springer Nature Switzerland AG 2018
M. Sun et al. (Eds.): CCL 2018/NLP-NABD 2018, LNAI 11221, pp. 398–409, 2018.
https://doi.org/10.1007/978-3-030-01716-3_33

natural language processing tasks, such as machine translation, semantic relation extraction, automatic summarization, and so on, have successfully outperformed conventional machine learning methods, and NER tasks are no exception.

Recently, attention mechanism was successfully applied to the machine translation model and achieved state-of-art result in many public data sets [2]. This made us recognize the importance of the attention mechanism. In the machine translation model, query vector was used to perform alignment with each encoding results, which was considered as an effective simulation of human reading and translating process. This method hinted that the attention mechanism would also be suitable for the sequence labeling model.

In a cross-disciplinary field of Biomedical Natural Language Processing (BioNLP), the mainstream text data comes from a huge publication repository PubMED (https://www.ncbi.nlm.nih.gov/pubmed/), run by U.S national library of medicine and national institutes of health. The dataset has been increasing dramatically, and the amount reached 28 million in 2018. PubMED was treated as a useful resource for Bioinformatics research to curate bio-related knowledge [3,4]. For example, 93,096 PubMED abstracts entries were found for pancreatic cancer at 6th, June, 2018, and the amount was still increasing. Since straightforward curation of mass data was hard for knowledge discovery, automatic NER of trigger words made it possible to increase the relevant entries filtering and knowledge inference [5–7].

In this paper, we modified the classic encoder/decoder model [8] for NER of trigger words, incorporated with Bengio's attention mechanisms [9]. In the encoding layer we used the usual Bidirectional Long Short Term Memory (Bi-LSTM) structure to capture the context information. In the decoding layer we used an unidirectional LSTM. For the LSTM unit, the output of encoding layer and the result of the attention mechanism were combined in each time step appropriately. Unlike the usual case that these two elements were connected in a straightforward manner, this is an effective modification taken in our proposed scheme. Actually, an illuminative trick-playing were carried on by replacing concatenation to element-wise product. Henceforth, a Bi-LSTM-Attention-ElmentwiseProduct (Bi-LSTM-AEP) algorithm was achieved. By applying this new model to a manually labeled dataset, the highest F-score was obtained by the proposed algorithm after comparing to other three baseline popular sequence labeling methods—two without attention [10,11], and one with attention [12]. As an application, a bioinformatics case study was carried on by using the proposed trigger word NER algorithm to increase the discovery of causal genes that affect pancreatic cancer.

2   Material and Method

2.1   Dataset

An under-developed corpus, Active Gene Annotation Corpus (AGAC) [13], was chosen as the training set in this research. AGAC corpus contained structured texts with semi-manual annotations, which included five trigger labels containing the biological concepts from molecular level to cell level, three regulatory labels representing the mutant directions and two kinds of semantic relations between trigger words.

To better understand the trigger word setting in this corpus, a snapshot of annotation example in PubAnnotation platform [14] for trigger words is shown at Fig. 1.

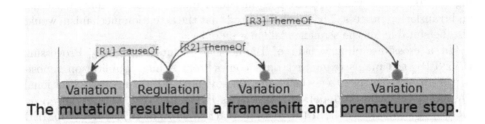

Fig. 1. An example for AGAC

In this work, we focused on 2 trigger labels: *Variation* and *Regulation*. Both of them are important for representing the function of genes, and these trigger labels were treated as the essential elements to present the causal gene information.

As a typical example shown in Fig. 1, a sentence "The mutation resulted in a frame shift and premature stop" was annotated. In this annotation manner, *Variation* label included the origin causality of all the other biological concept labels, and *Regulation* label was the regulatory label which was the center of a relation. In this sentence, "mutation", "result in", "frameshift", and "premature stop", were all treated as trigger words. Therefore, if a gene co-occurred with the above trigger words, this gene actually would play a so-called Loss-of-function role [15] in the molecular level, and that made this gene a causal gene in the context.

Since the purpose of this research is to compare the Bi-LSTM-AEP with other popular encode/decode-based mechanisms, AGAC corpus is an acceptable data resource for training and testing. In addition, it was assumed that trigger words in AGAC corpus represented the functioning effectiveness of curated genes, and this assumption made it possible to discriminate causal genes in a mass text data, i.e., PubMED.

2.2   Word Embeddings Library

It is widely accepted that the quality of the pre-trained word embeddings is a key factor for various NLP tasks, including NER [16]. Due to the differences in writing style and huge variation of terminology in biology, it is still a challenge to make a domain-free word embedding library to suit the biology-related application [17]. Therefore, a domain-specific word embeddings, BioASQ [18], was selected as our pre-trained embedding library. BioASQ was trained from a corpus of 10,876,004 English abstracts of biomedical articles from PubMed and contained 1,701,632 distinct words, thus that made it a proper one for embedding preprocessing in our method.

2.3 General Encode/Decode Framework Based on Recurrent Neural Network, and Two LSTM-Based Baseline Methods Without Attention Mechanism

As a general encode/decode framework for tackling NER or machine translation task [8], the input is a sequence of vectors $x = (x_1, \cdots, x_{T_x})$, and output is a sequence of label vectors or word vector $y = y_1, \cdots, y_{T_y}$. For tacking NER, in a general Encoder-Decoder framework, there is an encoder which reads the input sentence into a context vector c, and a decoder which predicts the next word y_t' given c and all the previously predicted words $\{y_1, \cdots, y_{t'-1}\}$.

In recent years, the most common approach is to use a recurrent neural network (RNN) and make decoder represent a probability over y by decomposing the joint probability into the ordered conditionals,

$$P(\{y_1, \cdots, y_{T_y}\}) = \prod_{t=1}^{T_y} P(\ y_t | \{y_1, \cdots, y_{t-1}\}, c\ ),\tag{1}$$

and models each conditional probability as

$$P(y_t | \{y_1, \cdots, y_{t-1}\}, c) = g(y_{t-1}, s_t, c),\tag{2}$$

where the context vector c is computed by using information of hidden layer of RNN, i.e., $h_t = RNN(x_t, h_{t-1})$.

The reason for using RNN in encode/decode frame work stems from the powerful modeling structure of recurrent network, generally speaking, is that RNN is regarded as being capable of capturing time dynamics via cycles in the graphs. Though RNN usually leads to gradient vanishing/exploding problems, some RNN variants perform better in practice, e.g. LSTM neural network.

LSTM is a set of special Recurrent neural networks, which could especially capture long-distance dependencies with the appropriate employment of gating functions at each time step i.e. input gate, forget gate, and output gate. These tricks are good solutions to the gradient vanishing for conventional RNNs. Formally, the formulas to update an LSTM unit at time t are:

$$\begin{cases} i_i = \sigma(W_{Ei}E_{xi} + U_{hi}h_{i-1} + b_i) \\ f_i = \sigma(W_{Ef}E_{xi} + U_{hf}h_{i-1} + b_f) \\ z_i = \tanh(W_{Ez}E_{xi} + U_{hz}h_{i-1} + b_z) \\ c_i = f_i * c_{i-1} + i_i * z_i \\ o_i = \sigma(W_{Eo}E_{xi} + U_{ho}h_{i-1} + b_o) \\ h_i = o_i * \tanh(c_i) \end{cases},\tag{3}$$

where σ is the element-wise sigmoid function and $*$ is the element-wise product. Here E_{xi} is the input vector at time i, usually represented by word embedding.

h_{i-1} is the hidden state vector of last time step $i - 1$. $W_{Ei}, W_{Ef}, W_{Ez}, W_{Eo}$ are the weight matrices of different gates for input E_{xi}, and $U_{hi}, U_{hf}, U_{hz}, U_{ho}$ are the weight matrices for hidden state h_t.

2.3.1 Baseline Method 1: Bi-LSTM-CRF Model

Bi-LSTM-CRF [10] is a typical neural network model used in sequence labeling tasks. It carried on LSTM training with the data for two times, and the only difference for each time was that the order of the two times input data was completely reversed, then the results of each LSTM layer were concatenated as an output of words encoding results. Thus, Bi-LSTM model captured both the past and the future information respectively. Then, the output vectors of Bi-LSTM were fed to the CRF layer to jointly decode the best label sequence. This model has been proved to be reasonable and has achieved state-of-art scores on many sequence labeling tasks.

2.3.2 Baseline Method 2: Bi-LSTM-ED Model

Bi-LSTM-ED [11] is another model used to carry on sequence labeling tasks. This model still used Bi-LSTM as encoding layers. Being different from other models, the decode layer in this model was a Variant LSTM. The units of the decoding LSTM were the same as the encoding LSTM except for the input gate, which was replaced by

$$\begin{cases} i_i = \sigma(W_{ii}E_{xi} + U_{ii}h_{i-1} + V_{ii}T_{i-1} + b_{ii}) \\ T_i = W_{is}h_i + b_{is} \end{cases}. \tag{4}$$

This model obtained good improvement in several sequence labeling tasks.

2.4 Proposed LSTM and Attention Neural Network for Trigger Words Recognition

In this section, Liu's model [12], a similar structure of Bi-LSTM-Attention [9], is listed as the 3rd baseline method. Subsequently, we propose an updated encode/decode scheme for NER by using Bi-LSTM and attention mechanism, while element product is used in decoding layer, and the abbreviation of this structure is Bi-LSTM-AEP.

In the conventional encode/decode attention-based models, encode layer is a bidirectional recurrent neural network (Bi-LSTM), which reads the words one by one in a sentence and then outputs metrics containing the forward and backward hidden states of each words. Attention mechanism acts as a weight calculator to change the importance of different inputs by taking both inputs and outputs into consideration. Decode layer is a unidirectional LSTM. There are many variants about this model. We notice that all of the variants only concerned how to optimize attention mechanism and design a appropriate score function, for instance, Luong et al. [19] presented a contend based and location based score functions, and Jean et al. [20] added the target word embedding as input for the score function.

2.4.1   Baseline Method 3: Bi-LSTM-Attention Model

In past few years, Bi-LSTM-Attention model [9] has been widely used in speech recognition, image caption generation, visual question answering, machine translation and other fields, while few people applied it to sequence labeling tasks. In 2016, Liu [12] converted Bi-LSTM-Attention model into a NER-purposed one, where s_i in encoder layer was computed by a GRU unit,

$$s_i = GRU(h_i, s_{i-1}, c_i), \tag{5}$$

where the detailed formulas are below:

$$\begin{cases} s_i = GRU(h_i, s_{i-1}, c_i) = (1 - z_i) \circ s_{i-1} + z_i \circ \tilde{s}_i \\ \tilde{s}_i = \tanh(W h_{i-1} + U[r_i \circ s_{i-1}] + C c_i) \\ z_i = \sigma(W_z h_i + U_z s_{i-1} + C_z c_i) \\ r_i = \sigma(W_r h_i + U_r s_{i-1} + C_r c_i) \end{cases} \tag{6}$$

2.4.2   Proposed Mechanism: Bi-LSTM-AEP Neural Network

In our work, we also applied Bahdanau et al's. model [9] in NER and sequence labeling tasks. Here attention mechanism was used to adjust the weight of input information. As introduced in the above section, the difference of Bi-LSTM-Attention model and our proposed model mainly exist at the Decoder-layer, see Eqs. (6) and (9), where GRU and element-wise LSTM were used separately.

In addition, we focus on how to combine the output of attention mechanism with the input of decoding layer. Here we propose an element-wise multiplication rather than conventional weight sum. The structure of the neural network is shown in Fig. 2, while the complete description of the framework is shown as below, which is a modification of Bi-LSTM-Attention model [9].

Algorithm of Bi-LSTM-AEP:

- Encode layer:
  Since we were desired to take both the last word and the next word into consideration, Bi-LSTM was employed as the encode layer, which contains a forward LSTM and a backward LSTM. At first, forward LSTM and backward LSTM read the words in a sentence x=$(x_1, x_2, \ldots, x_{Tx})$ respectively, and calculates the hidden states of each word: $(\overrightarrow{h_1}, \overrightarrow{h_2}, \ldots, \overrightarrow{h_{Tx}})$ and $(\overleftarrow{h_1}, \overleftarrow{h_2}, \ldots, \overleftarrow{h_{Tx}})$ [9].
  Then the forward hidden states and the backward hidden states were combined in the third dimension as the annotation for each word: $h_j = [\overrightarrow{h_j}, \overleftarrow{h_j}]$. Therefore, annotation h_j represents not only the x_j itself but also the context information around it.
- Attention-Mechanism:
  After encoding, the input words were transformed to annotation h. If attention mechanism was not taken into consideration, the annotation h for each input words (from h_1 to h_j) would be combined directly as a context vector which contained all the information in the sentence equally, and then the context vector was one of the inputs to decode layer. However, the importance of

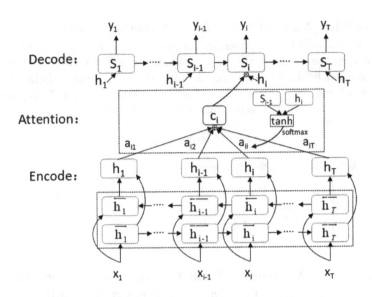

Fig. 2. Architecture of Bi-LSTM-AEP neural network

each annotation h should be different, so we introduced attention mechanism to calculate the different weight for each input by scoring to the alignment of input at position j and output at i [9].

At this part, the inputs is annotation h for each words in the sentence and the hidden state at last position of output sequence s_{i-1}, and the output is the context vector after considering the weight for each annotation h. The formulas of attention mechanism are shown below:

$$\begin{cases} c_i = \sum_{j=1}^{T_x} a_{ij} h_j \\ a_{ij} = \dfrac{exp(e_{ij})}{\sum_{k=1}^{T_x} exp(e_{ik})} \\ e_{ij} = V_a^T tanh(W_a s_{i-1} + U_a h_j) \end{cases} \qquad (7)$$

where a_{ij} is the weight for different annotation h, c_i is the context vector, and e_{ij} is the alignment scores.

- Decode-layer:
  The decode layer is a unidirectional LSTM. In previous models, the output of encode results $h = (h_1, h_2 \ldots hi, \ldots h_{Tx})$ and the context vector c_i from

attention mechanism were added at each time step $s_i = GRU(h_i, s_{i-1}, c_i)$ [12]. However, we proposed our decode-layer

$$s_i = LSTM(h_i * c_i, s_{i-1}), \qquad (8)$$

where $*$ is the element-wise product. The complete specific formulas are:

$$\begin{cases} i_i = \sigma(W_{ei}[h_i * c_i] + U_{si}s_{i-1}) \\ f_i = \sigma(W_{ef}[h_i * c_i] + U_{sf}s_{i-1}) \\ z_i = tanh(W_{ez}[h_i * c_i] + U_{sz}s_{i-1}) \\ \widetilde{z}_i = f_i * \widetilde{z}_{i-1} + i_i * z_i \\ o_i = \sigma(W_{eo}[h_i * c_i] + U_{so}s_{i-1}) \\ s_i = LSTM(h_i * c_i, s_{i-1}) = o_i tanh(\widetilde{z}_i) \end{cases} \qquad (9)$$

For simplicity, the bias terms were omitted in the above formulas.

3 Experimental Settings

The experiments are designed for two purposes: (1) to preliminary evaluate the significance of the database, which we designed for large-scale biological literature mining; (2) to test the performance of our newly designed model.

In this work, 28 labeled texts were selected from the AGAC data set. BioASQ was employed as pre-trained embeddings, and the dimension of words vector is 200. The hidden unites of Encode/Decode layer are set as 100. The model is trained with the RMSProp algorithm.

Precision, recall and F1-measure are taken as the evaluation criteria. The finally evaluation scores are computed by averaging all tags scores. Weights in all formulas are randomly initialized as uniform distribution with support $[-0.01, 0.01]$.

3.1 Models

In order to better evaluate the performance of the proposed models, we compare it with four different kinds of model. Bi-LSTM-CRF is a usual used model in sequence labeling tasks, while Bi-LSTM-ED performed better in some tasks. Besides, Bi-LSTM-Attention was also provided to evaluate the importance of attention mechanism in such tasks.

3.2 Case Study Design for Trigger Words Detection in Terms of Pancreatic Cancer Pathway Enrichment

In order to evaluate the application of our trigger word detection in mass text. A case study was performed.

As a target disease, Pancreatic cancer is a disease that attracts much attention in academia. Through keywords searching in the PubMed-wide scale, ninety hundreds abstracts were downloaded and all of the bio-entities, including gene mentions, and were retrieved by using Pubtator [21]. By counting the occurrence of the gene mentions, the rank of active gene related to pancreatic cancer was obtained. According to a practice of co-occurrence manner, we obtained a knowledge entry rank list for causal gene discrimination in terms of pancreatic cancer. In another words, the higher a gene ranked in the list, the higher chance the gene had to be relevant with the pancreatic cancer.

For the sake of performance evaluation for the application of trigger word detection in Re-ranking. Gene mentions were filtered by using Bi-LSTM-AEP Model. Here, a medium-sized training model of AGAC corpus were used to handle all of the Pancreatic-related PubMED abstracts. Subsequently, trigger words were detected for each abstract. In the re-ranked gene list, only genes co-occurred with these trigger words were treated causal and kept in the list. Thus a new gene list was obtained.

After using trigger words detection, the updated gene rank list were compared with the previous gene rank according its enrichment in pathway. Within domain knowledge, pathway is a directional map connected genes, proteins and metabolites, and the pathway enrichment analysis was generally an effective method to evaluate the accuracy of gene relevance. The design of the case study was shown in Fig. 3.

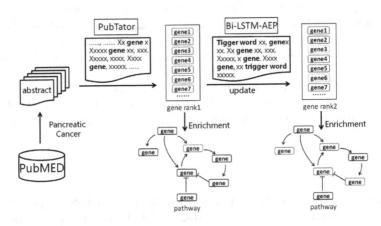

Fig. 3. A Case study in trigger words detection in terms of pancreatic cancer.

4   Result

4.1   Performance of Trigger Words Extraction Algorithm

The experimental results were based on a five-fold cross-validation and three replicates were performed. Finally, the results of the three experiments were averaged as the evaluation criteria.

Table 1 showed the results of the four different models: Bi-LSTM-Attention, Bi-LSTM-AEP, Bi-LSTM-CRF, Bi-LSTM-DE. In these four models, the first two were ones without attention mechanism, and the latter two were attention-based. A dramatic improvement in F value was achieved when using the attention mechanism.

In model with attention mechanism proposed by Liu [12], the F value is 0.4382. When using the model Bi-LSTM-AEP proposed by us, the F value further improved to 0.4951. As a result, the result showed that our proposed Bi-LSTM-AEP model outperformed other lSTM-based encoder/decoder models, and it was most suitable for our tasks.

Table 1. Experimental results on four different models.

Method	Precision	Recall rate	F1-measure
Bi-LSTM-AEP (ours)	0.7576	0.4171	0.5160
Bi-LSTM-Attention [12]	0.7092	0.3286	0.4368
Bi-LSTM-ED [11]	0.6604	0.3249	0.4263
Bi-LSTM-CRF [10]	0.5849	0.3051	0.3947

4.2   Case Study in PubMED-Wide Trigger Words Detection in Pancreatic Cancer

Disease-related pathways are always the focus of attention in disease and drug research. Here, Kyoto Encyclopedia of Genes and Genomes (KEGG) is an authoritative and commonly used database in biological research. KEGG pathway database contains a large number of manually curated pathway maps focusing on intermolecular interaction networks. A biological pathway is a series of molecular actions in a cell, which produces metabolites or generate changes in the cell. For example, Inactivation of the SMAD4 tumour suppressor gene leads to a loss of the inhibitory influence of the transforming growth factor-beta (TGF-Beta) signaling pathway and henceforth promotes the occurrence of cancer.

To compare the accuracy of the above mentioned two methods, Pubtator and Bi-LSTM-AEP. The result of the comparison is shown in Table 2.

Table 2. The result of the comparison

	Extracted terms	Extracted pathway genes	Accuracy
Method 1 (Pubtator)	28336	54	0.19%
Method 2 (Bi-LSTM-AEP)	11675	52	0.45%

Due to the rigorousness of KEGG database, the number of pathway genes in pancreatic cancer is very small compared to the results of text mining. Even

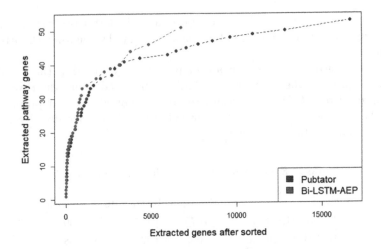

Fig. 4. Distribution of KEGG pathway genes in the results of two methods (Color figure online)

so, Bi-LSTM-AEP extracted 52 key genes out of 11,675 extracted ones, while the ratio for Pubtator was 54/28,336. The result showed that Bi-LSTM-AEP narrow down the gene searching scale from 28,336 to 11,675, which is a 58.80% reduce in amount, and the target genes remained in the shorter list.

It's noted that the accuracy of both methods were lower than 1%, i.e., 0.19% and 0.45%, respectively. However, the discovery of causal gene was actually an unsolved challenge. Fortunately, using a better text mining tool was capable of achieve a higher knowledge discovery rate.

As shown in Fig. 4, in order to explore the distribution of the KEGG pathway genes in two gene lists, we visualized the causal gene discovery rate by representing the appearing order of relevant genes. In detail, extracted genes were sorted by frequency, and the amount of accumulated target genes was marked with round dot at the corresponding sorted positions of the KEGG genes. In the Fig. 4, it clearly showed that the red line is at the top left of the blue one, which was sufficed to discover more target genes within less amount.

Acknowledgement. This work is funded by the Fundamental Research Funds for the Central Universities of China (Project No. 2662018PY096).

References

1. Nadeau, D., Sekine, S.: A survey of named entity recognition and classification. Lingvist. Investig. **30**(1), 326 (2007)
2. Vaswani, A., Shazeer, N., Parmar, N., et al.: Attention is all you need (2017)
3. Sintchenko, V., Anthony, S., Phan, X.H., Lin, F., Coiera, E.W.: A PubMed-wide associational study of infectious diseases. PLoS One **5**(3), e9535 (2010)

4. Allot, A., Peng, Y., Wei, C.H., Lee, K., Phan, L., Lu, Z.: LitVar: a semantic search engine for linking genomic variant data in PubMed and PMC. Nucl. Acids Res. **46**(W1), W530–W536 (2018)
5. Cohen, K.B., et al.: High-precision biological event extraction with a concept recognizer. In: Proceedings of the Workshop on Current Trends in Biomedical Natural Language Processing: Shared Task, 5 June 2009, pp. 50–58. Association for Computational Linguistics (2009)
6. Song, M., Kim, M., Kang, K., Kim, Y.H., Jeon, S.: Application of public knowledge discovery tool (PKDE4J) to represent biomedical scientific knowledge. Front. Res. Metr. Anal. **3**, 7 (2018)
7. Zhou, H., Yang, Y., Ning, S., Liu, Z., Lang, C., Lin, Y., Huang, D.: Combining context and knowledge representations for chemical-disease relation extraction. IEEE/ACM Trans. Comput. Biol. Bioinform. (2018). https://doi.org/10.1109/TCBB.2018.2838661
8. Cho, K., et al.: Learning phrase representations using RNN encoder-decoder for statistical machine translation. arXiv preprint arXiv:1406.1078 (2014)
9. Bahdanau, D., Cho, K., Bengio, Y.: Neural machine translation by jointly learning to align and translate. arXiv preprint arXiv:1409.0473 (2014)
10. Huang, Z., Xu, W., Yu, K.: Bidirectional LSTM-CRF models for sequence tagging. arXiv preprint arXiv:1508.01991 (2015)
11. Zheng, S., Hao, Y., Lu, D., et al.: Joint entity and relation extraction based on a hybrid neural network. Neurocomputing **257**, 1–8 (2017)
12. Liu, B., Lane, I.: Attention-based recurrent neural network models for joint intent detection and slot filling. arXiv preprint arXiv:1609.01454, 6 September 2016
13. Wang, Y., et al.: Guideline design of an active gene annotation corpus for the purpose of drug repurposing. In: OHDSI 2018 Workshop, July, Guangzhou (2018, submitted)
14. Kim, J.D., Wang, Y.: PubAnnotation: a persistent and sharable corpus and annotation repository. In: Proceedings of the 2012 Workshop on Biomedical Natural Language Processing, pp. 202–205. Association for Computational Linguistics (2012)
15. Wang, Z.Y., Zhang, H.Y.: Rational drug repositioning by medical genetics. Nat. Biotechnol. **31**(12), 1080–1082 (2013)
16. Ma, X., Hovy, E.: End-to-end sequence labeling via bi-directional LSTM-CNNs-CRF (2016)
17. Huang, E.H., Socher, R., Manning, C.D., Ng, A.Y.: Improving word representations via global context and multiple word prototypes. In: Proceedings of the 50th Annual Meeting of the Association for Computational Linguistics: Long Papers, vol. 1, pp. 873–882. Association for Computational Linguistics (2012)
18. Pavlopoulos, I., Kosmopoulos, A., Androutsopoulos, I.: Continuous space word vectors obtained by applying Word2Vec to abstracts of biomedical articles (2014)
19. Luong, M.-T., Pham, H., Manning, C.D.: Effective approaches to attention-based neural machine translation. arXiv preprint arXiv:1508.04025 (2015)
20. Jean, S., Cho, K., Memisevic, R., Bengio, Y.: On using very large target vocabulary for neural machine translation. arXiv preprint arXiv:1412.2007 (2014)
21. Wei, C.H., Kao, H.Y., Lu, Z.: PubTator: a web-based text mining tool for assisting biocuration. Nucl. Acids Res. **41**(W1), W518–W522 (2013)

Author Index

Printed in the United States
By Bookmasters